TEXT BOOK
OF
CIRCLES AND PARABOLA

DPH MATHEMATICS SERIES

TEXT BOOK OF CIRCLES AND PARABOLA

By

A.K. Sharma

DISCOVERY PUBLISHING HOUSE
NEW DELHI-110002

Published by:
Tilak Wasan

DISCOVERY PUBLISHING HOUSE PVT. LTD.
4383/4B, Ansari Road, Darya Ganj
New Delhi-110 002 (India)
Phone : +91-11-23279245, 23253475, 43596065
E-mail : discoverypublishinghouse@gmail.com
sales@discoverypublishinggroup.com
web : www.discoverypublishinggroup.com

***Edition:* 2020**

ISBN: 978-81-7141-995 1

Text Book of Circles and Parabola

Printed at:
Infinity Imaging Systems
Delhi

Preface

This book Text Book of Circle and Parabola has been written to meet the requirement of B.Sc. and engineering student of various universities in India. The subject matter has been discussed in such a simple way that the students will find no difficulty to understand it. The article have been explained in details in a nice manner and all the examples have been completely solved. We have tried to solve each problem in an elegant and more interesting way. The book contain almost all the questions set at the various examinations held by Indian universities and it cover the syllabi of all Indian universities.

The author will feel amply rewarded it the book serves the purpose for which it is meant suggestions for the empowerment of the book will be gratefully accepted.

A.K. Sharma

Contents

Equations of the Radical Axis and one Circle of the System are Given.

①

CIRCLES

Definition: *A circle is the set of all points in a plane which are at a constant distance from a fixed point in a plane. The fixed point is called the centre and the constant distance is called the radius of the circle. The circle with centre O and radius r, is denoted by C(O, r).*

Or

It is the locus of a point whose distance from the fixed point remains constant. The fixed point is called centre and the fixed distance is called radius of the circle.

EQUATION OF A CIRCLE

To find the equation of a circle when the radius and centre is given.

Let P(x, y) be the any point on the circle join the point P(x, y) with the centre C(h, k) using distance formula

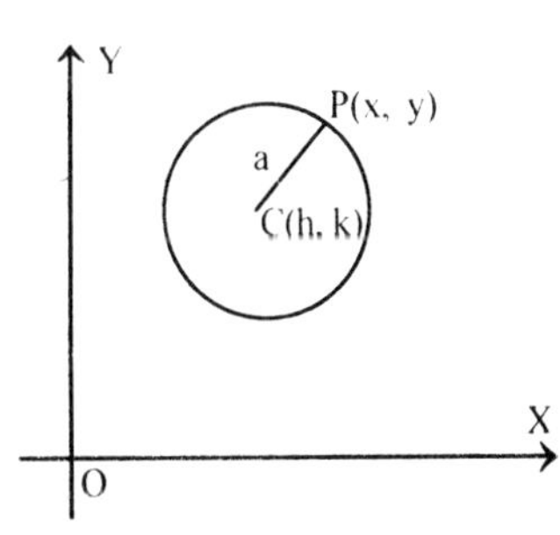

$$CP = \sqrt{(x - h)^2 + (y - k)^2}$$

$$r = \sqrt{(x-h)^2+(y-k)^2}$$

squaring both side, we get

$$r^2 = (x - h)^2 + (y + k)^2 \quad ...(i)$$

Conversely if a point O(x, y) satisfies the equation (i) *i.e.*, $(x - h)^2 + (y - k)^2$ then surely the distance between from $O(x_1 - y)$ to C(h, k) is r and so O is on the circle with centre C and radius r.

Hence equation (i) represent a circle with centre (h, k) and radius r.

Corollary : If the centre is origin then the equation of the circle is

$$x^2 + y^2 = a^2.$$

STANDARD FORM OF THE EQUATION OF A CIRCLE

Let that $x^2 + y^2 + 2gx + 2fy + c = 0$ represents a circle. Find its radius and centre.

Proof:

We know that the equation of the circle with centre (h, k) and radius a is

The given equation is $(x - h)^2 + (y - k)^2 = r^2$. ...(i)

$$x^2 + y^2 + 2gx + 2fy + c = 0 \quad ...(A)$$

$$\Rightarrow \quad x^2 + y^2 + 2gx + 2fy + g^2 + f^2 = g^2 + f^2 - c = 0$$

$$\Rightarrow \quad (x^2 + 2gx + g^2) + (y^2 + 2fy + y^2) = g^2 + f^2 - c$$

$$\Rightarrow \quad (x + g)^2 + (y + f)^2 = g^2 + f^2 - c$$

$$\Rightarrow \quad [x - (-g)]^2 + [y - (-f)]^2 = \sqrt{g^2 + f^2 - c} \quad ...(ii)$$

Comparing equation with $(x - h)^2 + (y - k)^2 = r^2$,

We have $h = g$, $k = -f$, r

Hence centre of the circle is $(-g, -f)$ and radius as $\sqrt{g^2 + f^2 - c}$. The equation (A) is known as the standard or canonical form of the equation of the circle.

POINT AND CIRCLE

Let the equation of circle $x^2 + y^2 + 2gx + 2fy + c = 0$ with centre $C(-g, -f)$ and $P(x, y)$ be any point then the point will be outside circle if the distance CP is greater then the radius of the circle and inside the circle if the corresponding distance is less than the radius. Mathematically, we can express that the point $P(x_1, y_1)$ lies outside, on or inside the circle if CP >

$\Rightarrow$ < Radius *i.e.*, $\sqrt{[x_1 - (-g)]^2 + [y_1 - (-f)]^2} >$ or = or $< \sqrt{(g^2 + f^2 - c)}$

which on simplification gives

$$x_1^2 + y_1^2 + 2gx_1 + 2f\,y_1 + c > \text{ or } = \text{ or } < 0$$

TO FIND THE EQUATION OF A CIRCLE DRAWN ON THE LINE JOINING (x_1, y_1) AND (x_2, y_2) AS DIAMETER

Let $A(x_1, y_1)$ and $B(x_2, y_2)$ be the end points of diameter AB of the circle. Take $P(x, y)$ be any point on the circle. Then by simple geometry property $\angle APB$ is a right angle. Then the product of the slopes of AP and BP should be –1. This gives the following results

$$m_1 = \frac{y - y_1}{x - x_1},$$

$$m_2 = \frac{y - y_2}{x - x_2}$$

$$m_1 \times m_2 = -1$$

$$\left(\frac{x - x_1}{y - y_1}\right) \times \left(\frac{x - x_2}{y - y_2}\right) = -1$$

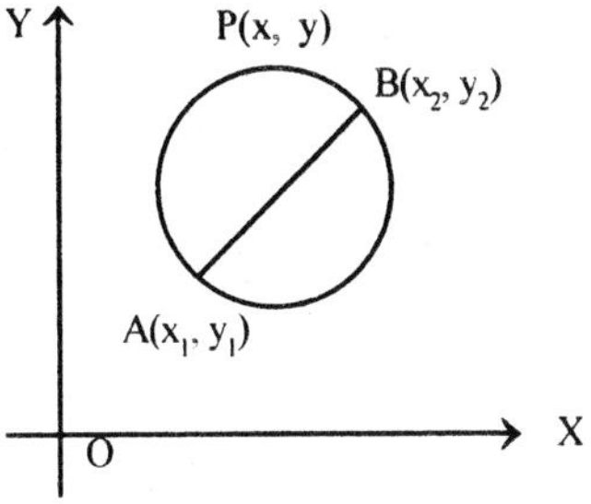

$\Rightarrow (x - x_1)(x - x_2) + (y - y_1)(y - y_2) = 0$, which is the required equation of the circle.

TO FIND THE EQUATION OF A CIRCLE THROUGH THREE GIVEN POINTS (x_1, y_1) (x_2, y_2) (x_3, y_3).

We know that the standard form of the circle is

$$x^2 + y^2\ 2gx + 2fy + c = 0. \quad ...(i)$$

It passes through (x_1, y_1), (x_2, y_2) and (x_3, y_3). This implies

$$x_1^2 + y_1^2\ 2gx_1 + 2fy_1 + c = 0. \quad ...(ii)$$

$$x_2^2 + y_2^2\ 2gx_2 + 2fy_2 + c = 0. \quad ...(iii)$$

$$x_3^2 + y_3^2\ 2gx_3 + 2fy_3 + c = 0. \quad ...(iv)$$

The required of the circle is obtained by eliminating g, f and c from the equations (i), (ii), (iii) and (iv).

We obtain the require equation by eliminating the determinant

$$\begin{vmatrix} x^2 + y^2 & x & y & 1 \\ x_1^2 + y_1^2 & x_1 & y_1 & 1 \\ x_2^2 + y_2^2 & x_2 & y_2 & 1 \\ x_3^2 + y_3^2 & x_3 & y_3 & 1 \end{vmatrix} = 0$$

INTERSECTION OF A LINE AND A CIRCLE

Let the equation of a circle be

$$x^2 + y^2 = r^2 \quad ...(i)$$

and the equation of straight line

$$y = mx + c. \quad ...(ii)$$

Solving (i) and (ii) simultaneously the value of y form (ii) to equation (i) we get

$$x^2 + (mx + c)^2 = r^2$$

$$x^2(1 + m^2) + 2mcx + c^2 - r^2 = 0. \qquad \text{...(iii)}$$

This equation being quadratic in x will give two values of x, *i.e.* There are two points of intersection. Equation (iii) may have real coincident or imaginary roots according as

$$4m^2c^2 - 4(1 + m^2)(c^2 - a^2) > \text{ or } = \text{ or } < 0$$

$$\Rightarrow \qquad r^2(1 + m^2) - c^2 > \text{ or } = \text{ or } < 0$$

If $r^2(1 + m^2) - c^2 > 0$ or $\dfrac{c}{\sqrt{(1+m)^2}} < r^2$, then the roots are real and the line cuts the circle at two points.

If $r^2(1 + m^2) - c^2 = 0$ *i.e.*, $c = \pm r\sqrt{1+m^2}$ or $\dfrac{c}{\sqrt{1+m^2}} = r$ then the roots are coincident *i.e.*, the line cuts the circle at one point *i.e.*, the line is tangent to the circle.

If $r^2(1 + m^2) - c^2 < 0$ or $\dfrac{c}{\sqrt{(1+m^2)}} > r$, the roots are imaginary *i.e.*, the lines do not meet the circle.

Also the perpendicular distance from (0, 0), the centre of the circle to the line $y = mx + c$ is $c/\sqrt{1+m^2}$. We can conclude that the line cuts the circle in two real points if the perpendicular distance is less than radius, in two coincident points if the perpendicular distance is equal to radius and in two imaginary points if the perpendicular distance is more than radius.

Example 1:

Find the equation to the circle whose radius is 5 and which touches the circle $x^2 + y^2 - 2x - 4y - 20 = 0$ at (5, 5)

Solution:

The centre of the given circle

$$x^2 + y^2 - 2x - 4y - 20 = 0 \text{ is } (1, 2)$$

and its radius is $\sqrt{1^2 + 2^2 + 20}$ or 5.

$$\therefore \qquad AP = 5.$$

Let the centre of the required circle be (a, b). Since its radius is

$$BP = 5 = AP,$$

P is the middle point of AB.

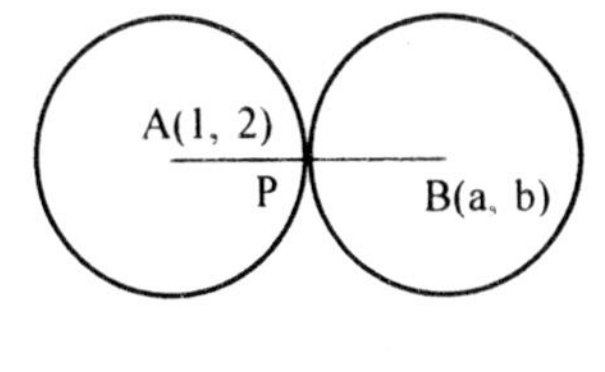

$\therefore \qquad 5 = \frac{a+1}{2} \Rightarrow a = 9$

$$5 = \frac{b+2}{2} \Rightarrow b = 8.$$

Hence the equation of the required circle is

$$(x - 9)^2 + (y - 8)^2 = (\text{radius})^2 = 5^2$$

$\Rightarrow \qquad x^2 + y^2 - 18x - 16y + 120 = 0.$

Example 2:

Find the equation to the circle which passes through the point (2, 0) and whose centre is the limit of the point of intersection of the lines

$$3x + 5y = 1, (2 + c)x + 5c^2y = 1 \text{ as } c \to 1$$

Solution:

The centre of the required circle will be the point of intersection of lines.

$$3x + 5y = 1 \qquad \text{...(i)}$$

$$(2 + c)x + 5c^2y = 1. \qquad \text{...(ii)}$$

Solving (i) and (ii), we get

$$x = \frac{c+1}{3c+2}$$

Taking limit as $c \to 1$, we get $x = \frac{2}{5}$. This gives $y = \frac{-1}{25}$.

$\therefore$ The centre of the circle is $\left(\frac{2}{5}, \frac{-1}{25}\right)$.

Let a be the radius.

$\therefore$ the equation of circle is

$$\left(x - \frac{2}{5}\right)^2 \left(y + \frac{1}{25}\right)^2 = a^2.$$

If passes through (2, 0),

$\therefore \qquad \left(2 - \frac{2}{5}\right)^2 + \left(0 + \frac{1}{25}\right)^2 = a^2$

$$\Rightarrow \qquad a^2 = \frac{1601}{625}$$

∴ The required equation is

$$\left(x - \frac{2}{5}\right)^2 + \left(y + \frac{1}{25}\right)^2 = \frac{\sqrt{1601}}{625}$$

Example 3:

Find the equation to the circle circumscribing the triangle formed by the lines $a_i x = b_i y + c_i = c$, $i = 1, 2, 3$, is

$$\begin{vmatrix} \dfrac{1}{a_1x+b_1y+c_1} & a_2a_3 - b_3b_2 & a_2b_3 + a_3b_2 \\ \dfrac{1}{a_2x+b_2y+c_2} & a_3a_1 - b_3b_1 & a_3b_1 + b_3a_1 \\ \dfrac{1}{a_3x+b_3y+c_3} & a_1a_2 - b_1b_2 & a_1b_2 + b_1a_2 \end{vmatrix} = 0$$

Solution:

The equation of the circle circumscribing the triangle formed by the given lines is given as

$$(a_1x + b_1y + c_1)(a_2x + b_2y + c_2) + \lambda(a_2x + b_2y + c_2)(a_3x + b_3y + c_3)$$
$$+ u(a_3x + b_3y + c_3)(a_1x + b_1y + c_1) = 0, \qquad \text{...(i)}$$

where λ and u are so chosen such that the coefficients if x^2 is equal to co-efficient of y^2 and the coefficient of xy is zero. This gives

$$a_1a_1 + \lambda a_2a_3 + ua_3a_1 = b_1b_2 + \lambda b_1b_3 + ub_3b_1$$

or $\qquad a_1a_1 + b_2b_3 + \lambda(a_2a_3 - b_2b_3) + u(a_3a_1 - b_3b_1) = 0 \qquad$ ...(ii)

and $\qquad (a_1b_2 - a_2b_1) + \lambda(a_2b_3 + a_3b_2) + u(a_3b_1 - a_1b_3) = 0 \qquad$ ...(ii)

The equation (i) can be re-written as

$$\frac{1}{a_3x+b_3y+c_3} + \frac{\lambda}{a_1x+b_1y+c_1} + \frac{u}{a_2x+b_2y+c_2} = 0. \qquad \text{...(iv)}$$

Eliminating l, u from (ii), (iii) and (iv) we have the required result.

Example 2:

Find the equation to the circles touching x-axis at (3, 0) and making an intercept of 8 units on the y-axis.

Solution:

Let r be the radius of the circle. There can be two circles satisfying the given conditions. Let A and B be the centres of these two circles. Since the circles touch x-axis at (3, 0) *i.e.*, OP = 3, the co-ordinates of A and B becomes (3, r) and (3, – r) respectively. As these circles made intercept of 8 units on the y-axis, we have

$$CD = 8 \text{ and } FF = 8$$

From A draw a segment AL perpendicular on y-axis, so that

$$CL = DL = 4$$

Also $\quad AL = PO = 3$

The right angled triangle CLA gives

$$r^2 = 3^2 + 4^2 = 25$$

$\Rightarrow \quad r = 5.$

$\therefore$ The equation of the circle with centre A is

$$(x - 3)^2 + (y - 5)^2 = 5^2$$

$$\Rightarrow \quad x^2 + y^2 - 6x - 10y + 9 = 0$$

Similarly the equation of circle with centre D is

$$x^2 + y^2 - 6x - 10y + 9 = 0.$$

TANGENT AND NORMALS

Definition : Let C be a curve and P and Q are any two points on a curve. Draw a line PQ. Now, as the point Q moves along the curve towards P. The line PQ turns about P and Q approaches P. The line PQ coincides with the limiting line PT. This limiting line PT is called the tangent line to the curve C at the point.

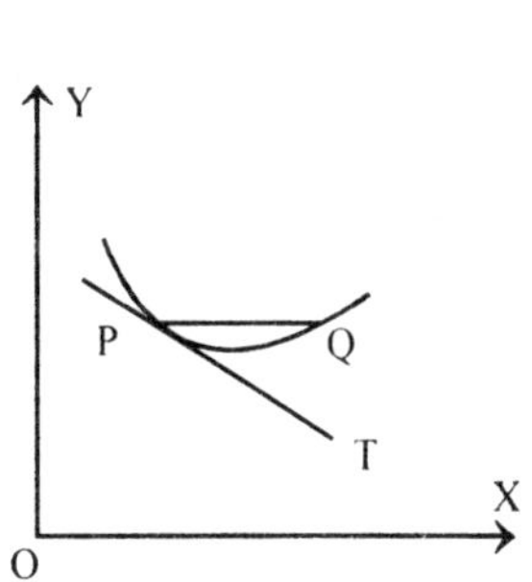

TO FIND THE EQUATION OF THE TANGENT AT ANY POINT ON THE CIRCLE

Let $P(x_1, y_1)$ and $Q(x_2, y_2)$ be any two points on the circle

$$x^2 + y^2 = r^2 \qquad \text{...(i)}$$

Then equation of the chord PQ by two point formula will be

$$y - y_2 = \frac{y_2 - y_1}{x_2 - x_1}(x - x_1) \qquad \text{...(ii)}$$

We obtain the value of the expression equivalent to $\frac{y_2 - y_1}{x_2 - x_1}$ as follows

Since the points lie on the circle, we get

$$x_1^2 + y_1^2 = r^2. \quad \text{...(iii)}$$

$$x_2^2 + y_2^2 = r^2. \quad \text{...(iv)}$$

On subtraction it gives

$$(x_1^2 + x_2^2) + (y_1^2 + y_2^2) = 0$$

$$\Rightarrow \quad \frac{y_2 - y_1}{x_2 - x_1} = -\frac{x_1 + x_2}{y_1 + y_2}$$

Equation (ii) now becomes

$$y - y_1 = -\frac{x_1 + x_2}{y_1 + y_2}(x - x_1). \quad \text{...(v)}$$

Now as Q approaches P, $x_2 \to x_1$ and $y_2 \to y_1$, the equation (v) becomes

$$y - y_1 = -\frac{x_1}{y_1}(x - x_1)$$

$$\Rightarrow \quad yy_1 - y_1^2 = -xx_1 + x_1^2$$

$$\Rightarrow \quad xx_1 + yy_1 = x_1^2 + y_1^2 = r^2 \text{ (due to (i))}$$

Hence the equation of the tangent to the circle $x^2 + y^2 = r^2$ at the point (x_1, y_1) on it is $xx_1 + yy_1 = r^2$.

EQUATION OF THE TANGENT AT ANY POINT (x_1, y_1) ON THE CIRCLE

$$x^2 + y^2 + 2gx + 2fy + c = 0$$

Let $P(x_1, y_1)$ and $Q(x_2, y_2)$ be two points on the circle. Equation of the line PQ by two point formula will be

$$y - y_1 = -\frac{y_2 - y_1}{x_2 - x_1}(x - x_1). \quad \text{...(i)}$$

Since PQ lies on the circle, then we have

$$x_1^2 + y_1^2 + 2gx_1 + 2fy_1 + c = 0. \quad \text{...(ii)}$$

and

$$x_2^2 + y_2^2 + 2gx_2 + 2fy_2 + c = 0. \quad \text{...(iii)}$$

Substracting and simplifying, we get

$$\frac{y_2 - y_1}{x_2 - x_1} = -\frac{x_1 + x_2 + 2g}{y_1 + y_2 + 2f}. \quad \text{...(iv)}$$

Substracting (iv) in (i) It becomes

$$y - y_1 = -\frac{x_1 + x_2 + 2g}{y_1 + y_2 + 2f}\ (x - x_1). \qquad ...(v)$$

For PQ to be tangent $x_2 \to x_1$ and $y_2 \to y_1$. This makes (v) as

$$y - y_1 = -\frac{x_1 + g}{y_1 + f}\ (x - x_1)$$

$$\Rightarrow \quad yy_1 + yf - y_1^2 - y_1 f = -xx_1 - gx + x_1^2 + gx_1$$

$$\Rightarrow \quad xx_1 + yy_1 + gx + fy = x_1^2 + y_1^2 + gx_1 + fy_1.$$

Adding $gx_1 + fy_1 + c$ to both sides, we have

$$xx_1 + yy_1 + g(x + x_1) + f(y + y_1) + c$$

$$= x_1^2 + y_1^2 + 2gx_1 + 2fy_1 + c = 0$$

Hence the equation of the tangent is

$$xx_1 + yy_1 + g(x + x_1) + f(y + y_1) + c = 0$$

TO FIND THE CONDITION THAT THE LINE $y = mx + c$ TOUCHES THE CIRCLE $x^2 + y^2 = r^2$.

The points of intersection of the line

$$y = mx + c \qquad ...(i)$$

and the circle

$$x^2 + y^2 = r^2 \qquad ...(ii)$$

are given by the equation

$$x^2 + (mx = c)^2 = r^2$$

$$\Rightarrow \quad x^2(1 + m^2) + 2mcx + c^2 - r^2 = 0. \qquad ...(iii)$$

This equation being quadratic in x gives two values of x.

Since $y = mx + c$ is to be tangent to the circle.

Equation (iii) must give two equal roots of x *i.e.*,

$$4m^2c^2 = 4(c^2 - r^2)(1 + m^2)$$

$$c = \pm a\sqrt{1 + m^2}.$$

Hence $y = mx + c$ will be tangent to the circle if

$$c = \pm a\sqrt{1 + m^2}$$

$$y = mx \pm a\sqrt{1 + m^2}$$

is the equation of a tangent to the given circle

$$x^2 + y^2 + r^2.$$

This is called equation to tangent to the circle in the slope form

$$c = \pm a\sqrt{1+m^2}.$$

EQUATION OF NORMAL TO A CIRCLE

Definition: The normal to the curve at any point is the straight line passing through that point and is perpendicular to the tangent at that point.

TO FIND THE EQUATION OF THE NORMAL TO A CIRCLE AT A GIVEN POINT (x_1, y_1) ON IT

(a) Equation of the circle is $x^2 + y^2 = r^2$

Equation of the tangent to the circle at (x_1, y_1) is given by

$$xx_1 + yy_1 = r^2 \qquad \text{...(i)}$$

Equation of the line perpendicular to (i) is given by

$$y_1x - x_1y + \lambda = 0. \qquad \text{...(ii)}$$

Since (ii) passes through the point (x_1, y_1) then we have

$$\lambda = 0.$$

Hence the normal to the circle $x^2 + y^2 + r^2$ at the point (x_1, y_1) given by

$$y_1x - x_1y = 0$$

(b) General Equation of the circle is

$$x^2 + y^2 + 2gx + 2fy + c = 0.$$

The equation of the normal at the point (x_1, y_1) on it is obtain similarly as

$$y(x_1 + g) - x(y_1 + f) + fx_1 - gy_1 = 0.$$

Example 1:

Let A be the centre of the circle $x^2 + y^2 - 2x - 4y - 20 = 0$. Suppose that the tangents at the points B(1, 7) and D(4, – 2) on the circle meet at the point C. Find the area of the quadrilateral ABCD.

Solution:

The equation of the given circle is

$$x^2 + y^2 - 2x - 4y - 20 = 0. \qquad \text{...(i)}$$

Equation of tangents at BC is $1.x + 7.y - (1 + x) - 2(y - 7) - 20 = 0$

$\Rightarrow \qquad y = 7.$ ...(ii)

Equation of the tangent at DC is

$$3x - 4y - 20 = 0. \qquad ...(iii)$$

Solving (ii) and (iii), we get the co-ordinate of C, the point of intersection of these two tangents as (16, 7)

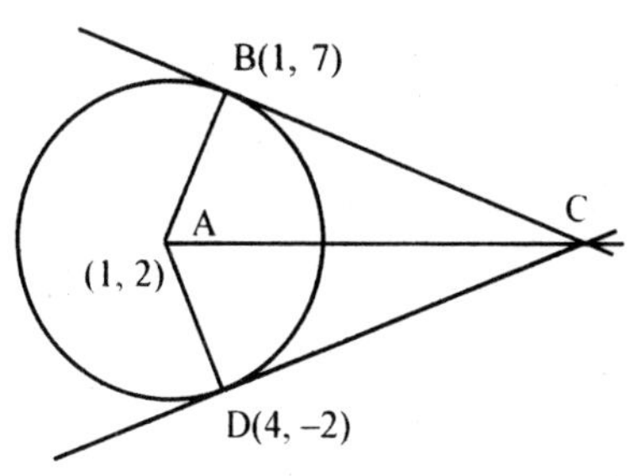

The area of quadrilateral

ABCD = ΔABD + ΔBDC

$$= \frac{15}{2} + \frac{135}{2} = 75.$$

Example 2:

Find the equation to the circle touching the y-axis and the straight line $3y = 4x$ at the point (3, 4).

Solution:

Let the equation of the circle be

$$x^2 + y^2 + 2gx + 2fy + c = 0. \qquad ...(i)$$

Since it touches y-axis, therefore, putting $x = 0$ in (i), we get

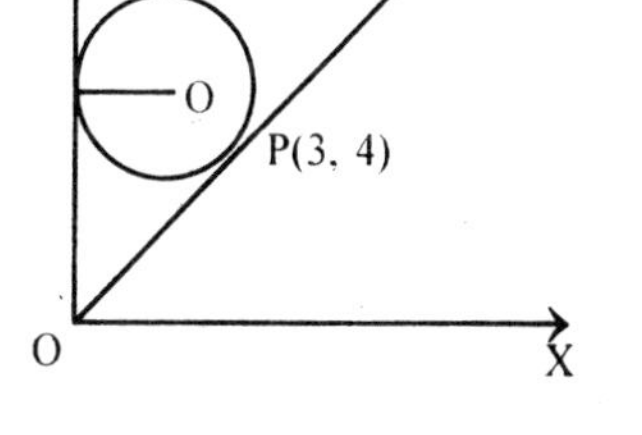

$$y^2 + 2fy + c = 0. \qquad ...(ii)$$

This must have equal roots

$\therefore 4f^2 = 4 \Rightarrow f^2 = c.$ ...(iii)

Now the circle (i) passes through the point (3, 4)

$\therefore 25 + 6g + 8f + c = 0.$ ...(iv)

Also the centre of the circle (i) is $(-g, -f)$ and radius = g. Since the line $4x = 3y = 0$ touches the circle (i),

$\therefore$ Perpendicular from the centre $(-g, -f)$ = Radius of the circle on the line.

i.e. $$\frac{-4g + 3f}{\sqrt{4^2 + 3^2}} = g \Rightarrow f = 3g. \qquad ...(v)$$

From (iii), (iv) and (v), we get

$$f^2 + 10f + 25 = 0$$

$\Rightarrow \qquad (f + 5)^2 = 0 \Rightarrow f = -5$

$\Rightarrow \quad g = -5/3 \Rightarrow c = 25.$

Substituting the values of g, f and c in (i) we get

$$3x^2 + 3y^2 - 10x - 30y + 75 = 0.$$

TWO TANGENTS TO A CIRCLE FROM A POINT OUTSIDE IT

Show that from a point two tangents can be drawn to a circle.

We have

$$y = mx \pm a\sqrt{1+m^2} \quad \text{...(i)}$$

is tangent to the circle

$$x^2 + y^2 = a^2 \quad \text{...(ii)}$$

for all values of m. Let the tangent passes through the point (x_1, y_1),

$$\therefore \quad y_1 = mx_1 \pm a\sqrt{1+m^2}$$

$$\Rightarrow \quad (y_1 - mx_1)^2 = a^2(1 + m^2)$$

$$\Rightarrow \quad m^2(x_1^2 - a^2) - 2mx_1y_1 + (y_1^2 - a^2) = 0. \quad \text{...(iii)}$$

This equation is quadratic in m, gives two values of m, m_1 and m_2 which given

$$y_1 = m_1x_1 \pm a\sqrt{1+m_1^2}$$

$$y_2 = m_2x_1 \pm a\sqrt{1+m_2^2}.$$

This proves that there exists two tangents $y = m_1x \pm a\sqrt{1+m_1^2}$ and $y = m_2x_1 \pm a\sqrt{1+m_2^2}$ both pass through (x_1, y_1). Hence from a point two tangents can be drawn to a circle. We also note the following.

(i) The tangents are real and distinct, coincident or imaginary according as the roots of (iii) are real and distinct, equal or imaginary according as

$$(-2x_1y_1)^2 - 4(x_1^2 - a^2)(y_1^2 - a^2) > = \text{ or } < 0$$

$$\Rightarrow \quad x_1^2y_1^2 - (x_1^2y_1^2 - a^2x_1^2 - a^2y_1^2 + a^4) > = \text{ or } < 0$$

$$\Rightarrow \quad x_1^2 + y_1^2 - a^2 > = \text{ or } < 0$$

i.e. according as the point lies outside, on or inside the circle.

(ii) Locus of the point of intersection of two perpendicular tangents to the circle $x^2 + y^2 = a^2$ is a concentric circle called the Director circle of the given circle.

Since the two tangents given by (iii) are perpendicular to each other, the product of their slopes must be -1. This gives

$$\frac{y_1^2 - a^2}{x_1^2 - a^2} = -1.$$

$\Rightarrow$ $x_1^2 - a^2 = -(y_1^2 - a^2)$

$\Rightarrow$ $x_1^2 + y_1^2 = 2a^2.$

Hence the locus is

$$x^2 + y^2 = 2a^2.$$

LENGTH OF THE TANGENT

Let equation of the givne circle is

Let $$x^2 + y^2 + 2gx + 2fy + c = 0 \quad \text{...(i)}$$

$P(x_1, y_1)$ be any point and PT is tangent to the circle at T. It is required to find the length PT.

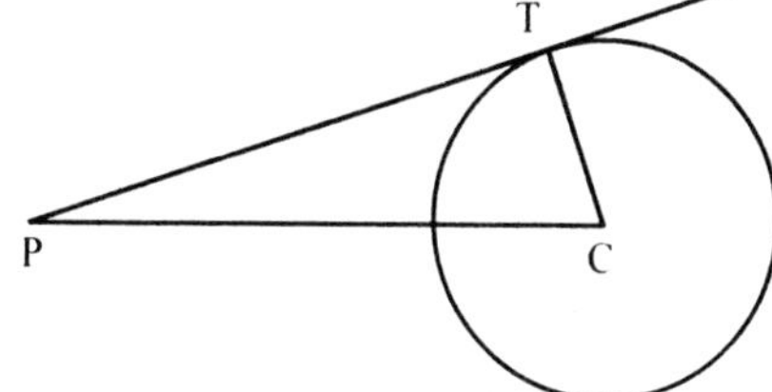

Since PT is the tangent and CT is the radius of circle, therefore PTC is a right angled triangle.

$\therefore$ $PT^2 = PC^2 - CT^2$

$= \{x_1 - (-g)\}^2 + \{y_1 - (-f)\}^2 - (g^2 + f^2 - c)$

$= x_1^2 + y_1^2 + 2gx_1 + 2fy_1 + c$

Hence the square of the length of tangent drawn from the point (x_1, y_1) to the circle

$x^2 + y^2 + 2gx + 2fy + c = 0$ is $x_1^2 + y_1^2 + 2gx_1 + 2fy_1 + c$

CHORD OF CONTACT

Let PQ and PR be two tangents drawn from P to the circle at Q and R respectively. The chord QR is called the chord of contact.

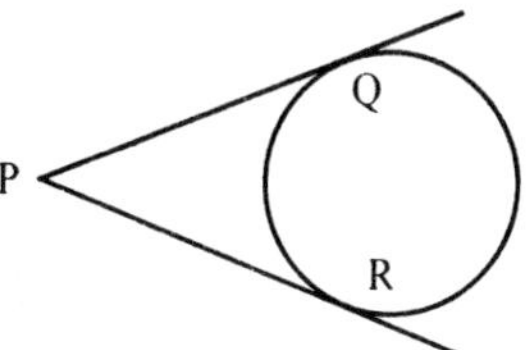

TO FIND THE EQUATION OF THE CHORD OF CONTACT

Let $x^2 + y^2 = a^2$ be the equation of circle and $P(x_1, y_1)$ be an external point. Let (x_2, y_2) and (x_3, y_3) be the co-ordinates of Q and R of the tangents PQ PR drawn to the circle from P

The equation of tangent at Q is given by

$$xx_2 + yy_2 = a^2. \qquad ...(i)$$

It passes through the point (x_1, y_1) then we have

$$x_1x_3 + y_1y_2 = a^2. \qquad ...(ii)$$

Similarly, the tangent at R passes through (x_1, y_1), is give by

$$x_1x_3 + y_1y_3 = a^2. \qquad ...(iii)$$

From (ii) and (iii), the equation of QR is $xx_1 + yy_1 = a^2$ as both points Q and R satisfy the equation. Hence the equation of the chord of contact of the tangents drawn from (x_1, y_1) to the circle $x^2 + y^2 = a^2$ is $xx_1 + yy_1 = a^2$. The equation of the chord of contact of tangents to the circle $x^2 + y^2 + 2gx = 2fy + c = 0$ can be similarly proved as $xx_1 + yy_1 + g(x + x_1) + f(y + y_1) + c = 0$.

POLE OF A LINE Ax + By + C = w.r.t. THE CIRCLE $x^2 + y^2 = a^2$

Let the pole be (h, k). Then the given straight line must be same as the equation of the polar *i.e.*, the straight line

$$Ax + By + C = 0$$

and

$$hx + ky - a^2 = 0$$

must be identical. Comparing the coefficients we have

$$\frac{h}{A} = \frac{k}{B} = \frac{-a^2}{C}$$

which gives $h = \dfrac{-Aa^2}{C}$ and $k = \dfrac{-Ba^2}{C}$

PROPERTIES OF POLE AND POLAR

(i) If the polar of a point P passes through a point Q then the polar of Q passes through P.

Let (x_1, y_1) and (x_2, y_2) be the co-ordinates of points P and Q and $x^2 + y^2 = a^2$ be the equation of circle. Then polar of P is given by

$$xx_1 + yy_1 = a^2 \qquad ...(i)$$

and polar of Q is $xx_2 + yy_2 = a^2$ ...(ii)

Since polar of P passes through point Q,

$$\therefore \quad x_1x_2 \; y_1y_2 = a^2 \qquad ...(iii)$$

Polar of Q passes through P if (x_1, y_1) satisfies the equation $xx_1 + yy_1 = a^2$ *i.e.* if $x_1x_2 + y_1y_2 = a$ which holds due to (iii). Hence the result.

(ii) If the pole of the line $lx + my + n = 0$ w.r.t. the circle $x^2 + y^2 = a^2$ lies on the line $l'x + m'y + n' = 0$, then the pole of the line $l'x + m'y + n' = 0$ lies on the line $lx + my + n = 0$.

The equation of the lines are

$$lx + my + n = 0 \qquad ...(i)$$

and $$l'x + m'y + n' = 0 \qquad ...(ii)$$

and circle of is $$x^2 + y^2 = a^2. \qquad ...(iii)$$

The pole of (i) is $\left(-\frac{a^2 l}{n}, \frac{-a^2 m}{n}\right)$. It lies on (ii) implies

$$l'\left(\frac{-a^2 l}{n}\right) + m'\left(\frac{-a^2 m}{n}\right) + n' = 0$$

or $$a^2(ll' + mm') - nn' = 0.$$

The symmetry of the result shows that the pole of (ii) also lies on (i). Lines (i) and (ii) are called conjugate lines.

(iii) If the polar of two points P and Q with respect to a circle meet in R, then R is the pole of the line PQ.

Let p and q be the polar of points P and Q, meeting at R. Since R lies on polar of P and Q, by reciprocal properties P and Q will lie on polar of R *i.e.*, PQ is the polar of R and R is the pole of the line PQ.

POLAR

Definition: *The polar of a point with respect to a circle is the locus of the point of intersection of tangents drawn at the extremities of chords through that point. In other words from a point P a straight line is drawn meeting the circle in two points Q and R. The tangents at Q and R meet in S. The locus of S is called the polar of P. P is called the pole.*

TO FIND THE EQUATION OF THE POLAR

Let equation of circle be $x^2 + y^2 = a^2$ and $P(x_1, y_1)$ be any point. Let the tangents at Q and R meet at S. If the co-ordinate of S be (h, k), then QR is interpreted as the chord of contact and its equation becomes

$$xh + yk = a^2.$$

As QR passes through $P(x_1, y_1)$, we have

$$x_1 h + y_1 k = a^2.$$

$\therefore$ locus of S is

$$xx_1 - yy_1 = a^2.$$

In a similar way it may be proved that the polar of point (x_1, y_1) with respect to the circle $x^2 + y^2 + 3gx = 2fy + c = 0$ is

$$xx_1 + yy_1 + g(x + x_1) + f(y + y_1) + c = 0.$$

EQUATION OF A PAIR OF TANGENTS

Let $P(x_1, y_1)$ be the external point to the given circle

$$x^2 + y^2 + 2gx + 2fy + c = 0 \qquad ...(i)$$

from which two tangents PA and PB are drawn to (i). AB is the chord of contact of (i). Therefore equation of AB is a

$$xx_1 + yy_1 + g(x + x_1) + f(y + y_1) + c = 0$$

or $x(x_1 + g) + y(y_1 + f) + (gx_1 + fy_1 + c) = 0.$

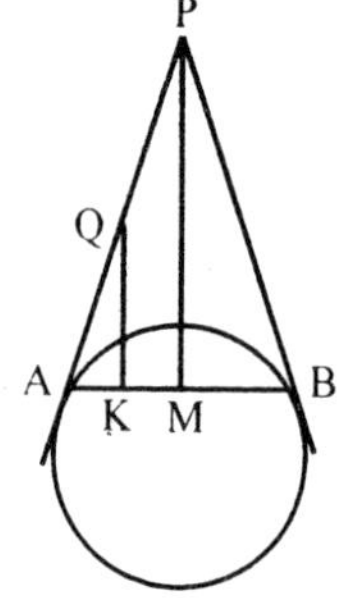

Take $Q(x_1, y_2)$ any point on PA. From P and Q draw PM and QK perpendiculars to AB. Now Δs QAK and PAM are similiar. Then we have

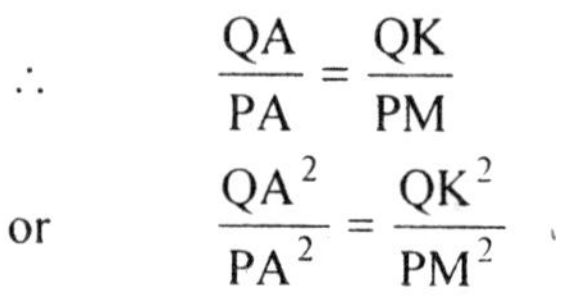

$$\therefore \qquad \frac{QA}{PA} = \frac{QK}{PM}$$

$$\text{or} \qquad \frac{QA^2}{PA^2} = \frac{QK^2}{PM^2}$$

Now we have $QA^2 = x_2^2 + y_2^2 + 2gx_2 + 2fy_2 + c$

$$PA^2 = x_2^2 + y_1^2 + 2gx_1 + 2fy_1 + c.$$

$$\text{Also} \quad QA = \frac{x_2(x_1+g)+y_2(y_1+f)+gx_1+fy_1+c}{\sqrt{\left[(x_1+g)^2+(y_1+f)^2\right]}}$$

$$\text{and} \quad PM = \frac{x_1(x_1+g)+y_1(y_1+f)+gx_1+fy_1+c}{\sqrt{\left[(x_1+g)^2+(y_1+f)^2\right]}}$$

Length of pre-pendicular

Substituting these values in (ii), we get

$$\frac{x_2^2+y_2^2+2gx_2+2fy_2+c}{x_1^2+y_1^2+2gx_1+2fy_1+c} = \frac{\left[x_2(x_1+g)+y_2(y_1+f)+gx_1+fy_1+c\right]^2}{\left[x_1(x_1+g)+y_1(y_1+f)+gx_1+fy_1+c\right]^2}$$

$$\frac{x_2^2+y_2^2+2gx_2+2fy_2+c}{x_1^2+y_1^2+2gx_1+2fy_1+c} = \frac{\left[x_2(x_1+g)+y_2(y_1+f)+gx_1+fy_1+c\right]^2}{\left\{x_1^2+y_1^2+2gx_1+2fy_1+c\right\}}$$

$$\text{or } (x_1^2 + y_1^2 + 2gx_1 + 2fy_1 + c)(x_2^2 + y_2^2 + 2gx_2 + 2fy_2 + c)$$
$$= [x_2(x_1 + g) + y_2(y_1 + f) + gx_1 + fy_1 + c]^2.$$

$\therefore$ locus of $Q(x_2, y_2)$ is

$$(x^2 + y^2 + 2gx + 2fy + c)(x_1^2 + y_1^2 + 2gx_1 + 2fy_1 + c)$$
$$= [x(x_1 + g) + y(y_1 + f) + gx_1 + fy_1 + c]^2$$
$$= [xx_1 + yy_1 + g(x + x_1) + f(y + y_1 + c]^2.$$

which is the equation of the pair of tangents. This equation can be easily remembered as

$$SS_1 = T^2$$

where $S = x^2 + y^2 + 2gx + 2fy + c$

$S_1 = x_1^2 + y_1^2 + 2gx_1 + 2fy_1 + c$

$T = xx_1 + yy_1 + g(x + x_1) + f(y + y_1) + c.$

EQUATION OF THE CHORD WHOSE MIDDLE POINT IS GIVEN

Let the equation of the circle be $x^2 + y^2 = a^2$ and AB be the chord whose middle point P(h, k) is known. The equation of AB is

$$y - k = m(x - h) \quad \text{...(i)}$$

Since P is the middle point of AB, OP is perpendicular to AB. This gives slope of OP as k/h. There fore, slope of AB is $\frac{-h}{k}$. Hence equation of AB becomes

$$y - k = \frac{-h}{k}(x - h)$$

$$xh + yk = h^2 + k^2$$

Which is required equ.

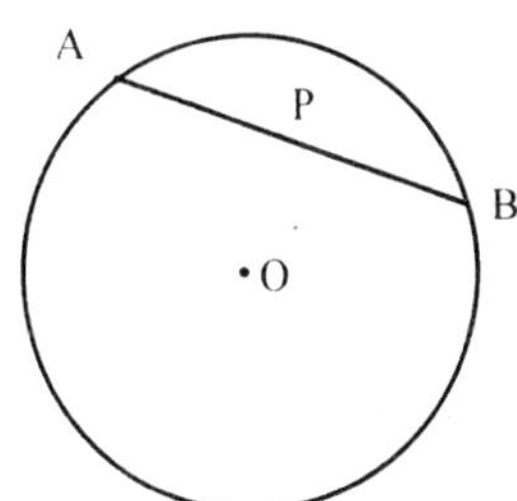

Example 1:

Prove that if the pole of a straight line with respect to a circle $x^2 + y^2 = a^2$ lies on the circle $x^2 + y^2 = 9a^2$, the line is a tangent to the circle $9x^2 + 9y^2 = a^2$

Solution:

Let the straight line be $lx + my + n = 0$ ...(i)

Let its pole be (x_1, y_1). Its polar w.r.t. to the circle

$$x^2 + y^2 = a^2 \quad \text{...(ii)}$$

is $xx_1 + yy_1 - a^2 = 0$ ...(iii)

But (i) and (iii) being the same straight line, we have on comparing the coefficients,

$$\frac{x_1}{l} = \frac{y_1}{m} = \frac{-a^2}{n}$$

$$\Rightarrow \quad x_1 = -\frac{l}{n}a^2 \text{ and } y_1 = \frac{-m}{n}a^2.$$

Since the pole lies on the circle $x^2 + y^2 = 9a^2$, we must have

$$l^2a^4 + m^2a^4 = 9a^2n^2$$

$$\Rightarrow \quad a^2l^2 + a^2m^2 = 9n^2. \quad \text{...(iv)}$$

The line (i) will touch the circle $9x^2 + 9y^2 = a^2$ if

$$\frac{n}{\sqrt{l^2 + m^2}} = \frac{a}{3}$$

or If $a^2l^2 + a^2m^2 = 9n^2$ which holds because of (iv). Hence the result follows.

Example 2:

Tangents are drawn from the point (h, k) to the circle $x^2 + y^2 = a^2$. Prove that the area of the triangle formed by them and the straight line joining their point of contact is

$$a(h^2 + k^2 - a^2)^{3/2}/(h^2 + k^2.)$$

Solution:

Equation of the chord of contact AB is given as

$$xh + yk = a^2. \quad \text{...(i)}$$

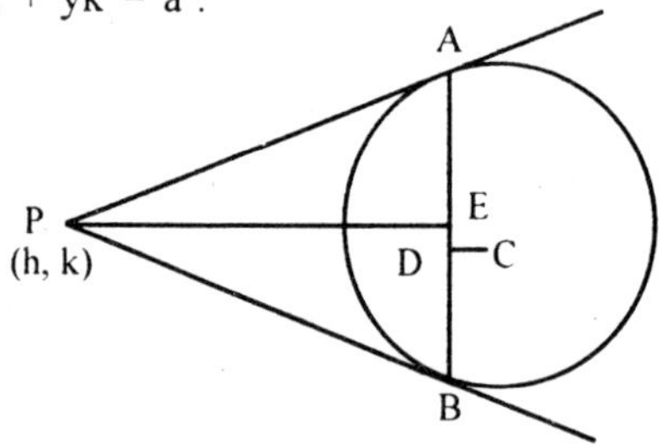

A
P
(h, k)
E
D
C
B

From C the centre of the circle. CD is drawn perpendicular to AB. This gives D as the middle point of AB.

Now $CD = \dfrac{a^2}{\sqrt{(h^2 + k^2)}}$ (by perpendicular distance formula)

From ΔCBD

$$BD^2 = CB^2 - CD^2$$

$$= a^2 - \frac{a^4}{h^2+k^2}$$

$$= \frac{a^2\left(h^2+k^2-a^2\right)}{h^2+k^2}$$

PE, the perpendicular distance from P on AB is given as

$$PE = \frac{h^2+k^2-a^2}{\sqrt{h^2+k^2}}.$$

$\therefore$ Area of the $\Delta APB = \frac{1}{2}$ AB.PE

$$= \frac{1}{2} \times 2BD.PE$$

$$= \frac{\sqrt{\left(h^2/k^2-a^2\right)}}{\sqrt{\left(h^2+k^2\right)}} = \frac{\left(h^2/k^2-a^2\right)}{\sqrt{\left(h^2+k^2\right)}}$$

$$= \frac{a\left(h^2+k^2-a^2\right)^{3/2}}{h^2+k^2}.$$

MISCELLANEOUS EXAMPLES

Example 1:

Through a fixed point O are drawn two straight lines OPQ and ORS to meet a circle in P and Q, and R and S, respectively. Prove that the locus of the point of intersection of PR and QS, is the polar of O with respect to the circle.

Solution:

Let the two lines OPQ and ORS be taken as axes of x and y respectively, O as origin, and $\angle ROP = \omega$.

Equation of the given circle may be take as

$$x^2 + y^2 + 2xy \cos \omega + 2gx + 2fy + c = 0. \qquad \text{...(1)}$$

Now equation of x-axis is $y = 0$. ...(2)

Putting the value of y from (2) in (1), we get

$$x^2 + 2gx = c = 0. \qquad ...(3)$$

Let the co-ordinates of P and Q be $(x_1, 0)$ and $(x_2, 0)$, x_1 and x_2 are given by; so

$$x_1 + x_2 = -2g, ...(4) \text{ and } x_1x_2 = c \qquad ...(5)$$

Similarly if the co-ordinates of R and S be $(0, y_1)$ and $(0, y_2)$

$$y_1 + y_2 = -2f, ...(6), \text{ and } y_1y_2 = c. \qquad ...(7)$$

Equation of PS will be $\frac{x}{x_1} + \frac{y}{y_1} = 1.$ ...(8)

Equation of QR will be $\frac{x}{x_2} + \frac{y}{y_2} = 1.$ ...(9)

The required locus will be found by eliminating y_1, y_2, x_1, x_2 from relations (8) upto (9).

Adding (8) and (9), $\frac{x}{x_1} + \frac{x}{x_2} + \frac{y}{y_1} + \frac{y}{y_2} = 2$

or $$x\left(\frac{x_2 + x_1}{x_2 x_1}\right) + y\left(\frac{y_2 + y_1}{y_2 y_1}\right) = 2.$$

Putting the values from (4) and (5) etc., we get

$$x\frac{(-2g)}{c} + y.\frac{(-2f)}{c} = 2 \text{ or } gx + fy + c = 0 \qquad ...(10)$$

Polar of (0, 0) with respect to (1), is

$$x.0 + y.0 + g(x + 2) + f(y + 0) + c = 0$$

or $$gx + fy + c = 0$$

which is same as (10).

Similarly we can prove that the locus of the point of intersection of PR and QS is the same.

Example 2:

Find the radical centre of the circle

$$(x - 2)^2 + (y - 3)^2 = 36, (x + 3)^2 + (y + 2)^2 = 49,$$

and $$(x - 4)^2 + (y + 5)^2 = 64.$$

Solution:

The circle are given as

$$(x - 2)^2 + (y - 3)^2 = 36 \text{ or } x^2 + y^2 - 4x - 6y - 23 = 0 \qquad ...(1)$$

$$(x + 3)^2 + (y + 2)^2 = 49 \text{ or } x^2 + y^2 + 6x + 4y - 36 = 0 \quad ...(2)$$

and $\quad (x - 4)^2 + (y + 5)^2 = 64 \text{ or } x^2 + y^2 - 8x + 10y - 23 = 0 \quad ...(3)$

The radical axis of (1) and (2) is

$$(x^2 + y^2 + 6x + 4y - 36) - (x^2 + y^2 - 4x - 6y - 23) = 0$$

or $\quad 10x + 10y - 13 = 0 \quad ...(4)$

The radical axis of (1) and (3) is

$$(x^2 + y^2 - 4x - 6x - 4y - 36) - (x^2 + y^2 - 4x - 6y - 23) = 0$$

or $\quad 10x + 10y - 13 = 0 \quad ...(4)$

Solving (4) and (5), $x = \frac{26}{25}$ and $y = \frac{13}{50}$

So the radical centre is $\left(\frac{26}{25}, \frac{13}{50}\right)$.

Example 3:

Find the radical axis of the pairs of circles:

$$x^2 + y^2 = 144 \text{ and } x^2 + y^2 - 15x + 11y = 0.$$

Solution:

The equation of the circles are given as

$$x^2 + y^2 = 144 \quad ...(1)$$

and $\quad x^2 + y^2 - 15x + 11y = 0 \quad ...(2)$

Substracting (2) from (1), the equation of the radical axis is

$$15x - 11y = 144.$$

Example 4:

$$x^2 + y^2 - 3x - 4y + 5 = 0 \text{ and } 3x^2 + 3y^2 - 7x + 8y + 11 = 0.$$

Solution:

The equations of the circles are given as

$$x^2 + y^2 - 3x - 4y + 5 = 0 \quad ...(1)$$

and $\quad 3x^2 + 3y^2 - 7x + 8y + 11 = 0. \quad ...(2)$

To make the coeffs. of x^2 same, multiplying (1) by 3, we get

$$3x^2 + 3y^2 - 9x - 12y + 15 = 0. \quad ...(3)$$

Subtracting (3) from (2), the equation of the radical axis is

$$2x + 20y - 4 = 0 \text{ or } x + 10y = 2.$$

Example 5:

Find the radical axis of the pair of circle

$x^2 + y^2 - xy + 6x - 7y + 8 = 0$ and $x^2 + y^2 - xy - 4 = 0$, the axex being inclined at 120°.

Solution:

The equations of the circles are given as

$$x^2 + y^2 - xy + 6x - 7y + 8 = 0 \quad ...(1)$$

and $$x^2 + y^2 - xy - 4 = 0 \quad ...(2)$$

Subtracting (2) from (1), the equation of the radical axis is

$$6x - 7y + 12 = 0.$$ **Ans.**

Example 6:

Find the radical centre of the sets of circles.

$x^2 + y^2 + x + 2y + 3 = 0$, $x^2 + y^2 + 2x + 4y + 5 = 0$,

and $x^2 + y^2 - 7x - 8y - 9 = 0$.

Solution:

The equations of the circles are

$$x^2 + y^2 + x + 2y + 3 = 0 \quad ...(1)$$

$$x^2 + y^2 + 2x + 4y + 5 = 0 \quad ...(2)$$

and $$x^2 + y^2 - 7x - 8y - 9 = 0 \quad ...(3)$$

The radical axis of (2) and (1) is

$$(x^2 + y^2 + 2x + 4y + 5) - (x^2 + y^2 + x + 2y + 3) = 0$$

or $$x + 2y + 2 = 0. \quad ...(4)$$

The radical axis of (2) and (3) is

$$(x^2 + y^2 + 2x + 4y + 5) - (x^2 + y^2 - 7x - 8y - 9 = 0$$

or $$9x + 12y + 14 = 0. \quad ...(5)$$

Solving (4) and (5), $x = -2/3$ and $y = -2/3$.

Hence the radical centre is $\left(-\frac{2}{3}, -\frac{2}{3}\right)$.

Example 7:

From any point on the circle $x^2 + y^2 + 2gx + 2fy + c = 0$ tangents are drawn to the circle

$$x^2 + y^2 + 2gx + 2fy + c \sin^2 \alpha + (g^2 + f^2) \cos^2 \alpha = 0;$$

prove that the angle between them is 2a.

Solution:

The given circle is

$$x^2 + y^2 + 2gx + 2fy + c = 0 \qquad ...(1)$$

Say P ≡ (h, k) is any point on (1), then

$$h^2 + k^2 + 2gh + 2fk + c = 0 \qquad ...(2)$$

The other circle is

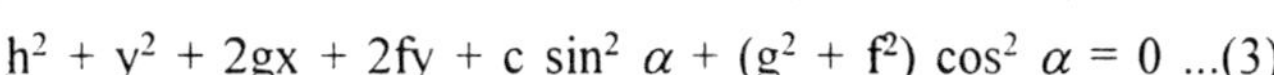

$$h^2 + y^2 + 2gx + 2fy + c \sin^2 \alpha + (g^2 + f^2) \cos^2 \alpha = 0 \ ...(3)$$

The radius of (3) is

$$CA = \sqrt{(g^2 + f^2} - \{c \sin^2\alpha + (g^2 + f^2) \cos^2 \alpha\}]$$

$$= \sqrt{[g^2} \ (1 - \cos^2\alpha) + f^2 (1 - \cos^2\alpha) - c \sin^2\alpha]$$

$$= \sqrt{(g^2} \sin^2\alpha + f^2 \sin^2\alpha - c \sin^2\alpha)$$

$$= \sin \alpha \ \sqrt{(g^2} + f^2 - c) \qquad ...(4)$$

Again

PA = length of the tangent drawn from the point P to the small circle given by (3)

$$\sqrt{\{h^2} + k^2 + 2gh + 2hg + 2fk + c \sin^2\alpha + (g^2 + f^2) \cos^2\alpha)$$

$$= \sqrt{\{h^2} + k^2 + 2gh + 2fk + c) - c \cos^2\alpha + (g^2 + f^2) \cos^2\alpha)$$

$$\{\because \sin^2\alpha = (1 - \cos^2\alpha)\}$$

$$= \sqrt{\{(g^2} + f^2 - c) \cos^2 \alpha\} = h^2 + k^2 + 2gh + 2fk + c = 0 \text{ by } (2)$$

$$= \cos \alpha \ \sqrt{(g^2} + f^2 - c).$$

Now ∠APB = 2∠APC

and $\qquad \tan APC = \dfrac{CA}{AP} = \dfrac{\sin \alpha \sqrt{(g^2 + f^2 - c)}}{\cos \alpha \left(g^2 + f^2 - a\right)} = \tan \alpha.$

$$= \angle APC = \alpha$$

So $\qquad \angle APB = 2 \angle APC = 2\alpha.$

Example 8:

Prove that the length of the common chord of the two circles equations are

$$(x - a)^2 + (y - b)^2 = c^2 \text{ and } (x - b)^2 + (y - a)^2 = c^2$$

is $\sqrt{\{4c^2} - 2(a - b)^2\}$.

Hence find the condition that the two circles may touch.

Solution:

Equations to the circles are given as

$$(x - a)^2 + (y - b)^2 = c^2$$

or $x^2 + y^2 - 2ax - 2by + a^2 + b^2 - c^2 = 0$ ...(1)

and $(x - b)^2 + (y - a)^2 = c^2$

or $x^2 + y^2 - 2bx - 2ay + a^2 + b^2 - c^2 = 0.$ ...(2)

Equation of the common chord is (1) – (2) = 0

or $2x(b - a) - 2y(b - a) = 0$ or $x - y = 0$ ...(3)

Let AB be a chord of a circle whose centre is C. If CN is perpendicular from the centre on the chord then AB = 2BN $= 2\sqrt{(CB^2 - CN^2)}$ or length of the chord

$= 2\sqrt{[(\text{radius})^2 - (\text{length of the perp. from the centre on the chord})^2]}$ .... (4)

Centre of (1) is (a, b) and its radius = c.

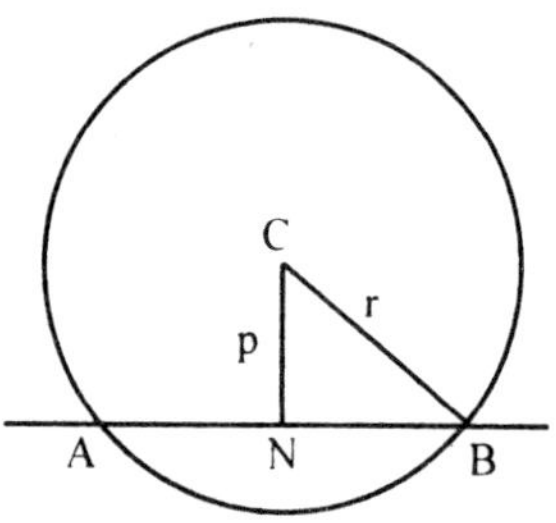

Perp. from (a, b) on the common chord (3) is $\frac{a - b}{\sqrt{(1+1)}}$.

Putting in (4), length of the chord

$$= 2\sqrt{[c^2 - \{(a-b)/\sqrt{2}\}^2]}$$

$$4\sqrt{[4c^2 - 2(a-b)^2]}.$$

If the circles touch other, the length of the common chord will be zero (as they simply touch each other); so

$$\sqrt{[4c^2 - 2(a-b)^2]} = 0 \text{ or } (a - b)^2 = 2c^2.$$

Example 9 :

A, B, C, and D are four points in a straight line; prove that the locus of a point P, such that the angles APB and CPD are equal, is a circle.

Solution :

Let A be the origin, the line along ABCD be the axis of x and the line perpendicular to it at A as y-axis. Say, P be any point satisfying the condition, having the co-ordinate as (h, k).

Do yourself.

$$m_1 = \frac{k}{h} \qquad \text{...(i)}$$

$$m_2 = \frac{k}{h-b} \qquad \text{...(ii)}$$

Example 10:

Find the equation to the circle passing through the points (0, a) and (b, h), and having its centre on the axis of x.

Solution:

As the centre C lies on x axis, let its co-ordinates be (g, 0). Again the circle passes through the points A ≡ (0, a) and B ≡ (b, h).

$\therefore \quad CA = CB \quad \text{or } CA^2 = CB^2$

Therefore $\quad [(g - 0)^2 + (0 - a)^2] = [(g - b)^2 + (0 - h^2)$

$\Rightarrow \quad g^2 + a^2 = g^2 + b^2 - 2bg + h^2$

$\Rightarrow \quad g = (h^2 + b^2 - a^2)/2b.$

The radius $CA = \sqrt{\left[\frac{(h^2+b^2-a^2)^2}{2b} + a^2\right]}$

Hence the equation of the circle is given as

$$x - \left(\frac{h^2+b^2-a^2}{2b}\right) + (y-0)^2 = \left(\frac{h^2+b^2-a^2}{2b}\right)^2 + a^2$$

$$\Rightarrow x^2 - 2x\frac{h^2+b^2-a^2}{2b} + \left(\frac{h^2+b^2-a^2}{2b}\right) + y^2 = \left(\frac{h^2+b^2-a^2}{2b}\right) + a^2$$

$$\Rightarrow x^2 - \frac{x}{b}\left(h^2+b^2-a^2\right) + y^2 = a^2$$

$\Rightarrow b\,(x^2 + y^2 - a^2) = x\,(h^2 + b^2 - a^2)$.

Example 11:

Find the centre and radius of the circle

$$\sqrt{(1+m^2)}\;(x^2 + y^2) - 2cx - 2mcy = 0.$$

Solution:

The equation of the circle is given as

$$\sqrt{(1+m^2)}\;(x^2 + y^2) - 2cx - 2mcy = 0.$$

$$\Rightarrow \qquad x^2 + y^2 - \frac{2c}{\sqrt{(1+m^2)}}\,x - \frac{2mc}{(1+m^2)}\,y = 0.$$

Half of the coefficients of x is $-\dfrac{c}{\sqrt{(1+m^2)}}$, half of the coefficient or

$$y \text{ is } -\frac{mc}{\sqrt{(1+m^2)}}.$$

Hence the centre is $\left(\dfrac{c}{\sqrt{(1+m^2)}}, \dfrac{mc}{\sqrt{(1+m^2)}}\right)$ and radius is

$$\sqrt{\left[\frac{c^2}{1+m^2} + \frac{m^2\,c^2}{1+m^2}\right]} = \sqrt{\left[\frac{c^2\,(1+m^2)}{1+m^2}\right]} = c$$

Example 12:

Find the equation to the circle whose radius is 3 and whose centre is (–1, 2).

Solution:

Centre is given as (– 1, 2) and radius is given as 3, hence the equation of the circle will be

$$[x - (-1)]^2 + (y - 2)^2 = (3)^2$$

$$\Rightarrow \qquad x^2 + y^2 + 2x - 4y = 4.$$

Example 13:

Find the equation of the circle. Whose radius is 10 and whose centre is (– 5, – 6).

Solution:

Centre is (–5, – 6) radius is 10. So the equation of the circle is

$$[x - (-5)]^2 + [y - (-6)]^2 = (10)^2$$

$$x^2 + 10x + y^2 + 12y = 39.$$

Example 14:

Find the equation of the circle. Whose radius is a + b and whose centre is (a, –b).

Solution:

Centre is (a, – b) and radius is (a + b). So equation of the circle is $(x - a)^2 + (y + b)^2 = (a + b)^2$

$$x^2 + y^2 - 2ax + 2by - 2ab = 0.$$

Example 15:

Find the equation of the circle. Whose radius is $\sqrt{(a^2 - b^2)}$ *and whose centre is (– a, – b).*

Solution:

Centre is (– a, – b) and radius $\sqrt{(a^2 - b^2)}$. Hence the equation of the circle is

$$(x + a)^2 + (y + b)^2 = \left[\sqrt{(a^2 - b^2)}\right]^2$$

or $$x^2 + y^2 + 2ax + 2by + 2b^2 = 0.$$

Example 16:

Find the centre and radius of the circle

$$3x^2 + 3y^2 - 5x - 6y + 4 = 0.$$

Solution:

The equation is

$$3x^2 + 3y^2 - 5x - 6y + 4 = 0$$

or $$x^2 + y^2 - 5/3\ x - 6/3\ y + 4/3 = 0$$

Half of the coefficient of x and y are – 5/6 and – 1.

Hence the centre will be (5/6, 1) and the radius will be

$$\sqrt{\left[\left(\frac{5}{6}\right)^2 + (1)^2 - \frac{4}{3}\right]} = \sqrt{\left(\frac{13}{36}\right)} = \frac{\sqrt{13}}{6}.$$

Example 17:

Find the centre and radius of the circle

$$x^2 + y^2 = k\ (x + k).$$

Solution:

$$x^2 + y^2 = k\ (x + k)$$

$$x^2 + y^2 - kx + 0.\ y - y\ k^2 = 0$$

Half of the coefficient of x is $-k/2$ and half of the coefficient of y is 0. So the centre is (k/2, 0) and radius is

$$\sqrt{\left[\left(\frac{k}{2}\right)^2 + 0 + k^2\right]} = \frac{k}{2}\sqrt{5}.$$

Example 18:

Find the centre and radius of the circle

$$x^2 + y^2 = 2gx - 2fy.$$

Solution:

The equation is given as

$$x^2 + y^2 = 2gx - 2fy$$

or $\quad x^2 + y^2 - 2gx + 2fy + 0.$

Example 19:

Find the centre and radius of the circle

$$x^2 + y^2 = 2ay.$$

center (g, f) radius $\sqrt{(g^2 + f^2)}$

Solution:

The equation of the circle is given as

$$x^2 + y^2 = 2ay$$

$\Rightarrow \quad x^2 + y^2 - 2ay = 0$

$\Rightarrow \quad x^2 + (y - a)^2 = a^2$

Hence the centre of the circle is (0, a) and radius = a.

Taking (0, a) as centre and a as radius draw the circle.

Example 20:

Find the centre and radius of the circle

$$3x^2 + 3y^2 = 4x.$$

Solution:

The equation is $3x^2 + 3y^2 = 4x$

$$x^2 + y^2 - \frac{4}{3}x = 0.$$

Hence the centre will be (2/3, 0) and radius will be 2/3.

Taking (2/3, 0) as centre and 2/3 as radius, draw the circle.

Example 21:

Find the centre and radius of the circle

$$5x^2 + 5y^2 = 2x + 3y.$$

Solution:

Equation is given as

$$5x^2 + 5y^2 = 2x + 3y$$

$$\Rightarrow \quad x^2 + y^2 - \frac{2}{5}x - \frac{3}{5}y = 0$$

Hence the centre of the circle is (1/5, 3/10) and radius is $\sqrt{(13)/10}$.

Taking (1/5, 3/10) as the centre and $\sqrt{(13)/10}$ as radius draw the circle.

Example 22:

Find the equation to the circle which passes through the points (1, –2) and (4, –3) and which has its centre on the straight line 3x + 4y = 7.

Solution:

Let the centre of the circle $C \equiv (h, k)$.

As the centre lines on the straight line $3x + 4y = 7$, its co-ordinates will satisfy $\quad 3x + 4y = 7$

Hence $\quad 3h + 4k = 7 \quad$ ...(1)

The circle passes through $A \equiv (1, -2)$ and $B \equiv (4, -3)$.

Hence $\quad CA = CB$ or $CA^2 = CB^2$

$$\Rightarrow \quad (h - 1)^2 + (k + 2)^2 = (h - 4)^2 + (k + 3)^2$$

$$\Rightarrow \quad 6h - 2k = 20$$

$$\Rightarrow \quad 3h - k = 10. \quad \text{...(2)}$$

Solving (1) and (2), we get $h = \frac{47}{15}, k = -\frac{3}{5}$.

So centre is $\left(\frac{47}{15}, -\frac{3}{5}\right)$.

$$\text{Radius} = CA = CA = \sqrt{\left\{\left(\frac{47}{15}-1\right)^2+\left(-\frac{3}{5}+2\right)^2\right\}} = \sqrt{\frac{1465}{225}}.$$

Therefore the equation of the circle is

$$\left(x-\frac{47}{15}\right)^2+\left(y+\frac{3}{5}\right)^2=\frac{1465}{225}$$

$$\Rightarrow \quad x^2-\frac{94}{15}x+\frac{2209}{225}+y^2+\frac{6}{5}y+\frac{9}{25}=\frac{1465}{225}$$

$$\Rightarrow \quad x^2-\frac{94}{15}x+y^2+\frac{6}{5}y-\left(\frac{1465}{225}-\frac{2209}{225}-\frac{9}{25}\right)=0$$

$$\Rightarrow \quad x^2-\frac{94}{15}x+y^2+\frac{6}{5}y+\frac{11}{3}=0$$

$$\Rightarrow \quad 15x^2+15y^2-94x+18y+55=0.$$

Example 23:

Find the equation of the circle which pass through (5, 7), (8, 1) and (1, 3).

Solution:

Let the equation of the circle be

$$x^2+y^2+2gx+2fy+c=0 \qquad ...(1)$$

The circle passes through (5, 7), (8, 1) and (1, 3).

Hence $5^2+7^2+2g.5+2.f.7=+c=0$

$$\Rightarrow \quad 10g+14f+c=-74 \qquad ...(2)$$

Again $(8)^2+(1)^2+2.g.8+2.f.1+c=0$

$$\Rightarrow \quad 16g+2f+c=-65. \qquad ...(3)$$

and $1^2+3^2+2.g.1+2.f.3+c=0$

$$\Rightarrow \quad 2g+6f+c=-10 \qquad ...(4)$$

Substracting (2) from (3) and (4) from (2), we get

$$6g-12f=9 \qquad ...(5)$$

and

$$8g+8f=-64. \qquad ...(6)$$

Solving (5) and (6), we get $g=-29/6$, $f=-19/6$.

Putting these values in (2), we get $c=56/3$.

Substituting the values of g, f and c in (1), we get

$$x^2 + y^2 + 2\left(-\frac{29}{6}\right)x + 2\left(-\frac{19}{6}\right)y + \frac{56}{3} - 0$$

or $\quad 3x^2 + 3y^2 - 29x - 19y + 56 = 0.$

Example 24:

Find the equation to the circle which goes through the origin and cuts off intercepts equal to h and k from the positive parts of the axes.

Solution:

As the circle passes through origin and cuts off an intercept of h from x-axis, it will pass through (h, 0), similarly if it cuts an intercept of k from y-axis and passes through origin, it will pass through (0, k).

Hence the circle passes through (0, 0), (h, 0) and (0, k).

Let its equation be $x^2 + y^2 + 2gx + 2fy + c = 0$. ...(1)

Substituting the co-ordinates of each point one by one in (1),

$0 + 0 + 0 + 0 + c = 0 \quad \Rightarrow \quad c = 0,$ ...(2)

$h^2 + 0 + 2gh + 0 + c = 0 \quad \Rightarrow \quad g = -\frac{h}{2},$ ...(3)

$(\because c = 0)$

$0 + k^2 = 0 + 2kf + c = 0 \quad \Rightarrow \quad f = -\frac{k}{2}.$ ...(4)

$(\because c = 0)$

Putting the values of e, g and f from (2), (3) and (4) in (1), we get

$x^2 + y^2 + 2gx + 2fy + c = 0$ ...(1)

The circle passes through (a, b), (a, –b), (a + b, a – b).

Hence $a^2 + b^2 + 2ga + 2fb + c = 0$ ...(2)

$a^2 + b^2 + 2ga - 2fb + c = 0$ ...(3)

$(a + b)^2 + (a - b)^2 + 2g(a + b) + 2f(a - b) + c = 0$

$\Rightarrow \quad 2a^2 + 2b^2 + 2g(a + b) + 2f(a - b) + c = 0$ ...(4)

Substracting (3) from (2), we have

$4fb = 0 \quad$ or $\quad f = 0$ ...(5)

Substracting (3) from (4), we have

$a^2 + b^2 + 2gb + 2fa = 0.$ ...(6)

Substituting the value of f from (5), $g = -(a^2 + b^2)/2b$.

Putting the values of g and f in (i)

$$c = \frac{(a^2 + b^2)(a - b)}{b}$$

Substituting the values of g.f and c in (i), the required equation will be

$$x^2 + y^2 - 2\frac{(a^2 + b^2)}{2b}x + 2.0\,y + \frac{(a^2 + b^2)(a - b)}{b} = 0$$

$\Rightarrow$ $bx^2 + by^2 - (a^2 + b^2)\,x + (a^2 + b^2)(a - b) = 0$

Example 25:

Find the equations to the circles which pass through the points (0, 0), (a, 0), and (0, b).

Solution:

Let the reqd. equation be

$$x^2 + y^2 + 2gx + 2fy + c = 0 \quad ...(1)$$

As the circle passes through (0, 0), (a, 0) and (0, b) each point will satisfy (1), Hence by (0, 0),

$$0 + 0 + 2g.0 + 2f.0 + c = 0 \quad \text{or} \quad c = 0 \quad ...(2)$$

By (a, 0), $a^2 + 0 + 2g.a + 2f.0 + c = 0$ or $g = -a/2$ ...(3)

(as $a \neq 0$ and $c = 0$)

By (0, b); $0 + b^2 + 2g.0 + 2f.b + c = 0$ or $f = -b/2$ ...(4)

(as $b \neq 0$ and $c = 0$).

Putting the values from (2), and (3) and (4) in (1), we get

$$x^2 + y^2 - ax - by = 0.$$

Example 26:

Find the equations to the circles which pass through the points. (1, 2), (3, – 4), and (5, –6).

Solution:

Let the equation be $x^2 + y^2 + 2gx + 2fy + c = 0$ ...(1)

The points are given as (1, 2), (3, –4), (5, –6)

As we know that

By (1, 2)' $1^2 + 2^2 + 2g.1 + 2f.2 + c = 0$

$\Rightarrow$ $2g + 4f + c = -5$ ...(2)

By (3, –4); $3^2 + (-4)^2 + 2g.3 + 2f(-4) + c = 0$

$\Rightarrow$ $6g - 8f + c = -25.$ ...(3)

By (5, –6); $5^2 + (-6)^2 + 2g.5 + 2f(-6) + c = 0$

$\Rightarrow \quad 10g - 12f + c = -61$ ...(4)

Subtracting (3) from (2), $-4g + 12f = 20$ ...(5)

Subtracting (4) from (3), $-4g + 4f = 36$ ...(6)

Subtracting (6) from (5), we get $8f = -16$ or $f = -2$.

Subtracting *this value* in (6), we get $c = 25$.

Putting these values of f, g, and c in (1), we get

$$x^2 + y^2 - 22x - 4y + 25 = 0.$$

Example 27:

Find the equation of the circle which pass through the point

(1, 1), (2, –1) and (3, 2).

Solution:

Let the equation of the circle be

$$x^2 + y^2 + 2gx + 2fy + c = 0 \quad ...(1)$$

The points are (1, 1), (2, –1) and (3, 2) lies on it, we have

$1^2 + 1^2 + 2g.1 + 2f.1 + c = 0$ or $2g + 2f + c = -2$ ...(2)

and $\quad 2^2 + (-1)^2 + 2g.2 + 2f(-) + c = 0$

or $\quad 4g - 2f + c = -5.$ ...(3)

and $\quad 3^2 + 2^2 + 2g.3 + 2f.2 + c = 0$ or $6g + 4f + c = -13$ ..(4)

Subtracting (3) from (2), we get $-2g + 4f = 3$. ...(5)

Subtracting (4) from (3), we get $-2g - 6f = 8$ ...(6)

Subtracting (6) from (5), we get $10f = -5$ or $f = -1/2$

Putting the values of g and f in (2), we get $c = 4$

Subtracting these values in (1) the reqd. equation is

$$x^2 + y^2 - 5x - y + 4 = 0.$$

Example 28:

ABC is square whose side is a; taking AB and AD as axes, prove that the equation to the circumscribing the square is

$$x^2 + y^2 = a(x + y).$$

Solution:

Taking AB and AD as axes, the co-ordinates of A, B, C and D respectively (0, 0), (a, 0), (a, a) and (0, a) as the side of the square is a by symmetry, the circle passing through A, B and D will also pass C.

Suppose that equation of the circle be

$$x^2 + y^2 + 2gx + 2fy + c = 0 \quad ...(1)$$

(0, 0) lies on it, hence

$$0 + 0 + 2g.0 + 2f.0 + c = 0 \quad \text{or} \quad c = 0 \quad ...(2)$$

(a, 0) lies on it, hence

$$a^2 + 0 + 2ga + 2f.0 + c = 0 \quad \text{or} \quad 2g = -a \quad ...(3)$$

$$(\because c = 0)$$

(0, a) lies on it, hence

$$0, a^3 + 2g.0 + 2f.a + 0 \quad \text{or} \quad 2f = -a \quad ...(4)$$

$$(\because c = 0)$$

Putting the values of g.f and c from (2), (3) and (4) in (1), we get

$$x^2 + y^2 - ax - ay + 0 = 0 \quad \text{or} \quad x^2 + y^2 = a(x + y).$$

Example 29:

Find the equation to the circle of radius a, which passes through the two points on the axis of x which are at a distance b from the origin.

Solution:

If the circle passes through the points on the x-axis at a distance of b from origin; say B and C, then their co-ordinates will be (–b, 0) and (b, 0) respectively.

By symmetry, it is clear the centre will be on the right bisector of BC, *i.e.* on y-axis. Let the co-ordinates of the centre C be (0, k).

Radius = BC; so $(0 + b)^2 + (k - 0)^2 = a^2$ (given)

then $\quad k^2 = a^2 - b^2$ or $k = \pm \sqrt{(a^2 - b^2)}$.

Hence the circles are

$$(x - 0)^2 + \left\{y \pm \sqrt{(a^2 - b^2)}\right\}^2 = a^2$$

$$\Rightarrow x^2 + y^2 \pm 2y\sqrt{[(a^2 - b^2)]} + a^2 - b^2 = a^2$$

$$\Rightarrow x^2 + y^2 \pm 2y\sqrt{[(a^2 - b^2)]} = b^2.$$

Example 30:

Find the equation to the circle passing through the origin and the points (a, b) and (b, a). Find the lengths of the chords that it cuts off from the axes.

Solution:

Let the equation of the circle passing through the origin (0, 0) and the points (a, b) and (b, a) be

$$x^2 + y^2 + 2gx + 2fy + c = 0 \qquad ...(1)$$

Putting the co-ordinates of each point in (1) one by one, we get

$$0 + 0 + 0 + 0 + c = 0 \quad \text{or } c = 0, \qquad ...(2)$$

$$a^2 + b^2 + 2ag + 2fb + c = 0, \qquad ...(3)$$

$$b^2 + a^2 + 2bg + 2ab + c = 0 \qquad ...(4)$$

Putting the value of c in (3) and (4) and substracting, we get

$$2g(a - b) + 2f(b - a) = 0 \qquad \text{or } g - f = 0 \qquad ...(5)$$

Putting g = f and c = 0 in (3), we get

$$a^2 + b^2 + 2af + 2b\,f = 0 \quad \text{or } 2f = \frac{a^2 + b^2}{a + b}$$

Similarly putting f = –g, we get $2g = -\frac{a^2 + b^2}{a + b}$

Substituting the values of 2g. 2f and c in (1), we get

$$x^2 + y^2 - \left(\frac{a^2 + b^2}{a + b}\right)x + \left(\frac{a^2 + b^2}{a + b}\right)y + 0 = 0$$

$$\Rightarrow x^2 + y^2 - \left(\frac{a^2 + b^2}{a + b}\right)x + y = 0 \qquad ...(6)$$

To get the intercept on x-axis, solve this equation with x-axis *i.e.*, y = 0. So *putting* y = 0 in (6), we get

$$x^2 - \frac{a^2 + b^2}{a + b}x = 0, \qquad \therefore x = 0 \text{ or } x = \frac{a^2 + b^2}{a + b}$$

Hence the intercept on x-axis is $\frac{a^2 + b^2}{a + b}$

Similarly on y-axis, the intercept will be $\frac{a^2 + b^2}{a + b}$.

Example 31:

Find the equation to the circle which passes through the origin and cuts off intercepts equal to 2 and 4 from the axes.

Solution:

As the circle passes through origin and cuts off the intercepts of 3 and 4 from axes of x and y respectively, the co-ordinates of the points of intersection with axes of x and y will be (3, 0) and (0, 4). So the circle passes through (0, 0), (3, 0) and (0, 4).

Let the equation of the circle be

$$x^2 + y^2 + 2gx + 2fy + c = 0 \qquad ...(1)$$

As (0, 0) lies on it, so

$$0 + 0 + 2g.0 + 2f.0 + c = 0 \qquad \text{or } c = 0 \qquad ...(2)$$

(3, 0) lies on it, so

$$9 + 0 + 2g.3 + 2f.0 + c = 0 \text{ or } 6g + c + 9 = 0 \qquad ...(3)$$

(0, 4) lies on it, so

$$0 + 16 + 2g.0 + 2f.4 + c = 0 \qquad \text{or } 8f + c + 16 = 0 \qquad ...(4)$$

From (3) and (4), we get $g = -3/2$ and $f = -2$ as $c = 0$.

Putting the values of c, f and g in (1), we get

$$x^2 + y^2 - 3x - 4y = 0.$$

Example 32:

Touches the axis of x and passes through the two points (1, –2) and (3, –4).

Solution:

If the circle touches axis of x, radius will be equal to the ordinate of the centre. Let the centre. Let the centre be (h, k) so radius will be k.

Distance of the point (1 – 2) from centre (h, k) must be equal to radius, so

$$(h - 1)^2 + (k + 2)^2 = k^2$$

or

$$h^2 - 2h + 4k + 5 = 0. \qquad ...(1)$$

Similarly, the distance of the point (3, –4) is equal to the radius, hence $(h - 3)^2 + k + 4)^2 = k^2$

or

$$h^2 - 6h + 8k + 25 = 0 \qquad ...(2)$$

Multiplying (1) by (2) and substracting (2) from it

$$h^2 + 2h - 15 = 0 \text{ or } (h + 5)(h - 3) = 0$$

Solving, we get either $h = -5$ or $h = +3$.

Putting $h = -5$ in (5), we get

$$25 + 10 + 4k + 5 = 0 \text{ or } k = -10.$$

If h = 3, putting h = 3 in (1), we get

$$9 - 6 + 4k + 5 = 0 \text{ or } k = -2.$$

Whence the centre is either (–5, – 10) or (3, – 2) and the corresponding radius is 10 or 2 units.

Therefore the equation of the circle will be either

$$(x + 5)^2 + (y + 10)^2 = (10)^2 \text{ or } x^2 + y^2 + 10x + 20y + 25 = 0$$

$$(x - 3)^2 + (y + 2)^2 = (2)^2 \text{ or } x^2 + y^2 - 6x + 4y + 9 = 0.$$

Example 33:

Touches both axes and passes through the point (–2, – 3).

Solution:

Let the radius of the circle be a. As it touches both the axes, the centre may be (± a, ± a).

Again the circle passes through (–2, –3), so it lies in the 3rd quadrant, hence the centre will be (–a, –a), and its distance from (–2, –3) will be equal to radius a. So

$$(-a + 2)^2 + (-a + 3)^2 = a^2$$

or $$a^2 + 4 - 4a + a^2 - 6a + 9 = a^2$$

or $$a^3 - 10a + 13 = 0 \quad \text{or} \quad a = \frac{10 \pm \sqrt{(100 - 52)}}{2}$$

or $$a = 5 \pm \sqrt{12}$$

Therefore the required equation is

$$\{x + (5 \pm \sqrt{12})\}^2 + \{y + (5 \pm \sqrt{12})\}^2 = \{5 \pm \sqrt{12}\}^2$$

or $$x^2 + y^2 + 2(5 \pm \sqrt{12})(x + y) = 37 \pm 19\sqrt{12}.$$

Example 34:

Find the equations to the circles in which the line joining the points, (a, b) and (b, –a) is a chord subtending an angle of 45° at any point on its circumference.

Solution:

In the Fig. of last question, let the co-ordinates of B and C be respectively (a, b) and (b, –a) and the co-ordinates of A be (h, k).

Slope of AB is $\dfrac{k - b}{h - a} = m_1$ (say)

Slope of AC is $\frac{k+b}{h-b} = m_2$ (say).

An angle ABC = 45°, we have

$$\tan 45^\circ = \pm \frac{m_1 - m_2}{1 + m_1 m_2} = \pm \frac{\frac{k-b}{h-a} - \frac{k+a}{h-b}}{1 + \frac{k-b}{h-a} \cdot \frac{k+a}{h-b}}$$

$$\Rightarrow 1 = \pm \frac{(k-b)(h-b) - (k+a)(h-a)}{(h-a)(h-b) + (k-b)(k+a)}$$

or $(h-a)(h-b) + (k-b)(k+a) = \pm [(k-b)(h-b) - (k+a)(h-a)]$

or $h^2 + k^2 - h(a+b) + k(a-b) = \pm [(k(a-b) - h(a+b) + b^2 + a^2]$

Taking +ve sign, we get

$$h^2 + k^2 = a^2 + b^2.$$

Generalising, we get

$$x^2 + y^2 = a^2 + b^2 \quad ...(1)$$

Taking – ve sign, we get

$$h^2 + k^2 - 2h(a+b) + 2k(a-b) + a^2 + b = 0$$

Generalising

$$x^2 + y^2 - 2x(a+b) + 2y(a-b) + a^2 + b^2 = 0. \quad ...(2)$$

(1) and (2) gives us the required equations.

Example 35:

Points (1, 0) and (2, 0) are taken on the axis of x, the axes being rectangular. On the line joining these points an equilateral triangle is described, its vertex being in the positive quadrant, find the equations to the circles described on its sides as diameters.

Solution:

Let the given points (1, 0) and (2, 0) be B and C respectively.

If M be the mid-point of BC its co-ordinates will be (3/2, 0).

Clearly the vertex of the equilateral triangle on BC will be on the perpendicular at M.

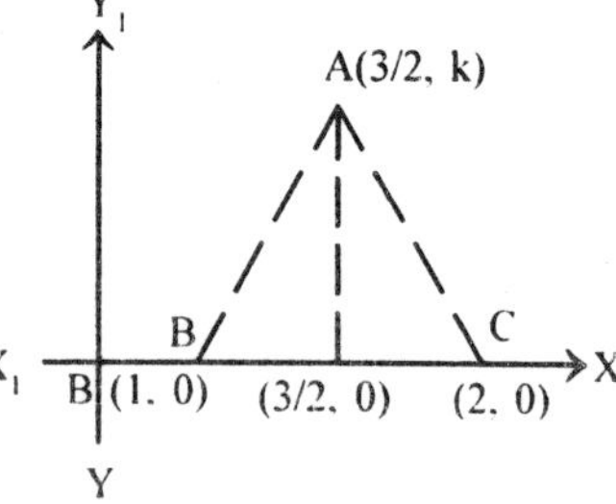

Let A be the third vertex. Its co-ordinates may be taken as (3/2, k).

Again as ABC is an equilateral triangle AB = BC = 1.

Hence

$$AB^2 = \left(\frac{3}{2} - 1\right)^2 + (k-0)^2 = (1)^2$$

$$\Rightarrow \qquad \frac{1}{4} + k^2 = 1 \text{ or } k = \pm\frac{\sqrt{3}}{2}.$$

As the vertex is in the first quadrant, the co-ordinates of A will be $\left(\frac{3}{2}, \frac{\sqrt{3}}{2}\right)$, neglecting the negative sign.

Now the co-ordinates A are $\left(\frac{3}{2}, \frac{\sqrt{3}}{2}\right)$, and of B are (1, 0).

Hence the equation of the circle drawn taking AB as the diameter will be

$$\left(x - \frac{3}{2}\right)(x-1) + \left(y - \frac{\sqrt{3}}{2}\right)(y-0) = 0$$

$$\Rightarrow \qquad (2x-3)(x-1) + (2y - \sqrt{3})(y) = 0$$

$$\Rightarrow \qquad 2x^2 + 2y^2 - 5x - \sqrt{3y} + 3 = 0.$$

Similarly the circle drawn as AC diameter will be

$$\left(x - \frac{3}{2}\right)(x-1) + \left(y - \frac{\sqrt{3}}{2}\right)(y-0) = 0$$

$$\Rightarrow \qquad (2x-3)(x-1) + (2y - \sqrt{3})\, y = 0$$

$$\Rightarrow \qquad 2x^2 + 2y^2 - 5x - \sqrt{3y} + 6 = 0.$$

And the circle described taking BC as diameter will be

$$(x-1)(x-2) + (y-0)(y-0) = 0$$

$$\Rightarrow \qquad x^2 + y^2 - 3x + 2 = 0.$$

Example 36:

If $y = mx$ be the equation of a chord of a circle whose radius is a, the origin of coordinates being one extremity of the chord and the axis of x being a diameter of the circle, prove that the equation of a circle of which this chord is the diameter is

$$(1 + m^2)(x^2 + y^2) - 2a(x + my) = 0.$$

Solution:

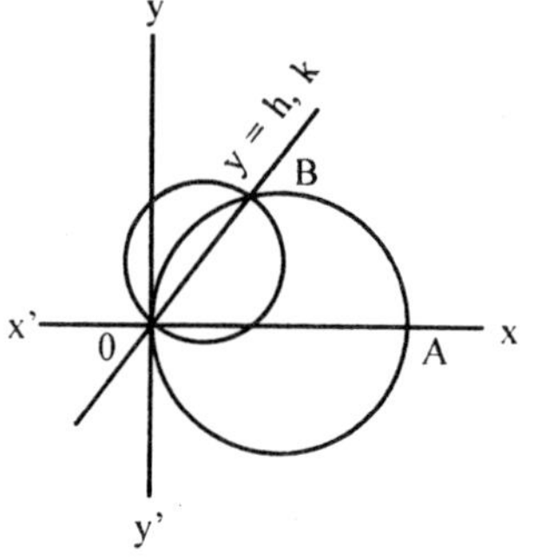

Equation of the circle passing through origin (0, 0) and of radius a, having its centre on the x-axis can be written as

$(x - a)^2 + (y - 0)^2 = a^2$

[as centre will be (a, 0) and radius = a]

or $x^2 + y^2 - 2ax = 0.$ ...(1)

Let the line $y = mx$ ...(2) cut the circle (i) at O and B.

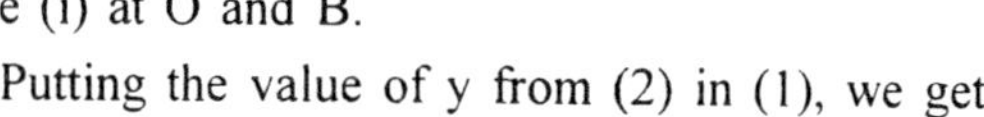

Putting the value of y from (2) in (1), we get

$$x^2 + m^2x^2 - 2ax = 0$$

or $$x^2 (1 + m^2) - 2ax = 0$$

or $$x [x (1 + m^2) - 2a] = 0$$

$$\therefore \quad x = 0 \quad \text{or} \quad x = \frac{2a}{1+m^2}$$

Putting these values of x in (ii)

if $x = 0,\ y = 0$. If $x = \dfrac{2a}{1+m^2},\ y = \dfrac{2am}{1+m^2}$.

Hence the point of intersection are O = (0, 0)

and $$B \equiv \left(\frac{2a}{1+m^2}, \frac{2am}{1+m^2}\right).$$

The equation of the circle drawn on OB as the diameter is

$$(x - 0)\left(x - \frac{2a}{1+m^2}\right) + (y - 0)\left(y - \frac{2am}{1+m^2}\right) = 0$$

$$\left(x^2 + y^2\right) - \frac{2a}{1+m^2}(x + my) = 0$$

$$(1 + m^2)(x^2 + y^2) - 2a(x + my) = 0$$

Example 37:

Touches the axis of x at a distance 3 from the origin and intercepts a distance 6 on the axis of y.

Solution:

Let the centre of the circle be (h, k). As it touches the x-axis at a distance of 3 from the origin, clearly

$$h = 3 \qquad ...(1)$$

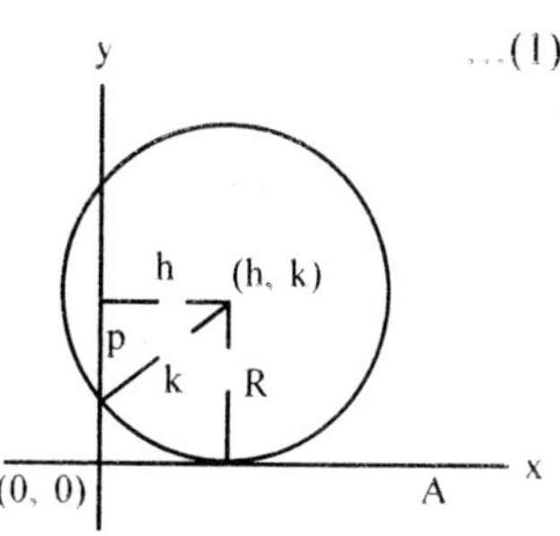

If l is the radius of the circle, clearly

$l^2 + h^2 +$ (half of the intercept on y-axis)$^2 = 3^2 + 3^2 = 18$

or $l = \pm\ 3\sqrt{2}$. ...(2)

Again radius will be equal to k, so $l = k = 3\sqrt{2}$.

Therefore the co-ordinates of the centre become $(3, \pm\ 3\sqrt{2})$ and radius as $3\sqrt{2}$ units.

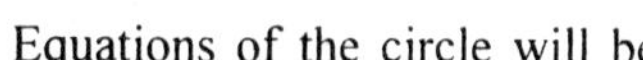

Equations of the circle will be

$$(x - 3)^2 + (y \pm 3\sqrt{2})^2 = (3\sqrt{2})^2$$

or $$x^2 + y^2 - 6x \pm 6\sqrt{2y} + 9 = 0.$$

xample 38:

Prove that the square of the tangent that can be drawn from any point on one circle to another circle is equal to twice the product of the perpendicular distance of the point from the radical axis of the two circles, and the distance between their centres.

Solution:

Let one circle be $\quad x^2 + y^2 = a^2 \qquad ...(1)$

and the other be $\quad x^2 + y^2 + 2gx + 2fy + c = 0. \qquad ...(2)$

If there be any point P whose co-ordinates are (h, k) on (1), then

$$h^2 + k^2 = a^2 \qquad ...(3)$$

If PT is the tangent from P on (2), then

$$PT^2 = h^2 + k^2 + 2gh + 2fk + c$$

$$= a^2 + 2gh + 2fk + c \text{[by (3)]} \qquad ...(4)$$

Again the equation of the radical axis of (1) and (2) is

$$(x^2 + y^2 + 2gx + 2fy + c) - (x^2 + y^2 - a^2) = 0$$

or $$2gx + 2fy + c + a^2 = 0 \qquad ...(5)$$

If PN be the perpendicular from P = (h, k) on the radical axis, then

$$PN = \frac{2gh + 2fg + c + a^2}{\sqrt{\{(2f)^2 + (2g)^2\}}} = \frac{2gh + 2fg + c + a^2}{2\sqrt{(g^2 + f^2)}} \quad ...(6)$$

Putting the values in (4) of last question, the length of the common chord

$$= 2\sqrt{\left[(3a)^2 - \left(\frac{19a}{\sqrt{65}}\right)^2\right]} = 2\sqrt{\left[9a^2 - \frac{361\,a^2}{65}\right]}$$

$$= 2\sqrt{\left[\frac{224}{65}\,.\,a^2\right]} = 8a\sqrt{\frac{14}{65}}\,.$$

Centre of (1) is (a, 2a) and its radius is

$$\sqrt{(a^2 + 4a^2 + 4a^2)} = 3a$$

Centre of (2) is (3a/2, – 2a) and (3a/2, – 2a) in the ratio of 3a and 5a/2 or 6: 5 externally,

$$x = \frac{9a = 5a}{6 - 5},\ y = \frac{-12a - 10a}{6 - 5},$$

i.e. (4a, – 22a) is

$$y + 22a = m\,(x - 4a). \quad ...(4)$$

If it is tangent to (1), the length of perpendicular from its centre will be equal to its radius; so

$$\frac{ma - 4am - 2a - 22a}{\sqrt{(1 + m^2)}} = 3a$$

$$\Rightarrow \quad -3m - 24 = 3\ 3\sqrt{(1 + m^2)}$$

$$\Rightarrow \quad -(m + 8) = \sqrt{(1 + m^2)}\,.$$

Squaring, we get $m^2 + 64 + 16m = 1 + m^2$

$$\Rightarrow \quad 0.m^2 + 16m + 63 = 0$$

As it is a quadratic in m^2, and the coeff. of m^2 is 0, so one root must be ¥, *i.e.* 1/m = 0 and other root is clearly – 63/16

Putting 1/m = 0 in (4), the equation becomes x – 4a = 0

Putting m = – 63/16 in (4), the equation becomes

$$y = 22a = -\frac{63}{16}\,(x - 4a)$$

$\Rightarrow \qquad 63x + 16y + 100a = 0.$

Hence the tangents are

$$x = 4a \text{ and } 63x + 16y + 100\,a = 0.$$

Again, the length of tangent from (4a, – 22a) upto (1)

$$= \sqrt{\{(4a)^2 + (-22a)^2 - a.\,4a - 4a\,(-22a) - 4a^2\}} = 24a$$

and the length of the tangent from (4a, – 22a) upto (2) is

∴ the length of the common tangent

$$= 24a - 20a = 4a.$$

Example 39:

Find the length of the common chord of the circles

$$x^2 + y^2 - 2ax - 4ay - 4a^2 = 0 \text{ and } x^2 + y^2 - 3ax + 4ay = 0.$$

Find also the equations of the common tangents and show that length of each is 4a.

Solution:

The circle are given as

$$x^2 + y^2 - 2ax - 4ay - 4a^2 \qquad ...(1)$$

and $$x^2 + y^2 - 3ax + 4ay = 0 \qquad ...(2)$$

The common chord is (1) – (2) = 0 ...(3)

$$ax - 8ya - 4a^2 = 0 \text{ or } \quad x - 8y - 4a = 0$$

Centre of (1) is (a. 2a) and radius of (1) is

$$\sqrt{(a^2 + 4a^2 + 4a^2)} = 3a.$$

Length of the perpendicular from the centre (a. 2a) upon the common chord (3)

$$= \frac{a - 8\,.\,2a - 4a}{\sqrt{[(1)^2 + (8)^2]}} = \frac{19a}{\sqrt{65}}$$

If (1) and (3) cut orthogonally, then

$$2g\,(-1) + 2f\,(-1) = 0 - 7.$$

Putting the value of g from (4) and solving, f = 29/6.

Substituting in (3), we get

$$x^2 + y^2 + 2\,.\left(-\frac{4}{3}\right)x + 2\,.\,\frac{29}{6}\,y = 0$$

or $$3x^2 + 3y^2 - 8x + 29y = 0.$$

Example 40:

Prove that the equation to the circle of which the points (x_1, y_1) and (x_2, y_2) are the ends of a chord of a segment containing an angle θ is

$$(x - x_1)(x - x_2) + (y - y_1)(y - y_2) \pm \cot\theta\,[(x - x_1)(y - y_2) - (x - x_2)(y - y_1)] = 0.$$

Solution:

Let the point B and C be respectively (x_1, y_1) and (x_2, y_2) and the co-ordinates of the moving point A be (h, k).

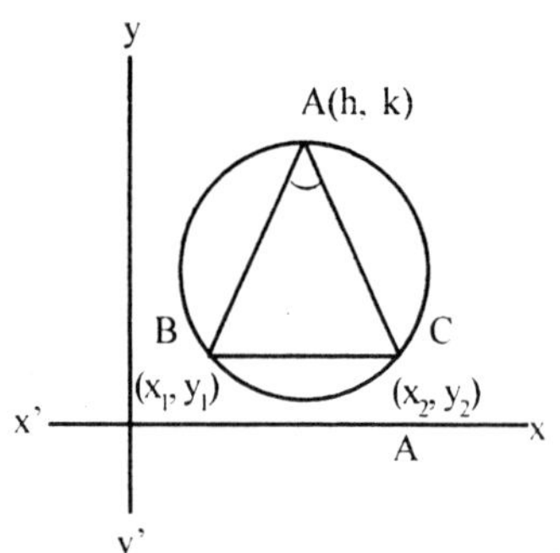

Slope AB and AC are respectively

$\dfrac{k-y_1}{h-x_1}$ and $\dfrac{k-y_2}{h-x_2}$ say m_1 and m_2.

If θ be the angle between AB and AC, then

$$\tan\theta = \pm\left(\frac{m_1 - m_2}{1 + m_1 m_2}\right) = \pm\frac{\dfrac{k-y_1}{h-x_1} - \dfrac{k-y_2}{h-x_2}}{1 + \dfrac{k-y_1}{h-x_1}\cdot\dfrac{k-y_2}{h-x_2}}$$

or $$\frac{1}{\cot\theta} = \pm\left[\frac{(k-y_1)(h-x_2) - (k-y_2)(h-x_1)}{(h-x_1)(h-x_2) + (k-y_1)(k-y_2)}\right]$$

or $$(h - x_1)(h - x_2) + (k - y_1) = \pm\cot\theta\,[(k - y_1)(h - x_2) - (k - y_2)(h - x_1)].$$

Generalising, we get

$$(x - x_1)(x - x_2) + (y - y_1)(y - y_2) = \pm\cot\theta\,[(y - y_1)(x - x_2) - (y - y_2)(x - x_1)].$$

Example 41:

Find the equation to the circle circumscribing the quadrilateral formed by the straight lines

$$2x + 3y = 2,\ 3x - 2y = 4,$$
$$x + 2y = 3 \text{ and } 2x - y = 3.$$

Solution:

The equations are given as $2x + 3y = 2$ ...(1)

$3x - 2y = 4.$ ...(2)

$x + 2y = 3$ ...(3)

and $2x - y = 3$ ...(4)

Let these lines represent the line AB, BC, CD and DA of a quadrilateral ABCD.

Solving (1) and (2), the co-ordinates of B are $(16/13, -2/13)$.

Solving (2) and (3), the co-ordinates of C are $(7/4, 5/8)$.

Solving (3) and (4), the co-ordinates of D are $(9/5, 3/5)$.

Solving (4) and (1), the co-ordinates of A are $(11/8, -1/4)$.

Let the equation of the circle passing through A, B and C be

$$x^2 + y^2 + 2gx + 2fy + c = 0 \quad \text{...(5)}$$

As A, B and C lies on the circle, the co-ordinates will satisfy (5) hence we get

$$\left(\frac{11}{8}\right)^2 + \left(-\frac{1}{4}\right)^2 + 2.g\left(\frac{11}{8}\right) + 2.f.\left(-\frac{1}{4}\right) + c = 0$$

$$\Rightarrow \quad \frac{121}{64} + \frac{1}{16} + \frac{11g}{4} - \frac{1}{2}f + c = 0 \quad \text{...(6)}$$

Again by B: $\left(\frac{16}{13}\right)^2 + \left(-\frac{2}{13}\right)^2 + 2g.\frac{16}{13} + 2f\left(-\frac{2}{13}\right) + c = 0$

of $$\frac{256}{169} + \frac{4}{169} + \frac{32}{13}g - \frac{4}{13}f + c = 0 \quad \text{...(7)}$$

and by C; $\left(\frac{7}{4}\right)^2 + \left(\frac{5}{8}\right)^2 + 2g\left(\frac{7}{4}\right) + 2f\left(\frac{5}{8}\right) + c = 0$

Solving (6), (7) and (8), we get

$$g = -\frac{25}{16} f = -\frac{3}{16} \text{ and } c = \frac{9}{4}$$

Substituting these values in (5); we get

$$x^2 + y^2 + 2x\left(-\frac{25}{16}\right) + 2.y\left(-\frac{3}{16}\right) + \frac{9}{4} = 0, \quad \text{...(8)}$$

or $8x^2 + 8y^2 - 25x - 3y + 18 = 0.$ ...(9)

This is the required equation. Substituting the co-ordinates of the point D on the L.H.S. of (9), we get

$$8\left(\frac{9}{5}\right)^2 + 8\left(\frac{3}{5}\right)^2 - 25\left(\frac{9}{5}\right) - 3\left(\frac{3}{5}\right) + 18 = 0$$

Hence the point lies on the circle, *i.e.* the circle passes through ABCD is given by (9).

Example 42:

Find the equation to the circle passing through the points (12, 43), (18, 39), and (42, 3) and prove that it also passes through the points (–54, – 69) and (–81, – 38).

Solution:

Let the equation of the circle is

$$x^2 + y^2 + 2gx + 2fy + c = 0 \quad ...(1)$$

As the circle passes through the points (12, 43), (18, 39) and (42, 3), these co-ordinates will satisfy (1). Substituting these values one by one, we get

$$(12)^2 + (43)^2 + 2g.12 + 2f.43 + c = 0$$

$\Rightarrow 144 + 1849 + 24g + 86f + c = 0$ or $24g + 86f + c = -1993.$ ...(2)

Similarly by (18, 39), we get

$$18^2 + 39^2 + 2g.18 + 2f.39 + c = 0$$

$\Rightarrow \quad 324 + 1521 + 36g + 78f + c = 0$

$\Rightarrow \quad 36g + 78f + c = -1845$ ...(3)

Similarly by (42, 3), we get

$$42^2 + 3^2 + 2.g.42 + 2.f.3 + c = 0$$

$\Rightarrow 1764 + 9 + 84g + 6f + c = 0$ or $84g + 6f + c = -1773$ ...(4)

Subtracting (2) from (3) and (3) from (4) respectively, we get

$12g - 8f = 148$ or $3g - 2f = 37$ ...(5)

and $48g - 72f = 72$ or $2g - 3f = 3$ ...(6)

Adding (5) and (6), we get

$5g - 5f = 40$ or $g - f = 8$ ...(7)

Subtracting (6) from (5), we get

$g + f = 34$ ...(8)

By addition and substracting of (7) and (8), we get

$$2g = 42 \qquad \text{and } 2f = 26$$

Substituting the values of 2g, 2f and c in (1), we get the required equation as

$$x^2 + y^2 + 42y + 26y - 3615 = 0$$

$$\Rightarrow \quad (x + 21)^2 - 441 + (y + 13)^2 - 169 - 3615 = 0$$

$$\Rightarrow \quad (x + 21)^2 + (y + 13)^2 = 4225$$

$$\Rightarrow \quad (x + 21)^2 + (y + 13)^2 = (65)^2.$$

Substituting the co-ordinates (–54, – 69) is equation we get

$$\text{L.H.S.} = (-54 + 21)^2 + (-69 + 13)^2$$

$$= (-33)^2 + (-56)^2 = 4225 = \text{R.H.S.}$$

Hence the point is on the circle.

Again, substituting the point (–81, – 38), we get

$$(-81 + 21)^2 + (-38 + 13)^2 = (60)^2 + (25)^2 = 4225 = \text{R.H.S.}$$

Hence this point also lies on the circle.

Example 43:

Find the equations of the common tangents of the circles

(i) $x^2 + y^2 - 2x - 6y + 9 = 0$ *and* $x^2 + y^2 + 6x - 2y + 1 = 0$,

(ii) $x^2 + y^2 = c^2$ *and* $(x - a)^2 + y^2 = b^2$.

Solution:

(i) The circles are given as

$$x^2 + y^2 - 2x - 6y + 9 = 0 \qquad ...(1)$$

and

$$x^2 + y^2 + 6x - 2y + 1 = 0 \qquad ...(2)$$

There centre of (1) is (1, 3) and radius is $\sqrt{(1+9-9)} = 1$.

The centre of (2) is (– 3, 1) and radius is 3.

Dividing the join of (1, 3) and (–3, 1) in the ratio of 1: 3 externally,

$$x = \frac{(1 \times -3) - (3 \times 1)}{1-3} \text{ and } y = \frac{(1 \times 1) - (3 \times 3)}{1-3} \text{ or } (3, 4)$$

Any line passing through (3, 4) is $y - 4 = m(x - 3)$ ...(3)

If (3) is the tangent to (1),

$$\frac{n(1-3)-(3-4)}{\sqrt{(1+m^2)}}=1 \quad \text{or } -2m+1=\sqrt{(1+m^2)}$$

Squaring both sides,

$$4m^2+1-4m+1+m^2$$

$\Rightarrow$ $3m^2-4m=0,$

whence, $m = 0$ or $m = 4/3$.

Putting in (3), we get $y - 4 = 0$ or $y = 4$

and $y-4=-\frac{4}{3}(x-3)$ or $3y + 4x = 0.$ ...(4)

Again, the co-ordinate of the point dividing the join of (1, 3) and (–3, 1) in the ratio of 1: 3 internally are

$$x=\frac{(1\times-3)+(3\times1)}{1+3} \text{ and } y=\frac{(1\times1)+(3\times3)}{1+3} \text{ or } \left(0,\frac{5}{2}\right)$$

Any line passing through $\left(0,\frac{5}{2}\right)$ is $y-\frac{5}{2}=mx$. ...(5)

If it is tangent to (1), we must have

$$\frac{m.1-3+\frac{5}{2}}{\sqrt{(1+m^2)}}=1 \quad \text{or} \quad 2m-1=2\sqrt{(1+m^2)}$$

Squaring, we get $4m^2 - 4m + 1 = 4(1 + m^2)$

$\Rightarrow$ $0.m^2-4m-3=0$

As the coeff of m^2 is zero, one root of this equation will be ∞ and other is clearly $-3/4$. Putting these values in (5), we get

either $x = 0$ or $y-\frac{5}{2}=-\frac{3}{4}x$ or $4y + 3x - 10 = 0.$

Hence the equations of the tangents are

$$x = 0,\ 4y + 3x + 10,\ y = 4 \text{ and } 3y = 4x.$$

(ii) The circles are given as

$x^2 + y^2 = c^2$ ...(1)

and $(x - a)^2 + y^2 = b^2$ ...(2)

The equation of any tangent to (1) may be given by

$y = mx + c\sqrt{(1+m^2)}$. ...(3)

If (3) is a tangent to (2), then the length of the perpendicular from the centre of (2), *i.e.* (a, 0) must be equal to its radius, *i.e.* = b; so

$$\frac{ma - 0 + c\sqrt{(1+m^2)}}{\sqrt{(1+m^2)}} = \pm\ b \qquad ...(4)$$

$\Rightarrow \qquad ma + c\ \sqrt{(1+m^2)} = b\sqrt{(1+m^2)}$ (taking +ve sign)

$\Rightarrow \qquad ma = (b - c)\ \sqrt{(1+m^2)}$.

Squaring, we get

$$(b - c)^2\ m^2 - m^2\ a^2 + (b - c)^2 = 0$$

$\Rightarrow \qquad (b - c)^2 - m^2\ [a^2 - (b - c)^2] = 0,$

whence $\qquad m = \pm \dfrac{b-c}{\sqrt{\{a^2-(b-c)^2\}}}$

Again, taking –ve in (4), we get

$$ma + c\ \sqrt{(1+m^2)} = -\ b\ \sqrt{(1+m^2)}.$$

Simplifying as before, we get

$$m = \pm \frac{b+c}{\sqrt{\{a^2-(b-c)^2\}}}.$$

Hence the equation of the four common tangents are given by (3), where either

$$m = \pm \frac{b-c}{\sqrt{\{a^2-(b-c)^2\}}}$$

or $\qquad m = \pm \dfrac{b+c}{\sqrt{\{a^2-(b-c)^2\}}}$.

Example 44:

Find the radical axis of the pairs of circles:

$$x^2 + y^2 = 144$$

and $\qquad x^2 + y^2 - 15x + 11y = 0.$

Solution:

The equation of the circles are given as

$$x^2 + y^2 = 144 \qquad ...(1)$$

and $\qquad x^2 + y^2 - 15x + 11y = 0 \qquad ...(2)$

Substracting (2) from (1), the equation of the radical axis is

$$15x - 11y = 144.$$

Example 45:

Find the Pole of the St. line $ax + by + 3a^2 + 3b^2 = 0$ *with respect to the circle* $x^2 + y^2 + 2ax + 2by = a^2 + b^2$.

Solution:

The equation of the line is given as

$$ax + by + 2ax + 2by = a^2 + b^2 \quad ...(2)$$

Let the pole of (1) with respect to (2) be (α, β). Then the polar of (α, β) with respect to (2) is

$$xa + yb + a(x + a) + b(y + b) - a^2 - b^2 = 0$$

$$\Rightarrow \quad x(\alpha + a) + y(\beta + b) + (a\alpha + b\beta - a^2 - b^2) = 0 \quad ...(3)$$

As (1) and (3) represent the same equation, hence comparing the co-efficients, we get

$$\frac{\alpha + a}{a} = \frac{\beta + b}{b} = \frac{a\,\alpha + b\,\beta - a^2 - b^2}{3a^2 + 3b^2}$$

As $\quad \dfrac{\alpha + a}{a} = \dfrac{\beta + b}{b}$ or $b\alpha - a\beta = 0 \quad ...(4)$

Again $\quad \dfrac{\alpha + a}{a} = \dfrac{a\,\alpha + b\,\beta - a^2 - b^2}{3a^2 + 3b^2}$

$$\Rightarrow \quad (2a^2 + 3b^2)\,\alpha - ab\beta + 4a(a^2 + b^2) = 0 \quad ...(5)$$

Substituting the value of b from (4) in (5), we get

$$\left(2a^2 + 3b^2\right)\alpha - a\,b.\frac{b}{a}.\alpha + 4a\left(a^2 + b^2\right) = 0$$

$$\Rightarrow \quad 2(a^2 + b^2)\,\alpha + 4a(a^2 + b^2) = 0 \text{ or } \quad a = -2a.$$

Substituting in (4), we get $b = -2b$.

∴ The co-ordinates of the required pole are

(– 2a, –2b). **Ans.**

Example 46:

Tangents are drawn from the point (h, k) to the circle $x^2 + y^2 = a^2$; *prove that the area of the triangle formed by them and the straight line joining their points of contact is*

$$\frac{a\left(h^2 + k^2 - a^2\right)^{3/2}}{h^2 + k^2}.$$

Solution:

The circle is given as

$$x^2 + y^2 = a^2 \quad ...(1)$$

and the point is given as (h, k).

If AB be the chord of contact for P with respect to (1), the equation of AB will be

$$xh + yk = a^2 \quad ...(2)$$

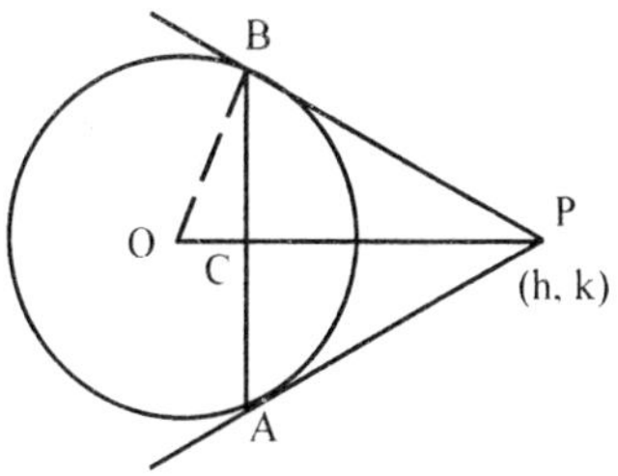

If OP cuts AB at C, then by geometry, we have

$$AC = \frac{1}{2} AB \text{ and } \perp AB$$

Hence the length of perpendicular from P on AB.

Hence the length of perpendicular from P on AB. Then we have

$$= \frac{h.h + k.k - a^2}{\sqrt{h^2 + k^2}}$$

or $$PC = \frac{h^2 + k^2 - a^2}{\sqrt{h^2 + k^2}} \quad ...(3)$$

The length of perpendicular from O (0, 0) upto AB

$$= OC = \frac{a^2}{\sqrt{h^2 + k^2}}$$

By triangle OBC, we have

$$BC^2 = OB^2 - OC^2$$

$$= a^2 - \frac{a^4}{h^2 + k^2} = \frac{(a^2 \; h^2 + k^2 - a^2)}{h^2 + k^2} \quad [\because OB = \text{radius} = a)$$

or $$BC = \frac{1}{2} AB = \frac{a\sqrt{(h^2 + k^2 - a^2)}}{\sqrt{(h^2 + k^2)}} \quad ...(4)$$

Area of the triangle. PAB

$$= \frac{1}{2} AB.PC$$

$$= \frac{a\sqrt{(h^2 + k^2 - a^2)}}{\sqrt{(h^2 + k^2)}} \cdot \frac{(h^2 + k^2 - a^2)}{\sqrt{(h^2 + k^2)}}$$

[putting the values from (3) and (4)] we get

$$= \frac{\left(h^2 + k^2 - a^2\right)^{3/2}}{\left(h^2 + k^2\right)}.$$ **Proved.**

Example 47:

Find the equation to the straight lines joining the origin to the points in which the straight line y = mx + c cuts the circle

$$x^2 + y^2 = 2ax + 2by.$$

Hence find the condition that these points may subtend a right angle at the origin.

Find also the condition that the straight line may touch the circle.

Solution:

The straight line is given as

$$y = mx + c \Rightarrow \frac{y}{c} - \frac{mx}{c} = 1. \qquad ...(1)$$

The equation of the circle is given as

$$x^2 + y^2 = 2ax + 2by \text{ or } x^2 + y^2 - 2ax - 2by = 0 \qquad ...(2)$$

Making the second equation homogeneous with the help of (1), we get the combined equation of the straight lines joining the points of intersection of (1) with (2) to the origin, hence the regd. equation is

$$x^2 + y^2 - 2ax\left(\frac{y}{c} - \frac{mx}{c}\right) - 2by\left(\frac{y}{c} - \frac{mx}{c}\right) = 0$$

$$\Rightarrow \quad cx^2 + cy^2 - 2ax(y - mx) - 2by(y - mx) = 0$$

$$\Rightarrow \quad x^2(c + 2am) + y^2(c - 2b) - 2xy(a - bm) = 0. \qquad ...(3)$$

If the lines represented by this equation are at right angles, the sum of the coefficients of x^2 and y^2 should be zero. Hence the reqd. condition is

$$c + 2am + c - 2b = 0 \quad \text{or} \quad c = b - am.$$

If the line (1) touches the circle, it is only possible when the line cuts the circle in two coincident points. Therefore the lines joining the origin to the two points will also coincide. Hence the equation (3) must represent one line only. It is only possible when it is a perfect square, so its discriminant must be zero.

So $$[2(a - bm)]^2 - 4(c + 2am)(c - 2b) = 0$$

$$\Rightarrow \quad a^2 + b^2m^2 - 2abm - c^2 + 2bc - 2amc + 4abm = 0$$

$$\Rightarrow \quad c^2 - 2c(b - am) - a^2 - b^2m^2 - 2abm = 0$$

$$\Rightarrow \quad c^2 - 2c(b - am) - (a + bm)^2 = 0$$

$$\therefore \qquad c = \frac{2(b-am) \pm \sqrt{\{4(b-am)^2 + 4(a+bm)^2\}}}{2}$$

$$= b - am \pm \sqrt{[(b^2+a^2)(1+m^2)]}$$ **Ans.**

Example 48:

Find the polar of the point (– 2, 3) with respect to the circle

$x^2 + y^2 - 4x - 6y + 5 = 0.$

Solution:

The point is (–2, 3) and the circle is

$x^2 + y^2 - 4x - 6y + 5 = 0.$

The equ. of the polar will be

$x(-2) + y \cdot 3 - 2(x - 2) - 3(y + 3) + 5 = 0$

$\Rightarrow \quad -4x = 0, x = 0$ **Ans.**

Example 49:

Find the polar of the point (5, –1/2) with respect to the circle

$$3x^2 + 3y^2 - 7x + 8y - 9 = 0$$

Solution:

The point is (5, –1/2) and the circle is

$$3x^2 + 3y^2 - 7x + 8y - 9 = 0.$$

The equation of the polar will be

$$3.x(5) + 3y\left(-\frac{1}{2}\right) - \frac{7}{2}(x + 5) + \frac{8}{2}\left(y - \frac{1}{2}\right) - 9 = 0.$$

Simplifying, we get 23x + 5y. **Ans.**

Example 50:

Find the poloar of the point (a, – b) with respective to the circle

$$x^2 + y^2 + 2ax - 2by + a^2 - b^2 = 0.$$

Solution:

The point is (a, – b) and the circle is

$$x^2 + y^2 + 2ax - 2by + a^2 - b^2 = 0.$$

The equation to the polar will be given as

$$x.a + y(-b) + a(x + a) - b(y - b) + a^2 - b^2 = 0$$

$\Rightarrow \quad 2ax - 2by + 2a^2 = 0$ or $by - ax = a^2.$ **Ans.**

Example 51:

Two circles are drawn through the points (a, 5a) and (4a, a) to touch the axis of y. Prove that they interest at an angle tan [1] 40/9.

Solution:

Let the centre of the circle be (h, k). As it touches y-axis, radius will be equal to abscissa of the centre = h. Hence the equation of the circle may be given by

$$(x - h)^2 + (y - k)^2 = h^2$$

$$\Rightarrow \quad x^2 + y^2 - 2hx - 2xy + k^2 = 0. \qquad ...(1)$$

As the circle passes through (a, 5a) and (4a, a), these points will satisfy the equation no. (1), hence

$$a^2 + 25a^2 - 2ah - 10ka + k^2 = 0 \qquad ...(2)$$

and

$$16a^2 + a^2 - 8ah - 2ka + k^2 = 0. \qquad ...(3)$$

Subtracting (3) from (2), we get

$$9a^2 + 6ha - 8ka = 0$$

or

$$2ah = \frac{1}{3}\left(8a\,k - 9a^2\right). \quad ...(4)$$

Substituting this value in (2), we get

$$a^2 + 25a^2 - \frac{1}{3}(8ak - 9a^2) - 10\,k\,a + k^2 = 0.$$

$$\Rightarrow \quad 3k^2 - 38ak + 87a^2 = 0$$

$$\Rightarrow \quad (k - 3a)(3k - 29a) = 0.$$

where $k = 3a$ or $k = \frac{29}{3}a$.

Substituting this value in (4), we get

$$h = \frac{5a}{2} \text{ and } \frac{205}{18}a.$$

Therefore the centre of the two circles will be respectively

$$\left(\frac{5a}{2}, 3a\right) \text{ and } \left(\frac{205}{18}a, \frac{29\,a}{3}\right).$$

Now the angle of intersection of two circles is the angle between the tangents drawn to these circles on the point of intersection which is same as the angle between the lines joining the two centres the point of intersection (as tangents are perpendicular to the radii).

The slope of the line joining the centre (5/2 a, 3a) to the point of intersection (4a. a) say m_1. will be

$$= \frac{3a - a}{\frac{5a}{2} - 4a} = -\frac{4}{3} = m_1.$$

Slope of the line joining $\left(\frac{205\,a}{18}, \frac{29}{3}a\right)$ to the point (4a, a)

$$= \frac{\frac{29}{3}a - a}{\frac{205}{18}a - 4a} = \frac{156}{133} = m_2 \quad \text{(say).}$$

If the angle between these two lines be θ, then

$$\tan\theta = \frac{m_1 - m_2}{1 + m_1\, m_2} = \frac{-\frac{4}{3} - \frac{156}{133}}{1 - \frac{4}{3}\cdot\frac{156}{133}} = \frac{40}{9}.$$

Therefore $\theta = \tan^{-1}\left(\frac{40}{9}\right)$. **Proved.**

Example 52:

Find the equation to the circle which has its centre at the point (3, 4) and touches the straight line

$$5x + 12y = 1.$$

Solution:

The centre of the circle is (3, 4).

If the circle touches the straight line

$$5x + 12y - 1 \qquad ...(1)$$

then the radius of the circle will be equal to the length of the perpendicular from centre (3, 4) upon straight line (1)

$$\text{So radius} = \frac{5.3 + 12.4 - 1}{\sqrt{[(5)^2 + (12)^2]}} = \frac{62}{13}.$$

So the required equation is $(x - 3)^2 + (y - 4)^2 = (62/13)^2$

or $$x^2 + y^2 - 6x - 8y + \frac{381}{169} = 0.$$ **Ans.**

Example 53:

Find the equation to the circle which touches the axes of coordinates and also the line

$$\frac{x}{a}+\frac{y}{b}=1$$

the centre being in the positive quadrant.

Solution:

As the circle touches both the axes, centre may be taken as (h, h) and its radius will also be h.

The given lines is $\frac{x}{a}+\frac{y}{b}=1$ or $bx + ay - ab = 0$ ...(1)

As the circle touches (1) also, the length of the perpendicular from (h, h) upon (1) must be equal to h.

Hence $$\frac{bh+ah-ab}{\sqrt{(a^2+b^2)}}=h$$

or $$bh + ah - ab = h\sqrt{(a^2+b^2)}\,.$$

Squaring, we get

$$b^2h^2 + a^2h^2 + a^2b^2 + 2abh^2 - 2h(ab^2 + ba^2) = h^2(a^2 + b^2)$$

$$\Rightarrow \quad 2h^2 - 2h(a + b) + ab = 0$$

$$\Rightarrow \quad h=\frac{2(a+b)\pm 4\sqrt{[(a+b)^2-8ab]}}{2}$$

$$\Rightarrow \quad h=\frac{(a+b)\pm\sqrt{(a^2+b^2)}}{2} \qquad \text{...(2)}$$

Therefore the required equation of the circle will be

$$(x - h)^2 + (y - h)^2 = h^2$$

$$\Rightarrow \quad x^2 + y^2 - 2hx - 2yh + h^2 = 0$$

where his given by (2).

Example 54:

Find the equation of the circle which has its centre at the point (1, – 3) and touches the straight line 2x – y – 4 = 0.

Solution:

As the centre is (1, – 3) and the circle touches the line $2x - y - 4 = 0$,

$$\text{radius} = \frac{2.1-(-3)-4}{\sqrt{(2^2+1)}}=\frac{1}{\sqrt{5}}$$

So the equation of the circle is

$$(x - 1)^2 + (y + 3)^2 = \left(\frac{1}{\sqrt{5}}\right)^2$$

$$\Rightarrow \quad 5x^2 + 5y^2 - 10x + 30y + 49 = 0.$$

Example 55:

A circle passes through the points (–1, 1), (0, 6) and (5, 5). Find the points on this circle the tangents at which are parallel to the straight line joining the origin to its centre.

Solution:

Let the equation of the circle be

$$x^2 + y^2 + 2gx + 2fy + c = 0. \quad ...(1)$$

As the points (–1, 1), (0, 6) and (5, 5) be on the circle hence each will satisfy (1), so

By (–1, 1), we have

$$1 + 1 - 2g + 2f + c = 0$$

or $$-2g + 2f + c = -2 \quad ...(2)$$

By (0, 6), we have

$$0 + 36 + 0.g + 12f + c = 0$$

or $$12f + c = -36 \quad ...(3)$$

By (5, 5), we have

$$25 + 25 + 10g + 10f + c = 0$$

or $$10g + 10f + c = -50. \quad ...(4)$$

Subtracting (3) from (2), we get $-2g - 10f = 34$ ...(5)

Subtracting (4) from (3), we get $-10g + 2f = 14$. ...(6)

Solving (5) and (6), we get $g = -2$, $f = -3$.

Putting the value of g and f in (2), we get $c = 0$.

Substituting these values of g, f and c in (1), we get the equation of the circle as

$$x^2 + y^2 - 4x - 6y = 0. \quad ...(7)$$

The centre of the circle is (2, 3) and radius is $\sqrt{13}$.

Slope of the line joining the centre to the origin is

$$\frac{3-0}{2-0} = \frac{3}{2}.$$

The line passing through the centre and perpendicular to the line joining the centre to the origin, is

$$y - 3 = \frac{2}{3}(x - 2)$$

or $$3y + 2x - 13 = 0 \quad \text{...(8)}$$

If the line (8) cuts the circle given by (7) at A and B, the tangents at A and B will be parallel to the line joining the centre and the origin.

To solve (7) and (8), putting the value of y from (8) in (7), we get

$$x^2 + \left(\frac{13-2x}{3}\right)^2 - 4x - 6\left(\frac{13-2x}{3}\right) = 0$$

$\Rightarrow$ $9x^2 + (13 - 2x)^2 - 36x - 18(13 - 2x) = 0$

$\Rightarrow$ $13x^2 - 52x - 65 = 0$

$\Rightarrow$ $(x - 5)(x + 1) = 0$ whence $x = 5$ or -1.

Therefore the required points are (5, 1) and (–1, 5). **Ans.**

Example 56:

Find the equation to a circle of radius r which touches the axis of y at a point distant h from the origin, the centre of the circle being in the positive quadrant.

Prove also that the equation to the other tangent which passes through the origin is

$$(r^2 - h^2)x + 2rhy = 0.$$

Solution:

The radius of the circle is given as r. As it touches y-axis, at a distance of h from the origin, it is clear by the figure that the co-ordinates of the centre C will be (r, h). Hence the equation of the circle will be

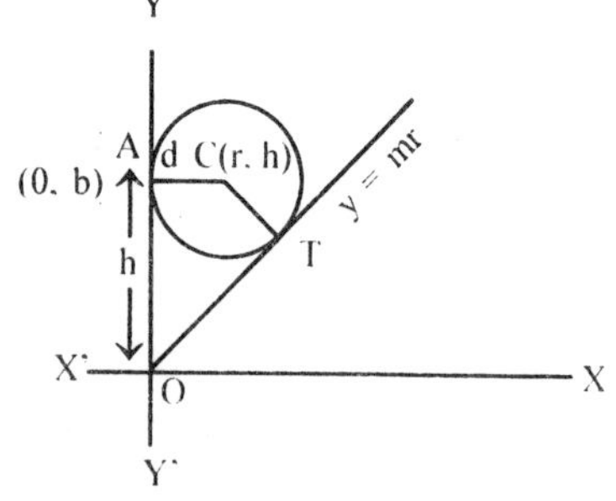

$$(x - r)^2 + (y - h)^2 = r^2$$

$$\Rightarrow x^2 + y^2 - 2rx - 2hy + h^2 = 0.$$

Let the equation of the other tangent OT be $y = mx$.

CT will be perpendicular to OT. Hence

$$CT = \frac{h - m}{\sqrt{(1+m^2)}} = r$$

or $(h - mr)^2 = r^2 (1 + m^2)$

whence $m = \dfrac{(h^2 - r^2)}{2hr}$

So the equation of the other tangent OT is $y = \dfrac{(h^2 - r^2)}{2hr}$

$\Rightarrow$ $(r^2 - h^2) x + 2hry = 0.$ **Hence Proved.**

Example 57:

Find the equation to the circles which pass through the origin and cut off equal chords a from the straight lines $y = x$ and $y = -x$.

Solution:

$$y = x \quad ...(i)$$

and $$y = -x \quad ...(ii)$$

By the figure it is clear that by symmetry, the one centre will be on the y-axis. Let the co-ordinates of the centre C be (0, b).

Draw CN perpendicular to AB.

Now $ON = \frac{1}{2} OL = \frac{1}{2} a.$

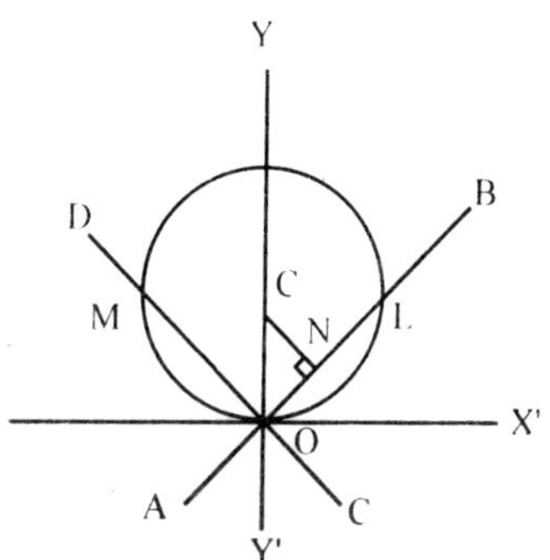

Length of the perpendicular from (0, b) on $y - x = 0$

$$= CN = \frac{b - 0}{\sqrt{(1+1)}} = \frac{b}{\sqrt{2}}.$$

In triangle CNO, $CO^2 = CN^2 + ON^2$

or $$b^2 = \frac{b^2}{2} + \left(\frac{a}{2}\right)^2$$

or $$b = \pm \frac{a}{\sqrt{2}}$$

$\therefore$ The co-ordinates of the centre will be $\left(0, \pm \frac{a}{\sqrt{2}}\right)$, and radius $= \frac{a}{\sqrt{2}}$.

Hence the equation of the circle is

$$(x - 0)^2 + \left(y \pm \frac{a}{\sqrt{2}}\right)^2 = \frac{a}{\sqrt{2}}\left(\frac{a}{\sqrt{2}}\right)^2$$

or $\quad x^2 + y^2 \pm \sqrt{2ax} = 0$

Similarly if the centre lies on x-axis the equation will be

$$x^2 + y^2 \pm \sqrt{2ax} = 0.$$ **Ans.**

Example 58:

Find the length of the chord joining the points in which the straight line

$$\frac{x}{a} + \frac{y}{b} = 1$$

meets the circle $x^2 + y^2 = r^2$.

Solution:

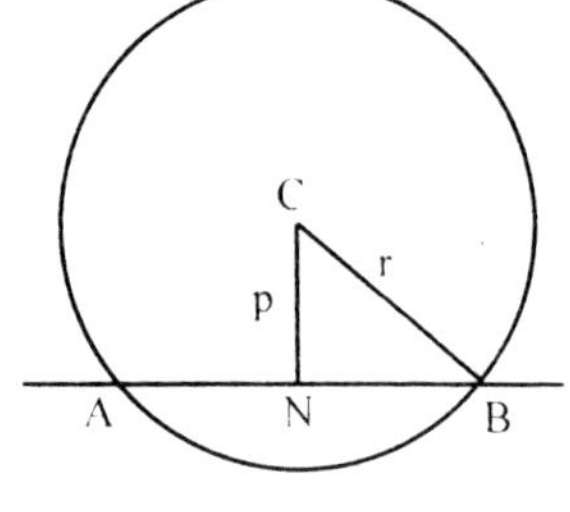

Let C be the centre of the circle

$$x^2 + y^2 = r^2 \quad ...(i)$$

Hence is centre C ≡ (0, 0)

and radius = r.

Again let AB be the line

$$\frac{x}{a} + \frac{y}{b} = 1 \quad ...(ii)$$

meeting the circle in A and B.

Draw CN perpendicular from the centre on the line. Clearly N will be the mid-point of AB, *i.e.*,

$$BN = \frac{1}{2} AB \quad ...(iii)$$

From the right angled triangle CNB,

$$NB^2 = CB^2 - CN^2 \quad ...(iv)$$

Length of the perpendicular from (0, 0) on AB

$$= CN = \frac{0 + 0 - ab}{\sqrt{(a^2 + b^2)}}$$

and $\quad$ CN = radius of the circle

Putting the values in (iv), $NB^2 = r^2 - \dfrac{a^2 b^2}{a^2 + b^2}$

$$\Rightarrow \quad NB = \sqrt{r^2 - \left(\frac{a^2 b^2}{a^2 + b^2}\right)}$$

$$\Rightarrow \qquad AB = 2NB = 2\sqrt{r^2 - \left(\frac{a^2\ b^2}{a^2 + b^2}\right)}.$$

Example 59:

Find whether the straight line $x + y = 2 + \sqrt{2}$ touches the circle

$$x^2 + y^2 - 2x - 2y + 1 = 0.$$

Solution:

The circle is given as $x^2 + y^2 - 2x - 2y + 1 = 0$, hence its centre is (1, 1) and radius $= \sqrt{(1^2 + 1^2 - 1)} = 1$.

The line is given as

$$x + y = 2 + \sqrt{2} \text{ or } x + y - 2 - \sqrt{2} = 0 \qquad ...(1)$$

If (1) touches the given circle, length of the perpendicular from the centre of the circle must be equal to its radius.

Length of the perpendicular form (1, 1) on (1) is

$$\frac{1 + 1 - 2 - \sqrt{2}}{\sqrt{(1+1)}} = -1 = 1 \text{ numerically}$$

As the perpendicular distance is equal to radius, hence the line touches the circle.

Example 60:

Find the condition that the straight line $3x + 4y = k$ may touch the circle $x^2 + y^2 = 10x$.

Solution:

The circle is given as

$$x^2 + y^2 = 10x$$

or $$x^2 + y^2 - 10x = 0.$$

hence its centre will be (5, 0) and radius = 5 units.

The line is given as $\quad 3x + 4y = k$

or $$3x + 4y - k = 0. \qquad ...(1)$$

If (1) is tangent to the circle, length of the perpendicular from centre on it must be equal to radius, hence,

$$\frac{3.5 + 4.0 - k}{\sqrt{(3^2 + 4^2)}} = \pm 5, \text{ whence } k = 40 \text{ or } -10. \qquad \textbf{Ans.}$$

Example 61:

Find the value of p so that the straight line

$$x \cos \alpha + y \sin \alpha - p = 0$$

may touch the circle

$$x^2 + y^2 - 2ax \cos \alpha - 2by \sin \alpha - a^2 \sin^2 \alpha = 0.$$

Solution:

The circle is given as

$$x^2 + y^2 - 2ax \cos \alpha - 2by \sin \alpha - a^2 \sin^2\alpha = 0.$$

Hence its centre is (a cos a, b sin a) and radius is

$$\{\sqrt{(a^2 + b^2 \sin^2 \alpha)}\}.$$

The line is given as

$$x \cos \alpha + y \sin \alpha - p = 0$$

For the line being a tangent to the given circle

$$\frac{a \cos \alpha \cos \alpha + b \sin \alpha \sin \alpha - p}{(\cos^2 \alpha + \sin^2 \alpha)} = \pm \sqrt{(a^2 + b^2 \sin^2 \alpha)}$$

$$p = a \cos^2 \alpha + b \sin^2 a \pm \sqrt{(a^2 + b^2 \sin^2 \alpha)}.$$ **Ans.**

Example 62:

Find the condition that the straight line $Ax + By + C = 0$ may touch the circle

$$(x - a)^2 + (y - b)^2 = c^2.$$

Solution:

The circle is given as $(x - a)^2 + (y - b)^2 = c^2$.

Clearly is centre is (a, b) and the radius is c.

The line is given as $Ax + By + C = 0$.

For the line being a tangent to the given circle.

$$\frac{A.a + B.b + C}{\sqrt{(A^2 + B^2)}} = \pm c$$

$$\Rightarrow \quad Aa + Bb + C = \pm c \sqrt{(A^2 + B^2)}.$$ **Ans.**

Example 63:

Find the equations to the tangents to the circle

$x^2 + y^2 = 4$ which are parallel to the line $x + 2y + 3 = 0$.

Solution:

The circle is $x^2 + y^2 = 4$, hence its centre is (0, 0) and radius = 2.

The tangent is parallel to the line $x + 2y + 3 = 0$ ...(1)

Any line parallel to (1) is $x + 2y - \lambda = 0$ ...(2)

Length of the perpendicular from the centre of the circle (0, 0) on (2)

$$= \frac{0+0-\lambda}{\sqrt{(1+4)}} = 2 \qquad \therefore \text{ radius} = 2$$

or $\quad l = \pm 2\sqrt{5}$.

Substituting in (2), we get the required equation as

$$x + 2y \pm 2\sqrt{5} = 0.$$

Example 64:

Find the equation to the tangents to the circle

$x^2 + y^2 + 2gx + 2fy + c = 0$ which are parallel to the line

$x + 2y - 6 = 0$.

Solution:

The circle is given as $x^2 + y^2 + 2gx + 2fy + c = 0$

Its centre is $(-g, -f)$ and radius is $\sqrt{(g + f^2 - c)}$

Any line parallel to the given line $x + 2y - 6 = 0$ is

$$x + 2y - \lambda = 0 \qquad ...(1)$$

The length of the perpendicular from centre $(-g, -f)$ on (1)

$$= \frac{-g-2f-\lambda}{\sqrt{(1+4)}} = \sqrt{(g^2+f^2-c)} \qquad \text{(radius of the circle)}$$

$\therefore \quad l = -g - 2f \pm \sqrt{5}\sqrt{(g^2 + f^2 - c)}$.

Putting this value in (1), we get the required equation as

$$x + 2h - [-g - 2f \pm \sqrt{5}\sqrt{(g^2+f^2-c)}]$$

or $x + 2y + g + 2f = \pm\sqrt{5}\sqrt{(g^2 + f^2 - c)}]$. **Ans.**

Example 65:

Find the equation to the tangent to the circle $x^2 + y^2 = a^2$ which

(i) is parallel to the straight line $y = mx + c$,

(ii) is perpendicular to the straight line $y = mx + c$,

(iii) passes through the point $(b, 0)$,

and (iv) makes with the axes a triangle whose area is a^2.

Solution:

The circle is $x^2 + y^2 = a^2$. So any tangent to it will be of the form

$$y = Mx \pm a\sqrt{(1+M^2)} \qquad ...(1)$$

(i) If the tangent is parallel to $y = mx + c$ ...(2)

The slope of (1) and (2) must be equal; hence $m = M$.

Substituting in (1), we get

$$y = mx \pm a\sqrt{(1+m^2)}$$

(ii) If (1) is perpendicular to $y = mx + c$, then

$$Mm = -1 \text{ or } M = -\frac{1}{m}$$

Putting in (1), we get $y = -\frac{1}{m}x \pm a\sqrt{\left(1+\frac{1}{m^2}\right)}$

Simplifying $\quad my + x = \pm a \sqrt{(1+m^2)}$.

(iii) If the tangent passes through the point (b, 0), these co-ordinates will satisfy (1), hence

$$0 = M.b \pm a\sqrt{(1+M^2)}$$

or $\quad Mb = \pm a\sqrt{(1+M^2)}$

Squaring $M^2b^2 = a^2\sqrt{(1+M^2)}$

Solving for M, we get $M = \pm a/\sqrt{(b^2-a^2)}$

Substituting in (1), we get

$$y = \pm\frac{ax}{\sqrt{(b^2-a^2)}} \pm a\sqrt{1+\frac{a^2}{(b^2-a^2)}}$$

Simplifying, we get $ax \pm y\sqrt{(b^2-a^2)} = ab$

(iv) Let (1) cut the axes at (h, 0) and (0, k); (h, 0) and (0, k) will satisfy (1), so, we get

$$0 = M.h \pm a\sqrt{(1+M^2)}$$

or $\quad h = \pm\frac{\sqrt{(1+M^2)}}{M} \qquad ...(2)$

and $\qquad k = 0 \pm a\sqrt{(1+M^2)}$

or $\qquad k = \pm a \{\sqrt{(1+M^2)}\}$ ...(3)

The area of the D between the axes and the line is 1/2 hk

$$= \frac{1}{2}\cdot\left\{\frac{a\sqrt{(1+M^2)}}{M}\times\sqrt{(1+M^2)}\right\} = \frac{a^2}{2M}\left(1+M^2\right) = a^2 \quad \text{(given)}$$

or $\qquad 1 + M^2 = 2M$

or $\qquad M^2 - 2M + 1 = 0$ or $(M - 1)^2 = 0$

$\therefore \qquad M = 1$

Putting in (1), we get

$$y = x \pm a\sqrt{(1+1)} \text{ or } y = x \pm a\sqrt{2}.$$

Example 66:

Prove that the straight line $y = x + c\sqrt{2}$ touches the circle $x^2 + y^2 = c^2$, and find its point of contact.

Solution:

Circle is given as $x^2 + y^2 = c^2$ ...(1)

and the line is $y = x + c\sqrt{2}$ ...(2)

Putting the value of y from (2) in (1), we get

$$x^2 + (x + c\sqrt{2})^2 = c^2$$

or $\qquad x^2 + x^2 + 2xc\sqrt{2} + 2c^2 - c^2 = 0$

or $\qquad (\sqrt{2}x + c)^2 = 0.$ ...(3)

This being a perfect square gives two coincident values of x, hence (2) is a tangent to (1)

Again by (3). $\sqrt{2}x + c = 0$

or $\qquad x = -\frac{c}{\sqrt{2}}$

Substituting the value of x in (2), $y = -\frac{c}{\sqrt{2}} + c\sqrt{2} = \frac{c}{\sqrt{2}}$

Hence point of contact is $\left(-\frac{c}{\sqrt{2}}, \frac{c}{\sqrt{2}}\right)$.

Example 67:

Find the condition that the straight line $cx - by + b^2 = 0$ may touch the circle $x^2 + y^2 = ax + by$ and find the point of contact.

Solution:

Circle is given as $x^2 + y^2 = ax + by$

$$\Rightarrow \qquad x^2 + y^2 - ax - by = 0 \qquad ...(1)$$

The line is given as $cx - by + b^2 = 0$

$$\Rightarrow \qquad y = \left(\frac{c}{b}x + b\right) \qquad ...(2)$$

Substituting the value of y from (2) in (1), we get

$$x^2 = \left(\frac{c}{b}x + b\right) - ax - b\left(\frac{c}{b}x + b\right) = 0$$

$$\Rightarrow \qquad x^2 + \frac{c^2}{b^2}x^2 + 2cx + b^2 - ax - b^2 = 0$$

$$\Rightarrow \qquad x^2\left(1 + \frac{c^2}{b^2}\right) + x\,(c - a) = 0 \qquad ...(3)$$

If it is a perfect square, then $c - a = 0$

or $\qquad c = a$

Now, if the line is the tangent, $c = a$,

hence the equation becomes

$$x^2\left(1 + \frac{c^2}{b^2}\right) = 0$$

or $\qquad x = 0$ as $1 + \dfrac{c^2}{b^2} \neq 0$

Substituting the value of x in (2), we get $y = b$

Therefore the point of contact is (0, b).

Example 68:

Find the equation to the circle whose centre is at the point (a, b) and which passes through the origin, and prove that the equation of the tangent at the origin is

$$\alpha x + \beta y = 0.$$

Solution:

As the centre is (α, β), say C and the circle passes through origin, O, hence the radius will be $CO = \sqrt{(\alpha^2 + \beta^2)}$.

Therefore the equation of the circle will be

$$(x - \alpha)^2 + (y - \beta)^2 = \alpha^2 + \beta^2$$

or $$x^2 + y^2 - 2\alpha x - 2\beta y = 0.$$

For tangent, any the passing through origin is

$$y = mx. \quad ...(1)$$

Slope of the radius CO is

$$\frac{\beta - 0}{\alpha - 0} = \frac{\beta}{\alpha}.$$

As (1) is tangent to the circle, it will be perpendicular to CO, hence the slope of tangent

$$= m = -\alpha/\beta.$$

Substituting in (1), we get $y = -\frac{\alpha}{\beta}x$

or $$\alpha x + \beta y = 0.$$

Example 69:

Find the Pole of the st. line $2x + y + 12 = 0$ with respect to the circle

$$x^2 + y^2 - 4x + 3y - 1 = 0.$$

Solution:

The line is given as $2x + y + 12 = 0$ ...(1)

The circle is given as $x^2 + y^2 - 4x + 3y - 1 = 0.$ ...(2)

Let the pole of (1) with respect to (2) be (α, β).

Polar (a, b) with respect to (2) is

$$x\,\alpha + y\,\beta - 2\,(x + \alpha) + 3/2\,(y + \beta) - 1 = 0$$

or $$x\,(2\alpha - 4) + y\,(2\beta + 3) - (4\alpha - 3\beta + 2) = 0 \quad ...(3)$$

As (1) and (3) represent the same straight lines, comparing the coefficients of x, y and the constant term, we get

$$\frac{2\alpha - 4}{2} = \frac{2\beta + 3}{1} = \frac{-(4\alpha - 3\beta + 2)}{12}$$

or $\alpha - 2\beta = 5$ ...(4) (by first two terms)

and $4\alpha + 21\beta = -38$ ...(5) (by last two terms)

Solving (4) and (5), we get a = 1 and b = – 2.

Hence the required pole is (1, –2).

Example 70:

Find the length of the common chord of the circles, whose equations are $(x - a)^2 + y^2 = a^2$ and $x^2 + (y - b)^2 = b^2$, and prove that the equation to the circle whose diameter is this common chord is

$$(a^2 + b^2)(x^2 + y^2) = 2ab(bx + ay).$$

Solution:

The two circles are given as

$$(x - a)^2 + y^2 = a^2$$

$$\Rightarrow x^2 + y^2 - 2ax = 0 \quad ...(1)$$

and $\quad x^2 + (y - b)^2 = b^2$

$$\Rightarrow x^2 + y^2 - 2by = 0. \quad ...(2)$$

(0, b)
A
B
O
C(a, 0)

The equation of the chord OA which is common to (1) and (2) is

$$(x^2 + y^2 - 2ax) - (x^2 + y^2 - 2by) = 0$$

$$\Rightarrow \quad by - ax = 0. \quad ...(3)$$

To get the co-ordinates of A, we solve (1) and (3); so putting the values of y from (3) in (1), we have

$$x^2 + (ax/b)^2 - 2ax = 0$$

$$\Rightarrow \quad b^2x^2 + a^2x^2 - 2ab^2x = 0$$

$$\Rightarrow \quad x\,[x\,\{a^2 + b^2) - 2ab^2] = 0.$$

So either $x = 0$ or $x = 2ab^2/(a^2 + b^2)$; the corresponding values of y are $y = 0$ and $y = 2a^2b/(a^2 + b^2)$.

So the co-ordinates of O are (0, 0) and of A are

$$\left(\frac{2ab^2}{a^2 + b^2}, \frac{2a^2b}{a^2 + b^2}\right).$$

Therefore the equation of the circle drawn on OA as diameter is

$$(x - a)\,x - \left(\frac{2ab^2}{a^2 + b^2}\right) + (y - 0)\left(y - \frac{2a^2b}{a^2 + b^2}\right) = 0$$

$$\Rightarrow \quad (a^2 + b^2)\,x^2 - 2ab^2x + (a^2 + b^2)\,y^2 - 2a^2by = 0$$

$$\Rightarrow \quad (x^2 + y^2)(a^2 + b^2) = 2ab(bx + ay).$$

Example 71:

The angular points of a triangle are the points

(a cos α, a sin α), (a cos β, a sin β), and (a cos γ, a sin γ);

prove that the coordinates of the orthocentre of the triangle are

a (cos α + cos β + cos γ) and a (sin α + sin β + sin γ).

Hence prove that if A, B, C and D be four points on a circle the orthocentre of the four triangles ABC, BCD, CDA and DAB lie on a circle.

Solution:

Let the point (a cos α, a sin α), (a cos β, a sin β) and (a cos γ, a sin γ) be A, B and C respectively, the the equation of the line AB is

$$y - a\sin\alpha = \frac{a\sin\beta - a\sin\alpha}{a\cos\beta - a\cos\alpha}(x - a\cos\alpha)$$

$$\Rightarrow \quad x\cos\frac{1}{2}(\alpha+\beta) + y\sin\frac{1}{2}(\alpha+\beta) = a\cos\frac{1}{2}(\alpha-\beta) \qquad ...(1)$$

Similarly equation of BC and CA are respectively

$$x\cos\frac{1}{2}(\beta+\gamma) + y\sin\frac{1}{2}(\beta+\gamma) = \cos\frac{1}{2}(\beta-\gamma) \qquad ...(2)$$

and
$$x\cos\frac{1}{2}(\gamma+\alpha) + y\sin\frac{1}{2}(\gamma+\alpha) = a\cos\frac{1}{2}(\gamma-\alpha) \qquad ...(3)$$

If AN and BM be the perpendiculars from A and B on BC and CA respectively, then the equation on AN is

$$y - a\sin\alpha = \frac{\sin\frac{1}{2}(\gamma+\beta)}{\cos\frac{1}{2}(\gamma+\beta)}(x - a\cos\alpha)$$

$$\Rightarrow x\sin\frac{1}{2}(\gamma+\beta) - y\cos\frac{1}{2}\cos\frac{1}{2}(\gamma+\beta) = a\sin\left(\frac{\beta+\gamma}{2} - \alpha\right) \qquad ...(4)$$

Similarly equation BM is

$$y - a\sin\beta = \frac{\sin\frac{1}{2}(\gamma+\alpha)}{\cos\frac{1}{2}(\gamma+\beta)}(x - a\cos\beta)$$

$$\Rightarrow x\sin\frac{1}{2}(\gamma+\alpha) - y\cos\frac{1}{2}(\gamma+\alpha) = a\sin\left(\frac{\gamma+\alpha}{2} - \beta\right) \qquad ...(5)$$

To solve (4) and (5) multiply (4) by cos 1/2 (γ + α) and (5) by cos 1/2 (γ + β) and subtracting, we have

$$x\left[\sin\frac{1}{2}(\gamma+\beta)\cos\frac{1}{2}(\gamma+\alpha) - \sin\frac{1}{2}(\gamma+\alpha)\cos\frac{1}{2}(\gamma+\beta)\right]$$

$$= a\left[\sin\left(\frac{\beta+\gamma}{2}-\alpha\right)\cos\frac{1}{2}(\gamma+\alpha)-\sin\left(\frac{\gamma+\alpha}{2}-\beta\right)\cos\frac{1}{2}(\gamma+\beta)\right]$$

or $x\sin\left[\frac{1}{2}(\gamma+\beta)\frac{1}{2}(\gamma+\alpha)\right]=\frac{a}{2}\left[2\sin\frac{1}{2}(\beta+\gamma-2\alpha)\cos\frac{1}{2}(\gamma+\alpha)\right.$

$$\left.-2\sin\frac{1}{2}(\gamma+\alpha-2\beta)\,.\,\cos\frac{1}{2}(\gamma+\beta)\right]$$

$$\Rightarrow x\sin\frac{1}{2}(\beta-\alpha)=\frac{a}{2}\sin\frac{1}{2}(\beta+2\gamma-\alpha)+\sin\frac{1}{2}(\beta-3\alpha)$$

$$\left.-\sin\frac{1}{2}(\gamma+\alpha-\beta)-\sin\frac{1}{2}(\alpha-3\beta)\right]$$

$$=\frac{a}{2}\left[\left\{\sin\frac{1}{2}(\beta+2\gamma-\alpha)-\sin\frac{1}{2}(2g+a-b)\right\}+\sin\frac{1}{2}(\beta-3\alpha)-\sin\frac{1}{2}\right](\alpha-3\beta)$$

$$\Rightarrow x\sin\frac{1}{2}(\beta-\alpha)=a\sin\frac{1}{2}(\beta-\alpha)\left[\cos\gamma+2.\cos\frac{1}{2}(\beta+\alpha)\cos\frac{1}{2}(\beta-\alpha)\right]$$

or $\quad x = [\cos\gamma + \cos\beta + \cos\alpha]$.

Similarly, $\quad y = a[\sin\gamma + \sin\beta + \sin\alpha]$.

∴ the co-ordinates of the orthocentre are

$$[a(\cos\alpha+\cos\beta+\cos\gamma),\ a(\sin\alpha+\sin\beta+\sin\gamma)]$$

Again. Clearly the three points A, B and C lie on a circle radius a and having centre as origin. Let D be another point on the same circle as $(a\cos\delta, a\sin\delta)$. Then orthocentre of ΔABC is

$$[a(\cos\alpha+\cos\beta+\cos\gamma),\ a(\sin\alpha+\sin\beta+\sin\gamma)]$$

Similarly the ortho-centres of the triangles ABD, ACD and BCD are respectively.

$$[a(\cos\alpha+\cos\beta+\cos\delta),\ a(\sin\alpha+\sin\beta+\sin\delta)].$$

$$[a(\cos\alpha+\cos\gamma+\cos\delta),\ a(\sin\alpha+\sin\gamma+\sin\delta)].$$

and $\quad [a(\cos\beta+\cos\gamma+\cos\delta),\ a(\sin\beta+\sin\gamma+\sin\delta)]$.

Clearly all these points lie on the circle

$$[x-a(\cos\alpha+\cos\beta+\cos\gamma+\cos\delta)]^2+[y-a(\sin\ +\sin\beta$$
$$+\sin\gamma+\sin\delta)]^2 = a$$

as they satisfy it. Hence all the ortho-centres lie on a circle with centre $(a\,\Sigma\cos\alpha,\ a\,\Sigma\sin\alpha)$ and radius a.

Example 72:

The polar of P with respect to the circle $x^2 + y^2 = a^2$ touches the circle $(x - \alpha)^2 + (y - \beta)^2 = b^2$; prove that its locus is the curve given by the equation $(\alpha x + \beta y - a^2)^2 = b^2 (x^2 + y^2)$.

Solution:

The circles are given as

$$x^2 + y^2 = a^2 \qquad ...(1)$$

and $$(x - \alpha)^2 + (y - \beta)^2 = b^2. \qquad ...(2)$$

Let the co-ordinates of the moving point P be (h, k).

Polar of (h, k) with respect to (1) is

$$xh + yk - a^2 = 0. \qquad ...(3)$$

If (3) is a tangent to (2), the length of the perpendicular from the centre of (2), *i.e.* (α, β) to (3) must be equal to be, the radius of (2). The length of perpendicular from (a, b) on (3) is

$$\frac{\alpha h + \beta k - a^2}{\sqrt{(h^2 + k^2)}} = b$$

or $$(ah - bk - a^2) = b \sqrt{(h^2 + k^2)}$$

Squaring and generalising, we get the required locus as

$$(\alpha x + \beta y - a^2) = b^2 (x^2 + y^2).$$

Example 73:

Find the locus of the foot of the perpendicular let fall from the origin upon any chord of the circle $x^2 + y^2 + 2gx + 2fy + c = 0$ which subtends a right angle at the origin.

Solution:

The circle is given as

$$x^2 + y^2 + 2gx + 2fy + c = 0 \qquad ...(1)$$

Let the equation of any chord be $lx + my = 1$ ...(2)

Making (1) homogeneous with the help of (2), we get

$$x^2 + y^2 + 2gx (lx + my) + 2fy (lx + my) + c (lx + my)^2 = 0.$$

Equation No. (3) represents two st. lines joining the origin to the point of intersection of (1) and (2). If the set two lines are at right angles, then

$$1 + 2gl + cl^2 + 1 + 2fm + cm^2 = 0$$

or $$2 + 2(gl + fm) + c(l^2 + m^2) = 0. \quad \text{...(3)}$$

(a) Let the perpendicular from the origin on (2) meet it at P whose co-ordinates are (h, k). Equation of this perpendicular will be

$$ly - mx = 0. \quad \text{...(4)}$$

As (h, k) is the point of intersection, it will satisfy (2) and (4) hence

$$lh + mk - 1 = 0 \quad \text{...(5)}$$

and $$lk - mh = 0. \quad \text{...(6)}$$

To get the required locus, we have to eliminate l and m from (3), (5) and (6).

Applying cross-multiplication for l and m in (5) and (6), we have

$$\frac{l}{0-h} = \frac{m}{-k-0} = \frac{1}{-h^2-k^2}$$

or $$l = \frac{h}{h^2+k^2} \text{ and } m = \frac{k}{h^2+k^2}; \qquad \therefore l^2 + m^2 = \frac{1}{h^2+k^2}.$$

Putting in (3), we get

$$2 + 2\left(g.\frac{h}{h^2+k^2} + f.\frac{k}{h^2+k^2}\right) + c.\frac{1}{h^2+k^2} = 0$$

or $$2h^2 + 2k^2 + 2gh + 2fk + c = 0.$$

Generalising, we get $2x^2 + 2y^2 + 2gx + 2fy + c = 0$

which is clearly a circle.

Similarly we can find the locus of the middle points of the chords.

Example 74:

Show that if the length of the tangent from a point P to the circle $x^2 + y^2 = a^2$ be four times the length of the tangent from it to the circle $(x - a)^2 + y^2 = a^2$, then P lies on the circle

$$15x^2 + 15y^2 - 32ax + a^2 = 0.$$

Prove also that these three circles pass through two points and that the distance between the centres of the first and third circles is sixteen times the distance between the centres of the second and third circles.

Solution:

The two circles are given as

$$x^2 + y^2 = a^2 \quad \text{...(1)}$$

and $$(x - a)^2 + y^2 = a^2$$

or $$x^2 + y^2 - 2ax = 0. \quad \text{...(2)}$$

If P be any point (h, k) satisfying the given condition, PT_1 and PT_2 be the lengths of tangents from P upon (1) and (2).

$$PT_1 = \sqrt{(h^2 + k^2 - a^2)} \text{ and } PT_2 = \sqrt{(h^2 + k^2 - 2ah)}$$

By hypothesis, $PT_1 = 4PT_2$ and $PT_1^2 = 16\ PT_2^2$

Putting the value, $h^2 + k^2 - a^2 = 16\ (h^2 + k^2 - 2ah)$

Simplifying and generalising, we get

$$15x^2 + 15y^2 - 32ax + a^2 = 0 \qquad ...(3)$$

which is the required locus.

Again, common chord of (1) and (2) is

$$2ax = a^2 \qquad \text{or} \qquad 2x = a$$

Common chord of (1) and (3) is obtained by multiplying (1) by 15 and subtracting (3) from it; so we get

$$32ax - a^2 = 15a^2 \qquad \text{or} \qquad 2x = a. \qquad ...(5)$$

As (4) and (5) are the same, the 3 circles have the same common chord meaning thereby that the three circles pass through 2 common points.

Again centre of (1) is (0, 0), centre of (2) is (a, 0) and centre of (3) (16/15 a, 0)

Distance between the centres of (1) and (3)

$$= \frac{16a}{15} - 0 = \frac{16a}{15}$$

Clearly the first distance is 16 times the second distance.

Example 75:

Two rods, of lengths a and b, slide along the axes, which are rectangular, in such a manner that their ends are always concyclic; prove that the locus of the centre of the circle passing through these ends is the curve $4(x^2 - y^2) = a^2 - b^2$.

Solution:

Let AB and CD be the two given rods which slide along the axes and are of length a and b respectively. A the four points A, B, C, D are concyclic, a circle will pass through them. Let the equation of the circle be

$$x^2 + y^2 - 2xh - 2yk + c = 0, \qquad ...(1)$$

so that the centre is (h, k).

Solving it with x axis, *i.e.* y = 0, we get

$$x^2 - 2xh + c = 0 \quad ...(2)$$

If A and B be $(x_1, 0)$ and (x_2, o) respectively, then x_1 and x_2 are given by (2).

Clearly AB = $x_2 - x_1 = \sqrt{\{(x_2 + x_2)^2 - 2x_2x_1\}} = 0$ (by hypothesis)

$\Rightarrow \quad (x_2 + x_1)^2 - 4x_2x_1 = a^2.$

By (2) $x_1 + x_2 = 2h$ and $x_1x_2 = c$, hence putting the values

$$4h_2 - 4c = a^2. \quad ...(3)$$

Similarly solving with y-axis, *i.e.* x = 0, we get

$$y^2 - 2yk + c = 0 \quad ...(4)$$

If C and D be $(0, y_1)$ and $(0, y_2)$, y_1 and y_2 are given by (4) and $y_1 + y_2 = 2k$ and $y_1y_2 = c$.

Again $CD^2 = (y_2 - y_1)^2 = (y_2 + y_1)^2 - 4y_1y_2 = b^2$ (by hypothesis)

$\Rightarrow \quad 4k^2 - 4c = b^2 \quad ...(5)$

(putting the values).

Subtracting (5) from (3), we get

$$4h^2 - 4k^2 = a^2 - b^2$$

Generalising, we get $(x^2 - y^2) = a^2 - b^2$ which is the required locus of the centre.

Example 76:

O is a fixed point and AP and PQ are two fixed parallel straight lines; BOA is perpendicular to both and POQ is a right angle. Prove that the locus of the foot of the perpendicular drawn from O upon PQ is the circle on AB as diameter.

Solution:

Let the fixed point O be the origin, x-axis along AOB and perpendicular at O to AB as y-axis, clearly by hypothesis, AP and BQ are parallel to y-axis. If OA and OB be respectively a and b, then equation of AP is

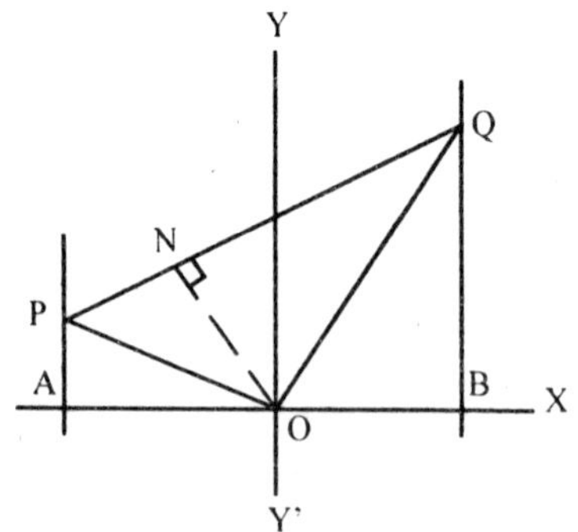

$$x = a$$

and of BQ is $x = -b$

Hence the combined equation is

$$(x - a)(x + b) = 0$$

$\Rightarrow \qquad x^2 - x\,(a - b) - ab = 0$ ...(1)

Again let PQ be any line $lx + my = 1$. ...(2)

Draw ON $\perp$ PQ and let the co-ordinates of N be (h, k)

As N lies on PQ, hence $lh + mk - 1 = 0$. ...(3)

Equation of ON, the line passing through O (0, 0) and perp. to PQ will be

$$mx - ly = 0$$

As N lies on ON,

$$mh - lk = 0,$$

$\Rightarrow \qquad lk - mh + 0 = 0.$ ...(4)

As P and Q are the points of intersection of (1) and (2), so the combined equation of OP and OQ can be found be making (1) homogeneous with the help of (2). So, we get

$$x^2 - (a - b)\,.\,x\,(lx + my) - ab\,(lx + my)^2 = 0$$

$\Rightarrow x^2\,(l - lx + lb - al^2) - bm^2y^2 - (a - b)\,mxy - 2ab/mxy = 0$

As OP and OQ are right angles, sum of coeffs of x^2 and y^2 must be zero.

$$1 - la + lb - abl^2 - abm^2 = 0$$

$\Rightarrow \qquad 1 - l\,(a - b) - ab\,(l^2 + m^2) = 0$ ...(5)

The required locus will be obtained by eliminating l and m from (3), (4) and (5). So by (3) and (4), by cross-multiplication for l and m, we get

$$\frac{l}{0 - h} = \frac{m}{-k - 0} = \frac{1}{-h^2 - k^2}$$

$$\Rightarrow \qquad l = \frac{h}{h^2 + k^2} \text{ and } m = \frac{k}{h^2 + k^2}$$

So $\quad l^2 + m^2 = \dfrac{h^2}{(h^2 + k^2)^2} + \dfrac{k^2}{(h^2 + k^2)^2} = \dfrac{h^2 + k^2}{(h^2 + k^2)^2} = \dfrac{1}{h^2 + k^2}$

Substituting in (5), we have

$$1 - \frac{h}{h^2 + k^2}(a - b) - ab\,.\,\frac{1}{h^2 + k^2} = 0$$

$\Rightarrow \qquad (h^2 + k^2) - h\,(a - b) - ab = 0$

$\Rightarrow \qquad h^2 - ha + hb - ab + k^2 = 0$

$\Rightarrow \qquad (h - a)\,(h + b) + k^2 = 0$

Generalising, the required locus is $(x - a)\,(x + b) + y^2 = 0$ which is a circle having AB as diameter.

Example 77:

Show that the locus of a point, which is such that the tangents from it two given concentric circles are inversely as the radii, is a concentric circle, the square of whose radius is equal to the sum of the squares of the radii of the given circles.

Solution:

Taking the common centre as the origin and the radii of the two circles as a and b, the equations of the circles will be

$$x^2 + y^2 = a^2 \qquad ...(1)$$

and $$x^2 + y^2 = b^2 \qquad ...(2)$$

Let P be any point such that PT_1 and PT_2 are lengths from it to (1) and (2) respectively, say its co-ordinates are (h, k).

Now $PT_1 = \sqrt{(h^2 + k^2 - a^2)}$, $PT_2 = \sqrt{(h^2 + k^2 - b^2)}$.

By hypothesis $\frac{PT_1}{PT_2} = \frac{b}{a}$ or $\frac{PT_1^2}{PT_2^2} = \frac{b^2}{a^2}$

Putting the values of PT_1 and PT_2, we get

$$\frac{h^2 + k^2 - a^2}{h^2 + k^2 - b^2} = \frac{b^2}{a^2}$$

$\Rightarrow \qquad a^2h^2 + a^2k^2 - b^2 h^2 + b^2 k^2 - b^4$

$\Rightarrow \qquad h^2 (a^2 - b^2) + k^2 (a^2 - b^2) = a^4 - b^4$

$\Rightarrow \qquad h^2 + k^2 = a^2 + b^2$

Generalising, we get the required locus as $x^2 + y^2 = a^2 + b^2$ which is clearly a circle with centre (0, 0) and radius

$$\sqrt{(a^2 + b^2)}.$$

Example 78:

A tangent is drawn to the circle $(x - a)^2 + y^2 = b^2$ and a perpendicular tangent to the circle $(x + a)^2 + y^2 = c^2$; find the locus of their point of intersection, and prove that the bisector of the angle between them always touches one or other of two fixed circles.

Solution:

The equations of the circles are given as

$$(x - 2)^2 + y^2 = b^2 \qquad ...(1)$$

$$(x + a)^2 + y^2 = c^2 \quad ...(2)$$

be any tangent to (1). Hence, the perpendicular from the centre of (1) must be equal to its radius; so

$$\frac{am + c}{\sqrt{(1 + m^2)}} = b \qquad \text{(as centre is (a, 0) and radius is b)}$$

$$\Rightarrow \qquad c = -am + b\sqrt{(1+m^2)}$$

$$\therefore \qquad y = m(x - a) + b\sqrt{(1+m^2)}. \quad ...(3)$$

Similarly the equation of any tangent to (2) will be

$$y = m'(x + a) + c\sqrt{(1+m^2)}. \quad ...(4)$$

As by hypothesis, the two tangents are at right angles,

$$mm' = -1. \quad ...(5)$$

The required locus will be obtained by eliminating m and m' from (3), (4) and (5).

$$y = -\frac{1}{m}(x+a) + c\left(1 + \frac{1}{m^2}\right)$$

$$\Rightarrow \qquad my = -(x + a) + c\sqrt{(1+m^2)}. \quad ...(6)$$

(3) is same as $b\sqrt{(1+m^2)} + m(x - a) - y = 0$

(6) is same as $c\sqrt{(1+m^2)} - my - (x + a) = 0$.

Applying cross multiplication, we get

$$\frac{\sqrt{(1+m^2)}}{-(x^2 - a^2) - y^2} = \frac{m}{-yc + b(x+a)} = \frac{1}{-by - c(x-a)}$$

$$= \frac{\sqrt{(1+m^2)}}{\sqrt{(-yc + bx + ba)^2 + (-by - cx + ac)^2}}$$

[by ratio and proportion].

Squaring and simplifying, we get

$$(a^2 - x^2 - y^2)^2 = (bx - cy + ab)^2 + (by + cx - ac)^2$$

which is the required locus.

Again. Equation of the bisector of (3) and (6) is given as

$$\frac{m(x-a) - y + b\sqrt{(1+m^2)}}{\sqrt{(1+m^2)}} = \pm\frac{(x+a) + my - c\sqrt{(1+m^2)}}{\sqrt{(1+m^2)}}$$

Taking +ve sign, we get

$$mx - ma - y + b\sqrt{(1+m^2)} = x + a + my - c\sqrt{(1+m^2)}$$

$$\Rightarrow x(m-1) - y(m+1) - a(m+1) + (b+c)\sqrt{(1+m^2)} = 0$$

$$\Rightarrow \quad (y+a)(m+1) = x(m-1) + (b+c)\sqrt{(1+m^2)}.$$

$$\Rightarrow \quad y + a = \frac{m-1}{m+1}.x + \frac{b+c}{m+1}\sqrt{(1+m^2)}.$$

Example 79:

A straight line moves so that the product or the perpendicular on it from fixed points is constant. Prove that the locus of the feet of the perpendiculars from each of these points upon the straight line is a circle, the same for each.

Solution:

Let A and B be the two fixed points. Taking axis of X along AB and the right bisector of AB as y-axis, the co-ordinates of A and B may be taken as (– a, 0) and (a, 0), then let the equation of the given line be

$$x\cos\alpha + y\sin\alpha = p. \qquad ...(1)$$

Equation of the line through (– a, 0) and perpendicular to (1) is

$$y - 0 = \frac{\sin\alpha}{\cos\alpha}(x+a)$$

$$\Rightarrow \quad y\cos\alpha - x\sin\alpha = a\sin\alpha. \qquad ...(2)$$

Again, if AN and BM be the perpendiculars from A and B respectively on the line (1), then by hypothesis

$$AN \times BM = \frac{-a\cos\alpha + 0.\sin\alpha - p}{\sqrt{(\cos^2\alpha + \sin^2\alpha)}} \times \frac{a\cos\alpha + 0.\sin\alpha - p}{\sqrt{(\cos^2\alpha + \sin^2\alpha)}}$$

$$= \lambda^2 \text{ (constt)}$$

$$\Rightarrow \quad p^2 - a^2\cos^2\alpha = \lambda^2$$

$$\Rightarrow \quad p^2 = \lambda^2 + a^2\cos^2\alpha \qquad ...(3)$$

To get the reqd. locus, we have to eliminate p and α from (1), (2) and (3); squaring (1) and (2) and then adding

$$x^2\cos^2\alpha + y^2\sin^2\alpha + 2xy\cos\alpha\sin\alpha + y^2\cos^2\alpha + x^2\sin^2\alpha$$

$$- 2xy\cos\alpha\sin\alpha = p^2 + a^2\sin^2\alpha$$

[putting the value of p^2 from (3)]

or $x^2 + y^2 = \lambda^2 + a^2$,

which is reqd. locus.

If we change a into – a, we do not get any change in the equation, so it is the locus for both the points, hence proved.

Example 80:

A circle touches the axis of x and cuts off a constant length 2l from the axis of y; prove that the equation of the locus of its centre is $y^2 - x^2 = l^2 \operatorname{cosec}^2 \omega$, the axes being inclined at an angle ω.

Solution:

Let C, the centre of the circle be (h, k) and the radius be a; then the equation of the circle will be

$$(x - h)^2 + (y - k)^2 + 2(x - h)(y - k)\cos\omega = a^2 \qquad ...(1)$$

Draw CN ⊥ OX, then by Δ CMN,

$$CN = CM \sin\omega$$

$$\Rightarrow \qquad a = k\sin\omega. \qquad ...(2)$$

Equation of the axis of y is $x = 0$. ...(3)

To get the point of intersection, we solve (1) and (3), so putting the value of x from (3) in (1), we get

$$h^2 + y^2 + k^2 - 2yk - (2hy - 2hk)\cos\omega = a^2$$

$$\Rightarrow y^2 - 2y(k + h\cos\omega) = h^2 + k^2 + 2hk\cos\omega - a^2 = 0 \quad ...(4)$$

If A and B be the points of intersection having co-ordinates as $(0, y_1)$ and $(0, y_2)$, then clearly

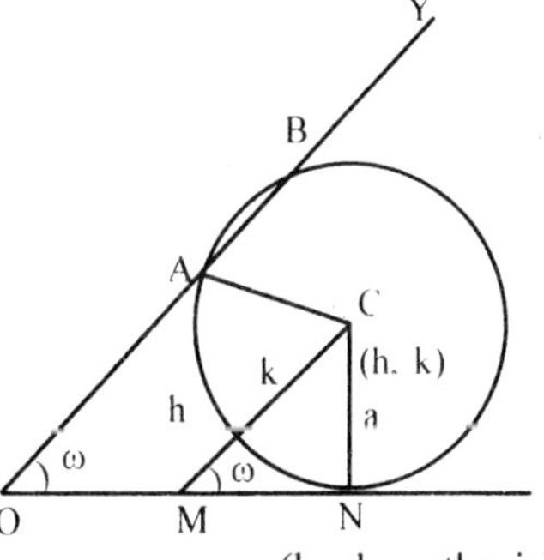

$$AB = y_2 - y_1 = 2l \qquad \text{(by hypothesis)}$$

$$\Rightarrow \qquad [(y_2 + y_1)^2 - 4y_1y_2] = (2l)^2 \qquad ...(5)$$

As (4) is a quadratic equation having roots as y_1 and y_2, so

$$y_2 + y_1 = -2(k + h\cos\omega)$$

and $$y_2y_1 = (h^2 + k^2 + 2hk\cos\omega - a^2)$$

Putting the values in (5), we get

$$4(k + h\cos\omega)^2 - 4(h^2 + k^2 + 2hk\cos\omega - a)^2 = 4l^2$$

[as $a^2 = k^2\sin^2\omega$ by (2)]

$$\Rightarrow \qquad k^2\sin^2\omega - h^2(1 - \cos^2\omega) = l^2$$

$\Rightarrow \qquad k^2 \sin^2 \omega - h^2 \sin^2 \omega = l^2$

$\Rightarrow \qquad k^2 - h^2 = l^2 \operatorname{cosec}^2 \omega.$

Generalising, we get $y^2 - x^2 = l^2 \operatorname{cosec}^2 \omega$.

Example 81:

O is any point in the plane of a circle, and OP_1P_2 any chord of the circle which passes through O and meets the circle in P_1 and P_2. On this chord is taken a point Q such that OQ is equal to (1) the arithmetic, (2) the geometric, and (3) the harmonic mean between OP_1 and OP_2, in each case find the equation to the locus of Q.

Solution:

Let O be the given fixed point, and A, the centre of the given circle.

Taking axis of x along OA and the line perpendicular to it through O as y-axis, the co-ordinates of O will be (0, θ) and the equation of the circle may be taken as

$$x^2 + y^2 - 2gx + c = 0 \qquad ...(1)$$

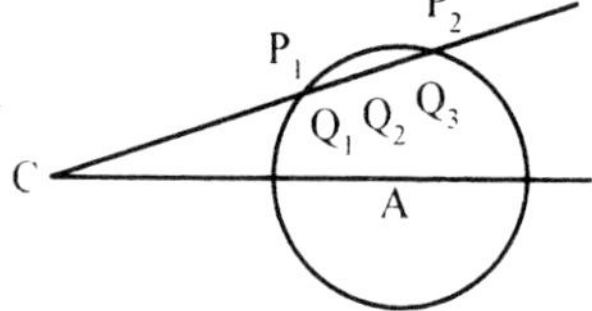

Suppose the equation of any line through O which cuts the circle at P_1 and P_2 be

$$y = mx. \qquad ...(2)$$

To solve (1) and (2), we put the value of y from (2) in (1), we get

$$x^2 + m^2x^2 - 2gx + c = 0 \qquad ...(3)$$

$$\Rightarrow \qquad x^2 (1 + m^2) - 2gx + c = 0$$

If the co-ordinates of P_1 and P_2 be (x_1, y_1) and (x_2, y_2) resp., then

$$x_1 + x_2 = 2g/(1+ m^2) \qquad ...(4)$$

and $\qquad x_1x_2 = c/(1 + m^2)$ by (3) $\qquad ...(5)$

Again $\quad OP_1^2 = x_1^2 + y_1^2 = x_1^2 + (mx_1)^2 = x_1^2\,(1+m^2) \qquad ...(6)$

Similarly $OP_2^2 = x_2^2 + y_2^2 = x_2^2 + m^2x_2^2\,(1+m^2) \qquad ...(7)$

(a) Let Q_1 be the point (x, y) such that OQ is the arith. mean of OP_1 and OP_2. Hence

$$OQ_1 = \frac{1}{2}(OP_1 + OP_2)$$

$$\Rightarrow \quad \sqrt{(x^2+y^2)} = \frac{1}{2}\left[\sqrt{\{x_1^2\,(1+m^2)\}} + \sqrt{\{x_2^2\,(1+m^2)\}}\right]$$

$$\Rightarrow x\sqrt{(1+m^2)} = \frac{1}{2}\sqrt{(1+m^2)}\,(x_1+x_2) \quad \text{(as Q lies on } OP_1P_2,\ y = mx)$$

$$\Rightarrow \qquad x = \frac{1}{2}(x_1+x_2) = \frac{1}{2}\,\frac{2g}{(1+m^2)} \qquad ...(8)$$

To get the required locus, we have to eliminate m between (2) and (8) So putting the value of m from (2) in (8), we get

$$g = x\left(1+\frac{y^2}{x^2}\right) \quad \text{or } x^2+y^2-gx = 0$$

It is the required locus and clearly is a circle with centre (g/2, 0) and radius g/2.

(b) Let Q_2 be the point (x, y) such that OQ_2 is the geometric mean between OP_1 and OP_2. Then

$$OQ_2^2 = OP_1 \,.\, OP_2$$

$$\Rightarrow \quad (x^2+y^2) = \{x_1\sqrt{(1+m^2)}\}\,\{x_2\sqrt{(1+m^2)}\}$$

$$\Rightarrow \quad x^2(1+m^2) = 1\,(1+m^2)\,.\,x_1\,x_2$$

$$\Rightarrow \quad x^2 = \frac{c}{1+m^2}\left(\text{as } x_1\,x_2 = \frac{c}{1+m^2} \text{ by (5)}\right)$$

$$\Rightarrow \quad (1+m^2)\,x^2 = c. \qquad ...(9)$$

To get the reqd. locus, we have to eliminate m between (2) and (9) so putting the value of m from (2) in (9), we get

$$x^2\left(1+\frac{y^2}{x^2}\right) = c \qquad \text{or} \quad x^2+y^2 = c$$

It is the reqd. locus and again is a circle with centre (0, 0) and radius $\sqrt{c}$.

(c) Let Q_3 be the point (x, y) such that OQ_3 is the harmonic mean between OP_1 and OP_2. So

$$OQ_3 = \frac{2OP_1\,.OP_2}{OP_1+OP_2}$$

$$\Rightarrow \quad \sqrt{(x^2+y^2)} = \frac{2x_1\sqrt{(1+m^2)}\,.x_2\sqrt{(1+m^2)}}{x_1\sqrt{(1+m^2)}+x_2\sqrt{(1+m^2)}}$$

$$\Rightarrow \quad x = \frac{2x_1\,x_2}{x_1+x_2},\ \text{As } x^2+y^2 = x^2(1+m^2)$$

$$\Rightarrow \quad x = \frac{2c/\sqrt{(1+m^2)}}{2g/\sqrt{(1+m^2)}} = \frac{c}{g}$$

$\Rightarrow \quad gx - c = 0$ ...(10)

It is the reqd. locus, which is clearly a st. line, parallel to y-axis at a distance of c/g.

Again, the polar of the origin O w.r.t. the circle (1) is $0 + 0 - g(x + 0) + c = 0$ or $gx - c = 0$ which is same as (10), hence the reqd. locus is the polar of O.

Example 82:

O is a fixed point and P any point on a given straight line; OP is joined and on it is taken a point Q such that OP.OQ = k^2; prove that the locus of Q, i.e. the inverse of the given straight line with respect to O, is a circle which passes through O.

Solution:

Let O be the fixed point and AB the straight line. Draw ON perpendicular on AB and OY parallel to AB. Taking the axes, along ON and OY, if ON = a, the co-ordinates of any point P on AB may be taken as (a, y_1). Join OP and take any point Q (α, β) on PO such that

$$OP.OQ = k^2 \quad ...(1)$$

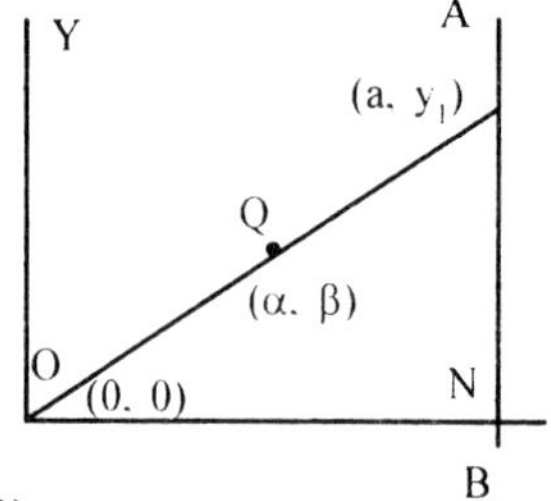

Now $OP = \sqrt{\{(a^2 + y_1^2)}$

and $OQ = \sqrt{\{(\alpha^2 + \beta^2)\}}$

Putting in (1), we get

$$\sqrt{\{(a^2 + y_1^2)\ (\alpha^2 + \beta^2)\}} = k^2. \quad ...(2)$$

Again equation to OP is $y - 0 = \dfrac{y_1 - 0}{a - 0}(x - 0)$

$\Rightarrow \quad ay = y_2 x.$

As Q ≡ (a, b) lies on OP, the co-ordinates will satisfy it so

$$a\beta = y_1\alpha \quad \text{or} \quad y_1 = a\beta/\alpha. \quad ...(3)$$

We have to eliminate y_1 from (2) and (3).

Putting this value in (2), we get

$$\sqrt{\{(a^2 + a^2\ \beta^2 / a^2)\ (\alpha^2 + b^2)\}} = k^2$$

$$\Rightarrow \quad \sqrt{\left\{\left(\frac{a^2\ \alpha^2 + a^2\ \beta^2}{\alpha^2}\right)\left(\alpha^2 + \beta^2\right)\right\}} = k^2$$

$$\Rightarrow \quad \frac{a}{\alpha}\left(\alpha^2 + \beta^2\right) = k^2$$

Generalising and simplifying for the variable (a, b), we get

$$a(x^2 + y^2) = k^2x$$

which is a circle.

Example 83:

Find the locus of the point of intersection of the tangent to a given circle and the perpendicular let fall on this tangent from a fixed point on the circle.

Solution:

Let C be the centre of the circle of radius a, and O be any point on its circumference. Take O as the pole and OC as the initial line.

Let AB be any tangent at the point T on the circle. Draw OP perpendicular to AB from O. Let the co-ordinates of P be (r, θ), so that

$$OP = r, \quad ...(1)$$

and $$\angle POC = \theta. \quad ...(2)$$

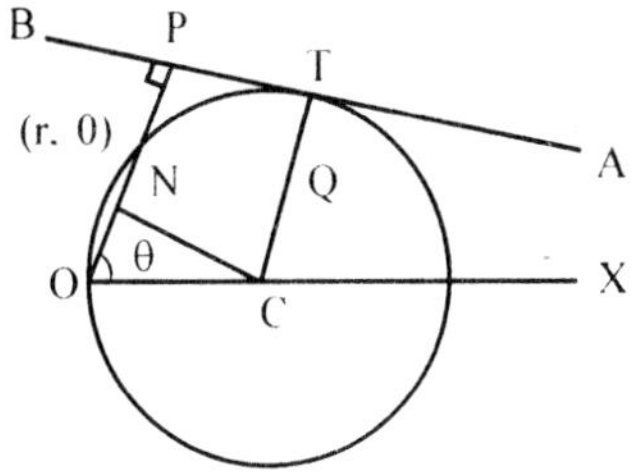

B P T (r, θ) N Q A O θ X C

If CN is perpendicular from C on OP, then

$$OP = ON + NP. \quad ...(3)$$

By Δ ONC, ON = OC

$$\cos NOC = a \cos \theta$$

and NP = CT = a (as OC = CT = NP = a, the radius).

Putting the value of OP, ON and NP in (3),

$r = a \cos \theta + a$ which is the reqd. locus.

Example 84:

In any circle prove that the perpendicular from any point of it on the line joining the points of contact of two tangents is a mean proportional between the perpendicular from the point upon the two tangents.

Solution:

Let the equation of the circle be $x^2 + y^2 = a^2$. ...(1)

Let there be any point P ≡ (a cos θ, b sin θ) on it. Again say T_1 and T_2 be any two other points as (a cos α, a sin α) and (a cos β, a sin β) on it.

Equation of the tangent at (a cos a, a sin a) to (1) is

$$x \cos \alpha + ya \sin \alpha - a^2 = 0$$

or $$x \cos \alpha + y \sin \alpha - a = 0 \quad ...(2)$$

Similarly equation of the tangent at (a cos β, a sin β) to (1) will be

$$x \cos \beta + y \sin \beta - a = 0 \quad \text{...(3)}$$

Equation of the chord of contact T_1 T_2 will be

$$y - a \sin \alpha = \frac{a \sin \beta - a \sin \alpha}{a \cos \beta - a \cos \alpha} (x - a \cos \alpha)$$

$$\Rightarrow \quad y - a \sin \alpha = \frac{a.2 \cos \frac{\beta+\alpha}{2} . \sin \frac{\beta-\alpha}{2}}{a.2 \sin \frac{\beta+\alpha}{2} . \sin \frac{\alpha-\beta}{2}} (x - \cos \alpha)$$

$$\Rightarrow \quad y \sin \frac{\beta+\alpha}{2} - a \sin \alpha . \sin \frac{\beta+\alpha}{2} = -x \cos \frac{\beta+\alpha}{2}$$

$$\Rightarrow \quad x \cos \frac{\alpha+\beta}{2} + y \sin \frac{\alpha+\beta}{2} = a \cos \frac{\alpha-\beta}{2}$$

If p_1, p_2 and p_2 be the length of perpendiculars from P ≡ (a cos θ, a sin θ) on (2), (3) and (4) respectively, then

$$p_1 = \frac{a \cos \alpha + a \sin \theta \sin \alpha - a}{\sqrt{(\cos^2 \alpha + \sin^2 \alpha)}} = a \ \{\cos (\theta - \alpha) - a\}$$

$$= 2a \sin \frac{2\theta - \alpha}{2} \text{ (omitting –ve sign).} \quad \text{...(5)}$$

Similarly, $p_2 = 2a \sin^2 \frac{\theta - \beta}{2}$, ...(6)

and $$p_3 = \frac{a \cos \theta \cos \frac{1}{2} (\alpha+\beta) + a \sin \theta . \sin \frac{1}{2} (\alpha+\beta) - a \cos \frac{1}{2} (\alpha-\beta)}{\sqrt{\left\{\cos^2 \frac{1}{2} (\alpha+\beta) + \sin^2 \frac{1}{2} (\alpha+\beta)\right\}}}$$

$$= a \left[\cos \left\{\theta - \frac{1}{2} (\alpha+\beta)\right\} - \cos \frac{1}{2} (\alpha-\beta)\right]$$

$$= 2a \sin \frac{1}{2} (\theta-\beta) \sin \frac{1}{2} (\theta-\alpha) \text{ (omitting – ve sign)} \quad \text{...(7)}$$

Now $$\sqrt{(p_1 p_2)} = \sqrt{\left\{2a \sin^2 \frac{1}{2} (\theta-\alpha)\right\} \left\{2a \sin \frac{1}{2} (\theta-\beta)\right\}}$$

$$= 2a \sin \frac{1}{2} (\theta-\alpha) . \sin \frac{1}{2} (\theta-\beta) = p_3$$

Hence p_3 is the geometric mean between p_1 and p_2.

Example 85:

A variable circle passes through the point of intersection O of any two straight lines and cuts off from them portions OP and OQ such that m. OP + n. OQ is equal to unity; prove that this circle always passes through a fixed point.

Solution:

Let the two given lines OX and OY be taken the axes, meeting at O, and the co-ordinates of the centre of any circle be (h, k). If the axes are at an angle, ω, then the equation of the circle may be given by

$$x^2 + y^2 + 2xy \cos \omega - 2x (h + k \cos \omega) - 2y (k + h \cos \omega)$$
$$+ h^2 + k^2 + 2hk \cos \omega - a^2 = 0$$

As the circle passes through origin, (0, 0) must satisfy the equation, therefore the constant term must be zero. So

$$x^2 + y^2 + 2xy \cos \omega - 2x (h + k \cos \omega)$$
$$- 2y (k + h \cos \omega) = 0 \quad ...(1)$$

If P be the other point (besides the origin) where circle cuts x-axis, then solving (1) with x-axis *i.e.* y = 0, we get

$$x^2 - 2x (h + k \cos \omega) = 0 \quad \text{or} \quad x [x - 2 (h + k \cos \omega)] = 0.$$

$\therefore$ either $x = 0$ or $x = 2 (h + k \cos \omega)$

Hence $OP = 2 (h + k \cos \omega)$. ...(2)

Similarly if Q be the point other than origin where the circle meets y axis. Solving with x = 0, we have

$$y^2 - 2y (k + h \cos \omega) = 0$$

Hence either $y = 0$ or $y = 2 (k + h \cos \omega)$.

So $OQ = 2 (k + h \cos \omega)$ ...(3)

By hypothesis, $m.OP + n.OQ = 1$. ...(4)

Putting the values from (3) and (2) in (4), we get

$$m.2 (h + k \cos \omega) + n.2 (k + h \cos \omega) = 1$$

Generalising, we get the locus of the centre as

$$2m (x + y \cos \omega) + 2n (y + x \cos \omega) = 1$$

or $2x (m + n \cos \omega) + y (n + m \cos \omega) = 1$. ...(5)

This is clearly a straight line say AB.

Let ON be the perpendicular to AB Produce ON upto O' such that ON = NO'. Clearly AB is the right bisector of OO'. So the circle having its centre ON AB and passing through the origin O must also pass through O' which is a fixed point.

Example 86:

Find the locus of the middle points of chords of the circle $x^2 + y^2 = a^2$ which subtend a right angle at the be point (c, 0).

Solution:

Let the equation of the circle be

$$x^2 + y^2 = a^2. \qquad ...(1)$$

Take any two points A and B on the circle as $(a \cos \alpha, a \sin \alpha)$ and $(a \cos \beta, a \sin \beta)$. If P be the middle point of AB as (h, k) then

$$h = \frac{a \cos \alpha + a \cos \beta}{2} = \frac{a}{2}(\cos \alpha + \cos \beta). \qquad ..(2)$$

and $$k = \frac{a \sin \alpha + a \sin \beta}{2} = \frac{a}{2}(\sin \alpha + \sin \beta) \qquad ...(3)$$

Again, if Q be the given point (c, 0), then

slope of AQ is $\dfrac{a \sin \alpha - 0}{a \cos \alpha - c} = \dfrac{a \sin \alpha}{a \cos \alpha - c}$

slope of BQ is $\dfrac{a \sin \beta - 0}{a \cos \beta - c} = \dfrac{a \sin \beta}{a \cos \beta - c}$

As angle AQB = 90°, hence AQ and BQ are perpendicular to each other, so

$$\frac{a \sin \alpha}{a \cos \alpha - c} \cdot \frac{a \sin \beta}{a \cos \beta - c} = -1$$

or $$a^2 \sin \alpha \sin \beta = -(a \cos \alpha - c)(a \cos \beta - c)$$

or $a^2 (\sin \alpha \sin \beta + \cos \alpha \cos \beta) - ac(\cos \alpha + \cos \beta) + c^2 = 0$ ...(4)

To eliminate α and β from (2), (3) and (4), squaring and adding (2) and (3), we get

$$h^2 + k^2 = \frac{a^2}{4}[\cos^2 \alpha + \cos^2 \beta + 2 \cos \alpha \cos \beta + \sin^2 \alpha + \sin^2 \beta + 2 \sin \alpha \sin \beta]$$

$$\Rightarrow h^2 + k^2 = \frac{a^2}{4}[2 + 2(\cos \alpha \cos \beta + \sin \alpha \sin \beta)]$$

$$\Rightarrow \cos\alpha\cos\beta + \sin\alpha\sin\beta = \frac{2\left(h^2+k^2\right)-a^2}{a^2} \qquad ...(5)$$

and by (2), $\cos\beta\alpha + \cos\beta = 2h\sqrt{a}$ ...(6)

Substituting the values from (5) and (6) in (4), we get

$$a^2\left[\frac{2\left(h^2+k^2\right)-a^2}{a^2}\right]-ac\frac{2h}{a}+c^2=0$$

$\Rightarrow$ $2\,(h^2 + k^2) - a^2 - 2ch + c^2 = 0.$

Generalising, we get $2x^2 + 2y^2 - 2cx + c^2 - a^2 = 0$

which is a circle. **Ans.**

Example 87:

Find the locus of the middle points of chords of the circle $x^2 + y^2 = a^2$ which pass through the fixed point (h, k).

Solution:

Any line passing through the given point (h, k) is

$$y - k = m\,(x - h) \qquad ...(1)$$

and the given circle is $x^2 + y^2 = a^2$ ...(2)

Let (α, β) be the co-ordinates of the middle point of the chord; then (a, b) will satisfy (1); so

$$\beta - k = m\,(\alpha - h) \qquad ...(3)$$

Again equation of the line perpendicular to the chord (1) and passing through the centre (0, 0) will be $my = x = 0$.

As it will cut the chord at the middle point, (a, b) will satisfy this equation also, that is $m\beta + \alpha = 0$.

To get the required locus we eliminate the variable m from (3) and (4).

So putting the value of m from (4) in (3), we get

$$(\beta - k) = -\frac{\alpha}{\beta}(\alpha - h)$$

or $\beta^2 - \beta k = -\alpha^2 + \alpha h.$

Generalising for the co-ordinates of the middle point (α, β), and rearranging, we get the required locus as

$x^2 + y^2 - xh - yk = 0$ which is a circle. **Ans.**

Example 88:

O is a fixed point and P any point circle; on OP is taken a point Q such that OQ is in a constant ratio to OP; prove that the locus of Q is a circle.

Solution:

Let the fixed point be the origin and the circle be

$$(x - a)^2 + (y - b)^2 = r^2.$$

The parametric co-ordinates of any point P on the circle may be taken as (a + r cos θ, b + r sin θ), where (α, β) is the centre.

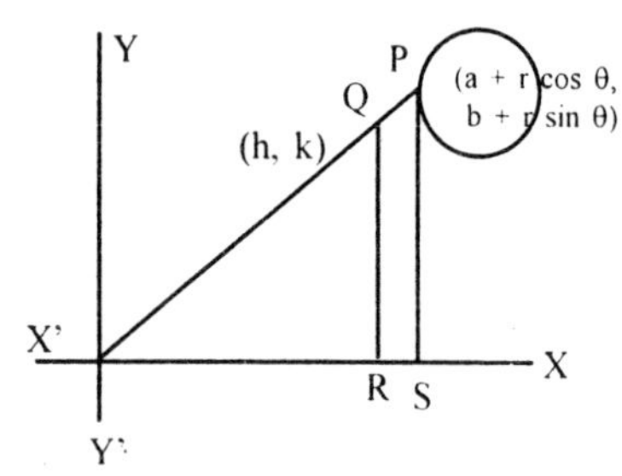

Suppose Q is the point satisfying the given condition

i.e. $\frac{OQ}{OP} = \lambda$

Draw PA and QB perpendicular to OX.

Then by similar triangles OQB and OPA,

$$\frac{OQ}{OP} = \frac{OB}{OA} = \frac{OB}{PA}$$

or $$\lambda = \frac{h}{a + r\cos\theta} = \frac{k}{b + r\sin\theta}$$

or $\quad la + \lambda r\cos\theta = h \quad$ or $\quad \lambda r\cos\theta = h - \lambda a \quad$ ...(1)

or $\quad \lambda b + \lambda r\sin\theta = k, \quad$ or $\quad \lambda r\sin\theta = k - \lambda b \quad$ ...(2)

The required locus will be found by generalising the eliminate for θ of (2) and (2); so to eliminate θ, squaring and adding (1) and (2), we get

$$\lambda^2 r^2 (\cos^2\theta + \sin^2\theta) = (h - \lambda a)^2 + (k - \lambda b)^2$$

or $$(h - \lambda a)^2 + (k - \lambda b)^2 = \lambda^2 r^2.$$

Generalising, we get $(x - \lambda a)^2 + (y - \lambda b)^2 = \lambda^2 r^2$

which is clearly a circle. **Ans.**

Example 89:

Whatever be the value of a, prove that the locus of the intersection of the straight lines

$$x\cos\alpha + y\sin\alpha = a \text{ and } x\sin\alpha - y\cos\alpha = b$$

is a circle.

Solution:

The two lines are given as

$$x \cos \alpha + y \sin \alpha = a \qquad ...(1)$$

and $$x \sin \alpha - y \cos \alpha = b \qquad ...(2)$$

If the point of intersection be P (h, k), line it will satisfy (1) and (2).

So $$h \cos \alpha + k \sin \alpha = c \qquad ...(3)$$

and $$h \sin \alpha - k \cos \alpha = b \qquad ...(4)$$

To eliminate the variable α, squaring (3) and (4) and adding, we get

$$h^2 (\cos^2 \alpha + \sin^2 \alpha) + k^2 (\sin^2 \alpha + \cos^2 \alpha) = a^2 + b^2$$

or $$h^2 + k^2 = a^2 + b^2.$$

Generalising, we get $x^2 + y^2 = a^2 + b^2$,

which is clearly a circle. **Ans.**

Example 90:

One vertex of a triangle of given species is fixed, and another moves along the circumference of a fixed circle; prove that the locus of the remaining vertex is a circle and find its radius.

Solution:

Let OAP be any triangle. Take O as the origin. A is any point on the circumference of a circle whose centre is C and radius a. Taking OC as the initial line, and OC = d, the polar co-ordinates of C will be (d, 0) and for A (r_1, θ_1), P be moving point (r, θ).

Let the angles POA, OAP and OPA be α, β and γ respectively, as the triangle is of given species.

Now in triangle AOC

$$AC^2 = OA^2 + OC^2 - 2.OA.OC \cos \theta$$

or $$a^2 = r_1^2 + d^2 - 2r_1 \cos \theta_1 \qquad ...(1)$$

Again from Δ OAP, by sine formula, $\dfrac{OA}{\sin \gamma} = \dfrac{OP}{\sin \beta}$

or $$\frac{r_1}{\sin \gamma} = \frac{r}{\sin \beta} \quad \text{or} \quad r_1 = \frac{r \sin \gamma}{\sin \beta} \qquad ...(23)$$

Putting the values of θ_1 and r_1 from (2) and (3) in (1), we get

$$a^2 = \left\{\frac{r \sin \gamma}{\sin \beta}\right\} + d^2 - 2\frac{r \sin \gamma}{\sin \beta} d \cos (\theta - \alpha)$$

This equation represents a circle with a radius sin β/sin γ. **Ans.**

Example 91:

From a point P on a circle perpendiculars PM and PN are drawn to two radii of the circle which are not at right angles; find the locus of the middle point of MN.

Solution:

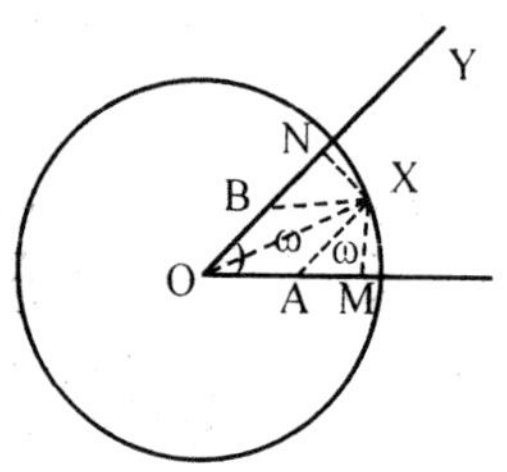

Taking the two lines as axes, the angle between them being w, let (h, k) be any point on the circle. Draw PA and PB parallel to y-axis and x-axis respectively, PM and PN are perpendiculars to the axis of x and y; then ∠OAP = (180° – ω) and OP is the radius say a

By ΔOAP, $OP^2 = OA^2 + AP^2 - 2OA.AP \cos(180° - \omega)$

or $\quad a^2 + h^2 + k^2 + 2.hk \cos w$ ...(1)

Again $\quad OM + OA + AM = h + k \cos \omega,$

$ON = OB + BN = k + h \cos \omega$

∴ Co-ordinates of M and N are respectively

$(h + k \cos \omega, 0)$ and $(0, k + h \cos \omega)$

If Q be the mid-point of MN, then if the co-ordinates of Q are (x, y), then

$$x = \frac{h + k \cos \omega}{2} \quad ...(2)$$

and
$$y = \frac{k + h \cos \omega}{2} \quad ...(3)$$

Solving (2) and (3), we get

$$h = \frac{2(x - y \cos \omega)}{\sin^2 \omega}, + \left\{\frac{2(y - x \cos \omega)}{\sin^2 \omega}\right\}^2$$

$$+ \frac{2.2(x - y \cos \omega)^2 (y - x \cos \omega)}{\sin^2 \omega \sin^2 \omega}.\cos \omega = a^2$$

or $\frac{4}{\sin^4 \omega}[(x^2 + y^2 \cos 2\omega - 2xy \cos \omega) + y^2 + x^2 \cos^2 \omega - 2yx \cos \omega)$

$$+ 2(xy + xy \cos^2 \omega - x^2 \cos \omega - y^2 \cos \omega) \cos \omega] = a^2$$

or $x^2(1 + \cos^2 \omega - 2\cos^2 \omega) + y^2(1 + \cos^2 \omega - 2\cos^2 \omega)$

$$- 2xy \cos \omega (1 - \cos^2 \omega) = \frac{a^2 \sin^4 \omega}{4}$$

or $$x^2 + y^2 - 2xy \cos \omega = \frac{a^2 \sin^2 \omega}{4}.$$

Example 92:

A point moves so that the sum of the squares of its distances from the four sides of a square is constant; prove that it always lies on a circle.

Solution :

Let the sides AB and AD of the square be taken as the axes, and the side of the square be a units, then the equation of the sides AB, BC, CD and DA are respectively.

$y = 0,$ ...(1) $x = a,$ ...(2)

$y = a,$ ...(3) and $x = 0$ ...(4).

If P be the moving point as (h, k) its distance from (1), (2), (3) and (4) are respectively

$$k, (a - h), (a - k) \text{ and } h.$$

By hypothesis $k^2 + (a - h)^2 + (a - k)^2 + h^2 = 2\lambda$ (constt.)

$\Rightarrow \quad 2h^2 + 2k^2 - 2ah - ak - 2\lambda = 0$

$\Rightarrow \quad h^2 + k^2 - ah - ak - \lambda = 0$

Generalising, we get $x^2 + y^2 - ax - ay - \lambda = 0$

which is clearly a circle. **Proved**

Example 93:

A point moves so that the sum of the squares of the perpendiculars let fall from it on the sides of an equilateral triangle is constant; prove that its locus is a circle.

Solution:

Let ABC be the given equilateral triangle. Take A as origin, AB and AC as the axis and 'a' as the side of the triangle. Clearly w = 60°, AB = AC a. Hence the equations of AB and BC are respectively

$y = 0,$ ...(1),

$x = 0$ ...(2)

and $\frac{x}{a} + \frac{y}{a} = 1$ or $x + y - a = 0$ ...(3)

Let the moving point P be (h, k).

Length of perpendicular from (h, k) on $x + y - a = 0$

$$= \frac{h+k-a}{\sqrt{(1+1-2.1.1\cos 60^\circ)}} \sin 60^\circ = (h + k - a).\ \frac{\sqrt{3}}{2} = PN \text{ (say)}$$

Similarly length of perpendicular from (h, k)

on $x = 0$ is $\quad h \sin 60^\circ = h.\sqrt{3/2} = PM$ (say)

and on $y = 0$ is $k \sin 60^\circ = k.\sqrt{3/2} = PL$ (say)

By hypothesis $PL^2 + PM^2 + PN^2 = 3/4\ \lambda$ (constt.)

$$\therefore \quad \left\{\frac{(h+k-a)\sqrt{3}}{2}\right\}^2 + \left(h.\frac{\sqrt{3}}{2}\right)^2 + \left(k.\frac{\sqrt{3}}{2}\right)^2 = \frac{3}{4}\lambda$$

$$\Rightarrow \quad \frac{3}{4}\left[(h+k-a)^2 + h^2 + k^2\right] = \frac{3}{4}\lambda$$

Generalising and simplifying, we get

$$2x^2 + 2y^2 + 2xy - 2ax - 2ay + a^2 - \lambda = 0$$

which is equation of a circle. **Proved.**

Example 94:

Tangents are drawn to a circle from a point which always lies on a given line; prove that the locus of the middle point of the chord of contact is another circle.

Solution:

Let the given line be $ax + by + c = 0$ ...(1)

and the circle be $\quad x^2 + y^2 = r^2$ ...(2)

Suppose (a, b) is any point on (1); then we have

$$a\alpha + b\beta + c = 0.$$

Chord of contact of (a, b) with respect to (2) is

$$\alpha x + \beta y = r^2. \quad ...(3)$$

If (h, k) be the middle point of this chord of contact, (h, k) will satisfy its equation; so $\alpha h + \beta k = r^2$. ...(4)

Again, if we draw a line perpendicular to the chord of contact through the centre of the circle (0, 0), it will cut it at the middle point.

Equation of the line perpendicular to the chord of contact through origin (0, 0) is $Bx - \alpha y = 0$.

As (h, k) lies on this line also, we have

$$\beta h - \alpha k = 0 \qquad ...(5)$$

To get the required locus, we have to eliminate the variables α and β from (3), (4) and (5). So solving (4) and (5) for α and β, we get

$$\alpha = \frac{xr^2}{x^2 + y^2} \text{ and } \beta = \frac{yr^2}{x^2 + y^2}.$$

Example 95:

The locus of a point whose distance from a fixed point is in a constant ratio to the tangent drawn from it in a given circle.

Solution:

Let Q be the fixed point as (α, β), circle be $x^2 + y^2 = a^2$ and the moving point P be (h, k)

Length of tangent PT from P to the circle

$$= \sqrt{(h^2 + k^2 - a^2)}$$

and distance PQ $= \sqrt{[(h - a)^2 + (k - b)^2]}$

By hypothesis, $\frac{PQ}{PT} = \lambda$ = constt. or $PQ^2 = \lambda^2 PT^2$

Substituting the values, we get

$$(h - a)^2 + (k - b)^2 = \lambda^2 (h^2 + k^2 - a^2)$$

$$\Rightarrow \quad h + k^2 - 2ha - 2kb + a^2 + b^2 = \lambda^2 h^2 + \lambda^2 k^2 - \lambda^2 a^2.$$

It is clearly a circle. **Ans.**

Example 96:

A point moves so that the sum of the squares of its distances from n fixed points is given. Prove that its locus is a circle.

Solution:

Let the n given fixed points A, B, C ... etc. bc (x_1, y_1), (x_2, y_2), (x_3, y_3), ... (x_n, y_n) and the moving point P be (h, k), then the hypothesis

$$PA^2 + PB^2 + PC^2 \ldots = \lambda \text{ (say)}$$

Putting the values,

$$\{(h - x_1)^2 (k - y_1)^2\} + \{(h - x_2)^2 + (k - y_2)^2\} + \ldots$$

$$\ldots + \{(h + x_n)^2 + (k - y_a)^2 = \lambda$$

$$\Rightarrow nh^2 + nk^2 - 2h(x_1 + x_2 + \ldots + x_n) - 2k(y_1 + y_2 + \ldots + y_n)$$

$$+ (x_1^2 + x_2^2 + \ldots + x_n^2) + (y_1^2 + y_2^2 + \ldots + y_n^2)\ \lambda$$

Generalising, we get

$$nx^2 + ny^2 - 2x.\Sigma x_1 - 2y\Sigma y_2 + \sum x_1^2 + \sum x_1^2 - \lambda = 0$$

which is a circle. **Proved.**

Example 97:

A point moves so that the sum of the squares of its distances from the angular points of a triangle is constant; prove that its locus is a circle.

Solution:

Let the triangle be ABC. Taking A as the origin, and axis of x along the side AB, the co-ordinates of B will be (a, 0) where AB = a Let c be (x_1, y_1) A is clearly (0, 0).

Taking P the moving point, as (h, k),

$$PA^2 = (h - 0)^2 + (k - 0)^2 = h^2 + k^2, \quad ...(1)$$

$$PB^2 = (h - a)^2 + (k - 0)^2 = h^2 + k^2 - 2ah + a^2 \quad ...(2)$$

$$PC^2 = (h - x_1)^2 + (k - y_1)^2$$

$$= h^2 + k^2 - 2hx_1 - 2ky_1 + x_1^2 + y_1^2. \quad ...(3)$$

By hypothesis, $PA^2 + PB^2 + PC^2 = \lambda$ (constt.)

Putting the value

$$(h^2 + k^2) + (h^2 + k^2 - 2ah + a^2) + (h^2 + k^2 - 2hx_1 - 2ky_1 + x_1^2 + y_1^2) = \lambda.$$

Generalising and simplifying, we get

$$3x^2 + 3y^2 - 2x(a + x_1) - 2yy_1 + (x_1^2 + y_1^2 + a^2 - \lambda) = 0$$

which is clearly a circle. **Proved.**

Example 98:

Find the locus of a point which moves so that the square of the tangent drawn it to the circle $x^2 + y^2 = a^2$ is equal to c times its distance from the straight line $lx + my + n = 0$.

Solution:

Let the moving point P be (h, k).

Length of tangent PT, from P upto the circle $x^2 + y^2 - a^2 = 0$ is given as

$$\sqrt{(h^2 + k^2 - a^2)}$$

PN, the length of perpendicular from P on $lx + my + n = 0$ is

$$= \frac{l\text{h} + \text{mk} + \text{n}}{\sqrt{(\lambda^2 + \text{m}^2)}}$$

By hypothesis $\text{PT}^2 = \text{c.PN}$

Hence $(\text{h}^2 + \text{k}^2 - \text{a}^2) = \text{c. (lh + mk + n)} \sqrt{(\lambda^2 + \text{m}^2)}$

$\Rightarrow \quad \sqrt{(l^2 + \text{m}^2)}\ (\text{h}^2 + \text{k}^2 - \text{a}^2) = \text{c}\ (l\text{h} + \text{mk} + \text{n})$

Simplifying and generalising, we get

$$\sqrt{(l^2 + \text{m}^2)}\ (\text{x}^2 + \text{y}^2 - \text{a}^2) - (l\text{x} + \text{my} + \text{n}) = 0$$

which is clearly a circle. **Ans.**

Example 99:

Prove that the straight line $y = mx$ will touch the circle

$$x^2 + 2xy \cos \omega + y^2 + 2gx + 2fy + c = 0$$

if $(g + fm)^2 = c\ (1 + 2m \cos \omega + m^2)$.

Solution:

The circle is given as

$$\text{x}^2 + \text{y}^2 + 2\text{xy} \cos \omega + 2\text{gx} + 2\text{fy} + \text{c} = 0 \quad ...(1)$$

and the line is $\text{y} = \text{mx}$. ...(2)

If (2) is a tangent to (1), the two points of intersection should coincide. Putting the value of y from (2) in (1), we get

$$\text{x}^2 + \text{m}^2\text{x}^2 + 2\text{x.mx} \cos \omega + 2\text{gx} + 2\text{fmx} + \text{c} = 0 \quad ...(3)$$

If the points of intersection coincide (3), must be a perfect square. So discriminant should be zero *i.e.*,

i.e. $\{2\ (\text{g} + \text{fm})^2 - 4\ (1 + \text{m}^2 + 2\text{m.}\cos \omega)\ \text{c} = 0$

$\Rightarrow \quad (\text{g} + \text{fm})^2\ (1 + 2\text{m} \cos \omega + \text{m}^2)$

which is the reqd. condition.

Example 100:

Find the centre and radius of the circle

$$\text{x}^2 + \sqrt{3\text{xy}} + \text{y}^2 - 4\text{x} - 6\text{y} + 5 = 0.$$

Solution:

Equation on as the given circle

$$x^2 + \sqrt{3xy} + y^2 - 4x - 6y + 5 = 0$$

Comparing coeffs. of xy etc.

$$2 \cos \sqrt{3} = \sqrt{3} \quad ...(1)$$

$$2 (h + k \cos \omega) = 4 \quad ...(2)$$

$$2 (k + h \cos \omega) = 6, \quad ...(3)$$

$$2 + k^2 + 2hk \cos \omega - a^2 = 5 \quad ...(4)$$

By (1), $\cos \omega = \sqrt{3/2}$ or $\omega = 30^o$.

Putting the values of cos w in (2) and (3), we get respectively

$$2h + \sqrt{3k} = 4 \quad ...(5)$$

and $$2k + \sqrt{3h} = 6 \quad ...(6)$$

Solving (5) and (6), we get $h = 8 - 6\sqrt{3}$ and $k = 12 - 4\sqrt{3}$.

Putting the values of h, k and cos w in (4), we have

$$\left(8-6\sqrt{3}\right)^2 + \left(12-4\sqrt{3}\right)^2 + 2.\left(8-6\sqrt{3}\right)\left(12-4\sqrt{3}\right)\left(\frac{\sqrt{3}}{2}\right) - a^2 = 5$$

or $$a^2 + 47 - 24\sqrt{3}$$

Hence $w = 30^o$, centre is $(8 - 6\sqrt{3}, 12 - 4\sqrt{3})$ and the radius is $\left\{\sqrt{47 - 24\sqrt{3}}\right\}$.

Example 101:

Find the locus of the vertex of a triangle, given (1) its base and the sum of the squares of its sides, (2) its base and the sum of m times the square of one side and n times the square of the other.

Solution:

Let BC be the base of the DABC. Taking BC as the axis of x and its right bisector as the axis of y, BC = 2a, the co-ordinates of B and C may be taken as (–a, 0) and (a, 0) respectively.

Suppose the moving vertex A be (h, k)

(i) By hypothesis, $AB^2 + AC^2 = 2\lambda$ (constant)

$$\Rightarrow \quad \{(h + a)^2 + k^2\} + \{(h - a)^2 + k^2\} = 2\lambda.$$

Simplifying and generalising, we get

$$x^2 + y^2 + a^2 - \lambda = 0,$$

which is clearly a circle.

(ii) By hypothesis, $m.AB^2 + n.AC^2 = \lambda$ (constant)

$\Rightarrow \quad m\{(h + a)^2 + k^2\} + n(h - a)^2 + k^2) = \lambda$

Simplifying and generalising, we get

$$x^2(m + n) + y^2(m + n) + 2xa(m - n) + a^2(m + n) - \lambda = 0$$

which is a circle. **Ans.**

Example 102:

Prove that the equation to a circle whose radius is a and which touches the axes of coordinates, which are inclined at an angle ω, is

$$x^2 + 2xy\cos\omega + y^2 - 2a(x + y)\cot\frac{\omega}{2} + a^2\cot^2\frac{\omega}{2} = 0.$$

Solution:

If the circle touches both the axes, the centre may be taken as (h, h). Again, if a be the radius of the circle, clearly a = h sin w or h = a cosec ω: hence centre becomes (a cosec ω, a cosec ω) and radius a

Putting these values in the general equation, we get

$$x^2 + y^2 + 2xy\cos\omega - 2.x(a\operatorname{cosec}\omega + a\operatorname{cosec}\omega\cos\omega)$$
$$-2y(a\operatorname{cosec}\omega + a\operatorname{cosec}\omega\cos\omega) + a^2\operatorname{cosec}^2\omega + a^2\operatorname{cosec}^2\omega$$
$$+ 2a\operatorname{cosec}\omega\, a\operatorname{cosec}\omega\cos\omega - a^2 = 0$$

$$\Rightarrow x^2 + y^2 + 2xy\cos\omega - 2xa\operatorname{cosec}\omega(1 + \cos\omega)$$
$$- 2ya\operatorname{cosec}\omega(1 + \cos\omega) + 2a^2\operatorname{cosec}^2\omega(1 + \cos\omega) - a^2 = 0 \quad ...(1)$$

Now $\operatorname{cosec}\omega(1 + \cos\omega) = \dfrac{2\cos^2(\omega/2)}{2\sin(\omega/2)\cos(\omega/2)} = \cot\dfrac{\omega}{2}$

and $\quad 2a^2\operatorname{cosec}^2\omega(1 + \cos\omega) - a^2$

$$= a^2\left[2\frac{2\cos^2(\omega/2)}{4\cos^2(\omega/2)\sin^2(\omega/2)} - 1\right]$$

$$= a^2\left[\operatorname{cosec}^2\left(\frac{\omega}{2}\right) - 1\right] = a^2\cot^2\frac{\omega}{2}$$

Putting the value in (1), we get the reqd. equation as

$$x^2 + y^2 + 2xy\cos w - 2xa\cot\frac{\omega}{2} - 2ya\cot\frac{\omega}{2} + a^2\cot^2\frac{\omega}{2} = 0$$

$$\Rightarrow x^2 + 2xy\cos w + y^2 - 2a(x + a)\cot\frac{\omega}{2} + a^2\cot^2\frac{\omega}{2} = 0.$$

Example 103:

The axes being inclined at an angle ω, find the equation to the circle whose diameter is the straight lin· ·oining the points

(x', y') and (x", y").

Solution:

Let the given points (x', y'a) and (x", y") be A and B respectively.

If P be any point (h, k) on the circle, as AB is the diameter ∠ABP = 90°.

$$\text{Slope of } PA = \frac{k-y'}{h-x'} = m_1 \text{ (say)}.$$

$$\text{Slope of } PB = \frac{k-y'}{h-x'} = m_2 \text{ (say)}.$$

As the angle between PA and PB is 90°, then

$$m_1 m_2 + (m_1 + m_2) \cos \omega + 1 = 0$$

$$\Rightarrow \quad \frac{k-y'}{h-x'} \cdot \frac{k-y''}{h-x''} + \left\{\frac{k-y'}{h-x'} + \frac{k-y''}{h-x''}\right\} \cos \omega + 1 = 0$$

$$\Rightarrow \quad (k - y')(k - y') + \{(y - y') + (y - y')(x - x'') + (y - y'')(x - x) \cos \omega = 0$$

Example 104:

The axis being inclined at 45°, find the equation to the circle whose centre is the point (2, 3) and whose radius is 4.

Solution:

When the axes are inclined at an angle of ω, the general equation of the circle with centre (h, k), radius a is

$$x^2 + y^2 + 2xy \cos \omega - 2(h + k \cos \omega) x - 2(k + h \cos \omega) y + h^2 + k^2 + 2hk \cos \omega - a^2 = 0 \quad ...(1)$$

If centre is (2, 3), radius is 4, and ω be 45°, substituting the values in (1) we get the reqd. equation as

$$x^2 + y^2 + 2xy \cos 45^\circ - 2(2 + 3 \cos 45^\circ) x - 2(3 + 2 \cos 45^\circ) y + 2^2 + 3^2 + 2.2.3 \cos 45^\circ - 4^2 = 0$$

$$\text{or } x^2 + y^2 + xy\sqrt{2} - x(2\sqrt{2}+3)\sqrt{2} - y(2+3\sqrt{2})\sqrt{2} + 6\sqrt{2} = 0.$$

Example 105:

The axes being inclined at 60°, find the equation to the circle whose centre is the point (– 3, – 5) and whose radius is 6.

Solution:

Here $\omega = 60^\circ$, centre is $(-3, -5)$ and radius is 6.

Putting these values in the general equation, we get

$$x^2 + y^2 + 2.xy \cos 60^\circ - 2(-3 - 5\cos 60^\circ) - 2(-5 - 3\cos 60^\circ)\,y$$
$$+ (-3)^2 + (-5)^2 + 2(-3)(-5)\cos 60^\circ - (6)^2 = 0$$

$\Rightarrow x^2 + y^2 + xy + 11x + 13y + 13 = 0.$ **Ans.**

Example 106:

The axes being inclined at an angle ω, find the centre and radius of the circle

$$x^2 + 2xy \cos \omega + y^2 - 2gx - 2fy = 0$$

Solution:

The given equation is

$$x^2 + 2xy \cos \omega + y^2 - 2gx - 2fy = 0$$

Comparing the coefficients of x, y and constant term with the standard eq. *i.e.*, we get

$$h + \cos \omega = g, \quad \text{...(1)}$$

$$k + h \cos \omega = f \quad \text{...(2)}$$

and $$h^2 + k^2 + 2hk \cos \omega - a^2 = 0 \quad \text{...(3)}$$

Solving (1) and (2), we get

$$h = (g - f\cos \omega)/\sin^2 \omega \text{ and } k = (f - g\cos \omega)/\sin^2 \omega.$$

Again, multiplying (1) by h, and (2) by k and adding, we get

$$h^2 + k^2 + 2hk.\cos \omega = gh + fk.$$

Putting the values in (3)

$$a^2 + h^2 + k^2 + 2hk \cos \omega = gh + fk$$

$$= \frac{g.(g - f\cos \omega)}{\sin^2 \omega} + \frac{f(f - g\cos \omega)}{\sin^2 \omega} = \frac{g + f^2 - 2fg\cos \omega}{\sin^2 \omega}$$

Hence the centre of the circle is given as

$$\left[\frac{(g - f\cos \omega)}{\sin^2 \omega}, \frac{f(f - g\cos \omega)}{\sin^2 \omega}\right]$$

and radius $\dfrac{\sqrt{(g^2 + f^2 - 2fg\,\omega)}}{\sin \omega}$. **Ans.**

Example 107:

Find the inclinations of the axes so that the following equations may represent circles, and in each case find the radius and centre;

$$x^2 - xy + y^2 - 2gx - 2fy = 0.$$

Solution:

The general equation of the circle of radius a, having the centre (h, k) and the axes being inclined at an angle of ω is given by

$$x^2 + y^2 + 2xy \cos \omega - 2x (h + k \cos \omega) - 2y (k + h \cos \omega)$$
$$+ h + k^2 + 2hk \cos \omega - a^2 = 0 \quad ...(1)$$

The given equation is

$$x^2 + y^2 - xy - 2gx + 2fy = 0. \quad ...(2)$$

Equating the coeffs. of xy, x, y and const. term in (1) and (2), as the coeffs. of x^2 and y^2 are equal, we get

$$2 \cos \omega = -1 \quad ...(3)$$

$$h + k \cos \omega = g \quad ...(4)$$

$$k + h \cos \omega = f \quad ...(5)$$

and $$h^2 + k^2 + 2hk \cos \omega - a^2 = 0 \quad ...(6)$$

By (3), $\cos \omega = -1/2$ or $\omega = 120^\circ$

Putting the value of cos ω in (4) and (5), we get respectively

$$h - \frac{k}{2} = g \quad ...(7)$$

and $$k - \frac{h}{2} = f$$

...(8)

Solving (7) and (8), we get

$$h = \frac{4g+2f}{3} \quad \text{and} \quad k = \frac{4f+2g}{3}$$

Putting the values of h, k and cos ω in (6), we get

$$\left(\frac{4g+2f}{3}\right)^2 + \left(\frac{4f+2g}{3}\right)^2 + 2\left(\frac{4g+2f}{3}\right)\left(\frac{4f+2g}{3}\right)\left(-\frac{1}{2}\right) - a^2 = 0$$

or $a^2 = \frac{4}{3}\left(g^2 + f^2 + gf\right)$.

Hence the angle between the axes is 120°.

The centre of the circle is $[(4g + 2f)/3, (4f + 2g)/3]$

and radius is $\frac{2\sqrt{3}}{3}\left\{\sqrt{(g^2+f^2+gf)}\right\}$. **Ans.**

Example 108:

Prove that the equation $r^2 \cos\theta - ar \cos 2\theta - 2a^2 \cos\theta = 0$ represents a straight line and a circle.

Solution:

The equation is given as

$$r^2 \cos\theta - ar \cos 2\theta - 2a^2 \cos\theta = 0 \quad ...(1)$$

or $r^2 \cos\theta - 2ar \cos^2\theta + ar - 2a^2 \cos\theta = 0$

$(\because \cos 2\theta = 2\cos^2\theta - 1)$

or $r\cos\theta\,(r - 2a\cos\theta) + a\,(r - 2a\cos\theta) = 0$

or $(r\cos\theta + a)\,(r - 2a\cos\theta) = 0$

Hence if $r\cos\theta + a = 0$, the $r = -a\sec\theta$ ...(2)

of if $r - 2a\cos\theta = 0$, then $r = 2a\cos\theta$. ...(3)

So (1) represents 2 curves given by (2) and (3) out of which (2) represents a straight line and (3) a circle.

Example 109:

Find the polar equation to the circle described on the straight line joining the points (a, α) and (b, β) as diameter.

Solution:

Let P and Q be the given points (a, α) and (b, β) respectively as the ends of the diameter.

If A be any point on the circle as (r, θ), clearly

$\angle PAQ = 90^\circ$; hence

$$PA^2 + AQ^2 = PQ^2 \quad ...(1)$$

In $\angle OAP$, $PA^2 = OP^2 + OA^2 - 2OP.OA \cos POA$

or $PA^2 = r^2 + a^2 - 2r.a.\cos(a - \theta)$ ...(2)

Similarly by ΔOPB, $AQ^2 = r^2 + b^2 - 2rb\cos(\theta - b)$ ...(3)

and ΔOAB, $PO^2 = a^2 + b^2 - 2ab\cos(\alpha - \beta)$ ...(4)

Substituting the values from (2), (3) and (4) in (1), we get

$\{r^2 + a^2 - 2r\,a\cos(\alpha - \theta)\} + \{r^2 + b^2 - 2rb\cos(\theta - \beta)\}$. **Ans.**

Example 110:

Prove that the equation to the circle described on the straight line joining the points (1, 60°) and (2, 30°) as diameter is

$$r^2 - r\,[\cos(\theta - 60^\circ) + 2\cos(\theta - 30^\circ)] + \sqrt{3} = 0.$$

Solution:

If (a, a) and (b, b) be the extremities of the diameter the equation of the circle is we know that

$$r^2 - r\,\{a\cos(\theta - \alpha) + b\cos(\alpha - \beta) + ab\cos(\alpha - \beta) = \quad ...(1)$$

Here the extremities are given as (1, 60°), (2, 30°). So putting the values of a, b, α and β in (1), we get

$$r^2 - r\,\{1\,.\cos(\theta - 60^\circ) + 2\cos(\theta - 30^\circ)\} + 1.2.\cos(60^\circ - 30^\circ) = 0$$

$$\Rightarrow \qquad r^2 - r\,\{\cos(\theta - 60^\circ) + 2\cos(\theta - 30^\circ)\} + \sqrt{3} = 0.$$

Example 111:

Draw the loci

$$(1)\ r = a;\ (2)\ r = a\sin\theta;\ (3)\ r = a\cos\theta;\ (4)\ r = a\sec\theta;$$
$$(5)\ r = a\cos(\theta - \alpha);\ (6)\ r = a\sec(\theta - \alpha).$$

Solution:

(1) r = a.

It is clearly a circle, centre being the pole and radius equal to a.

(2) $r = a\sin\theta$

I will be a circle passing through the pole, initial line being tangent to it and radius to a/2.

(3) $r = a\cos\theta$.

It will be a circle passing through the pole, centre being on the initial line and radius equal to a/2.

(4) $r = a\sec\theta$

It is a straight line perpendicular to the initial line at a distance of a from the pole.

(5) $r = a\cos(\theta - \alpha)$

It will be a circle passing through the pole, the centre being $(a/2, \alpha)$.

(6) $r = a\sec(\theta - \alpha)$

It is clearly a straight line at a distance of a from the pole, the perpendicular from the pole on the line being inclined at an angle of α to the initial line.

Example 112:

Prove that the equations $r = a \cos(\theta - \alpha)$ and $r = b \sin(\theta - \alpha)$ represent two circles which cut at right angles.

Solution:

The equations are given as

$$r = a \cos(\theta - \alpha) \qquad ...(1)$$

and

$$r = b \sin(\theta - \alpha)$$

$$= b \cos\{(\theta - \alpha) - 90^\circ\}$$

$$= b \cos\{\theta - (\alpha + 90^\circ)] \qquad ...(1)$$

As both of the equations represents circles passing through the origin, hence pole is one of the points of intersection. The angle of intersection of the circles is same as the angle between the tangent of the point of intersection which is equal to the angle between the radii of the two circles passing through the point of intersection.

So if O is the pole, OX the initial line, A and B are the centres of the circles (1) and (2) respectively, then

$\angle AOX = a$ as co-ordinates of A will be (a, α)

and $\angle BOX = \alpha + 90^\circ$ as co-ordinates of B will be $\{b, (\alpha, + 90)\}$.

$\therefore$ AOB will be 90°.

Therefore the angle between the radii is 90°. **Hence proved.**

Example 113:

Find the condition that the straight line

$$\frac{1}{r} = a \cos\theta + b \sin\theta$$

may touch the circle $r = 2c \cos\theta$.

Solution:

The circle is $r = 2c \cos\theta$ ...(1)

The lines is $\dfrac{1}{r} = a \cos\theta + b \sin\theta$ ...(2)

If (1) is a tangent to (2), solving the two we must get one and only one value of intersection.

By (1), $\cos\theta = r/2c$; hence $\sin\theta = \sqrt{1 - \dfrac{r^2}{4c^2}} = \dfrac{\sqrt{(4c^2 - r^2)}}{2c}$

Putting the values of $\cos\theta$ and $\sin\theta$ in (2), we get

$$\frac{1}{r} = a.\frac{r}{2c} + b.\left\{\frac{\sqrt{(4c^2 - r^2)}}{2c}\right\}$$

$\Rightarrow \quad \left(\frac{2c}{r} - ar\right)^2 = b^2\,(4c^2 - r^2).$

$\Rightarrow \quad r^4\,(a^2 + b^2) - 4c\,(a + b^2c)\,r^2 + 4c^2 = 0.$

This being quadratic in r^2; the values will coincide if discriminant is zero

i.e., $\quad \{4c\,(a + b^2c)\}^2 - 4\,(a^2 + b^2)\,.\,4c^2 = 0$

$\Rightarrow \quad a^2 + b^4c^2 + 2ab^2c - a^2 - b^2 = 0$

$\Rightarrow \quad b^2\,(b^2c^2 + 2ac - 1) = 0$

$\Rightarrow \quad b^2c^2 + 2ac - 1 = 0$ as $b \neq 0$

$\Rightarrow \quad b^2c^2 + 2ac = 1.$ **Ans.**

Example 114:

Find the polar equation of a circle, the initial line being a tangent. What does it become if the origin be on the circumference ?

Solution:

The equation of the circle having radius a and centre as (R, α) is given by

$$r^2 - 2Rr\cos(\theta - \alpha) + R^2 - a^2 = 0 \qquad ...(1)$$

If the initial line be the tangent, clearly $R = a\,\text{cosec}\,\alpha$.

Substituting in (1), we get

$$r^2 - 2.a\,\text{cosec}\,\alpha.r\cos(\theta - \alpha) + a^2\,\text{cosec}^2\,\alpha - a^2 = 0$$

$\Rightarrow \quad r^2 - 2ar.\,\text{cosec}\,\alpha\cos(\theta - \alpha) = a^2\,(\text{cosec}^2\,\alpha - 1) = 0$

$\Rightarrow \quad r^2 - 2ar.\text{cosec}\,\alpha\cos(\theta - \alpha) + a^2\cot^2\alpha = 0.$

If the origin is on the circumference and the tangent is the initial line, clearly $a = 90°$

Hence $\quad \text{cosec}\,a = 1$ and $\cot\alpha = 0.$

Putting these values in previous equation, we get

$$r^2 - 2ar\cos(q - 90°) = 0$$

$\Rightarrow \quad r^2 - 2ar\sin\theta = 0 \quad \{\because \cos(\theta - 90°) = \cos(90° - \theta) = \sin\theta\}$

$r = 2a\sin\theta.$ **Ans.**

Example 115:

The distances from the origin of the centres of three circles $x^2 + y^2 - 2\lambda x = c^2$ (where c is a constant and λ a variable) are in geometrical progression; prove that the lengths of the tangents drawn to them from any point on the circle $x^2 + y^2 = c^2$ are also in geometrical profession.

Solution:

The equations of the circles are given as

$$x^2 + y^2 - 2\lambda x = c^2 \quad \text{...(1)}$$

A all the circles given by (1) have their centres on the axis of x, let there by any three centres A, B and C having the co-ordinates as $(\lambda_1, 0)$, $(\lambda_2, 0)$ and $(\lambda_3, 0)$ respectively.

Then the circles will be

$$x^2 + y^2 - 2\lambda_1 x - c^2 = 0 \quad \text{...(1)}$$

$$x^2 + y^2 - 2\lambda_2 x - c^2 = 0 \quad \text{...(2)}$$

and $$x^2 + y^2 - 2\lambda_3 x - c^2 = 0 \quad \text{...(3)}$$

Distances of the centres A, B and C from origin O are as

$$OA = \lambda_1,\ OB = \lambda_2 \text{ and } OC = \lambda_3$$

As OA, OB and OC are in G.P., hence $\lambda_1\lambda_3 = \lambda_2^2$ ...(4)

Again let there be any other point $p \equiv (h, k)$ on the given circle $x^2 + y^2 = c^2$. As P lies on this circle, hence

$$h^2 + k^2 = c^2 \text{ or } h^2 + k^2 - c^2 = 0 \quad \text{...(5)}$$

Length of the tangent from P upto (1)

$$= \sqrt{(h^2 + k^2 - 2\lambda_1 h - c^2)} = PT_1 \text{ (say)}$$

As $h^2 + k^2 - c^2 = 0$, hence $PT_1^2 = -2\lambda_1 h$ ...(6)

Length of the tangent from T upto (2)

$$= \sqrt{(h^2 + k^2 - 2\lambda_2 h - c^2)} = PT_2 \text{ (say)}$$

Again as $h^2 + k^2 - c^2 = 0$, by (5), hence $PT_2^2 = -2\lambda_2 h$ ...(7)

Similarly if PT_3 be the tangent on the 3rd circle,

$$PT_3^2 = -2\lambda_2 h \quad \text{...(8)}$$

PT_1, PT_2 and PT_3 are in G P if

$$PT_1 \times PT_3 = (PT_2)^2$$

or $$(-2h\lambda_1) \times (2h\lambda_3) = (-2\lambda_2 h)^2$$

or $\lambda_1\lambda_3 = \lambda_2^2$.

This relation is given by (4) as OA, OB and OC are in G.P.

Example 116:

Given the three circles

$$x^2 + y^2 - 16x + 60 = 0,$$

$$3x^2 + 3y^2 - 36x + 81 = 0,$$

and $$x^2 + y^2 - 16x - 12y + 84 = 0$$

find (1) the point from which the tangents to them are equal in length, and (2) this length.

Solution:

The three circles are given as

$$x^2 + y^2 - 16x + 60 = 0 \qquad ...(1)$$

$$3x^2 + 3y^2 - 36x + 81 = 0$$

or $$x^2 + y^2 - 12x + 27 = 0 \qquad ...(2)$$

and $$x^2 + y^2 - 16x - 12y + 84 = 0. \qquad ...(3)$$

Let the co-ordinates of the point P from which the tangents to these circles are equal be (h, k). Then the length of the tangent from (h, k) to (1)

$$= \sqrt{(h^2 + k^2 - 16h + 60)} = PT_1 \text{ (say)}.$$

Length of the tangent from (h, k) to (2)

$$\sqrt{(h^2 + k^2 - 12k + 27)} = PT_2 \text{ (say)}.$$

Length of tangent from (h, k) to (3)

$$= \sqrt{(h^2 + k^2 - 16h - 12k + 84)} = PT_3 \text{ (say)}.$$

As the tangents are equal, hence by $PT_1 = PT_2$

$$\sqrt{(h^2 + k^2 - 16h + 60)} = \sqrt{(h^2 + k^2 - 12h + 27)}$$

or $h^2 + k^2 - 16h + 60 = h^2 + k^2 - 12h + 27$ (on squaring)

$$-4h = -3 \quad \therefore \; h = \frac{33}{4}.$$

Again, $PT_1 = PT_3$ also; hence

$$\sqrt{(h^2 + k^2 - 16h + 60)} = \sqrt{(h^2 + k^2 - 16h - 12k + 84)}$$

Solving, we get k = 2.

Hence the reqd. point is $\left(\frac{33}{4}, 2\right)$

Length of the tangent from $\left(\frac{33}{4}, 2\right)$ on (1).

$$= \sqrt{\left[\left\{\left(\frac{33}{4}\right)^2 + (2)^2 - 16 \times \frac{33}{4} + 60\right\}\right]} = \sqrt{\left(\frac{1}{16}\right)} = \frac{1}{4}.$$ **Ans.**

Example 117:

Prove that the distances of two points, P and Q, each from the polar of the other with respect to a circle, are to one another as the distances of the points from the centre of the circle.

Solution:

Let the circle $x^2 + y^2 = a^2$ ...(1)

having O as its centre.

Let there be two points P and Q whose co-ordinates are respectively (x_1, y_1) and (x_2, y_2).

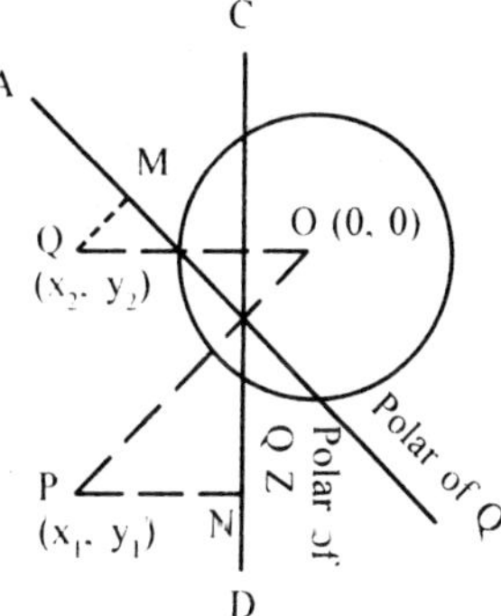

If AB be the polar of P with respect to (1), its equation will be

$$xx_1 + yy_1 = a^2. \quad ...(2)$$

The equation of CD, the polar of Q with respect to (1), will be

$$xx_2 + yy_2 = a^2 \quad ...(3)$$

If PN be the perpendicular distance from P upto CD then

$$PN = \frac{(x_1 x_2 + y_1 y_2)}{(x_1^2 + y_1^2)} \quad ...(4)$$

Similarly if QM be the perpendicular from Q upto AB then

$$QM = \frac{x_2 x_1 + y_2 y_1 - a^2}{\sqrt{(x_1^2 + y_1^2)}} \quad ...(5)$$

By (4) and (5)

$$\frac{PN}{QM}=\frac{x_1x_2+y_1y_2-a^2}{\sqrt{(x_2^2+y_2^2)}}/\frac{x_1x_2+y_1y_2-a^2}{(x_1^2+y_1^2)}$$

$$\frac{PN}{QM}=\frac{\sqrt{(x_1^2+y_1^2)}}{\sqrt{(x_2^2+y_2^2)}} \quad ...(6)$$

Again PO = $\sqrt{(x_1^2+y_1^2)}$ and QO = $\sqrt{(x_2^2+y_2^2)}$.

Hence $$\frac{PO}{QO}=\frac{\sqrt{(x_1^2+y_1^2)}}{\sqrt{(x_2^2+y_2^2)}} \quad ...(7)$$

By (6), (7), we get $\frac{PN}{QM}=\frac{PO}{QO}$. **Proved.**

Example 118:

Find the condition that the chord of contact of tangents from the point (x', y') to the circle $x^2 + y^2 = a^2$ should subtend a right angle at the centre.

Solution:

The circle is given as $x^2 + y^2 = a^2$. ...(1)

The point P is given as (x', y').

The equation of the chord of contact of (x', y') with respect to (1), *i.e.* AB will b

$$x\,x' + y\,y' = a^2 \text{ or } \frac{x\,x'}{a^2}+\frac{y\,y'}{a^2}=1. \quad ...(2)$$

To get the combined equation of the lines which join the origin O, the points of intersection of (1) and (2), A and B, we make (1) homogeneous with the help of (2).

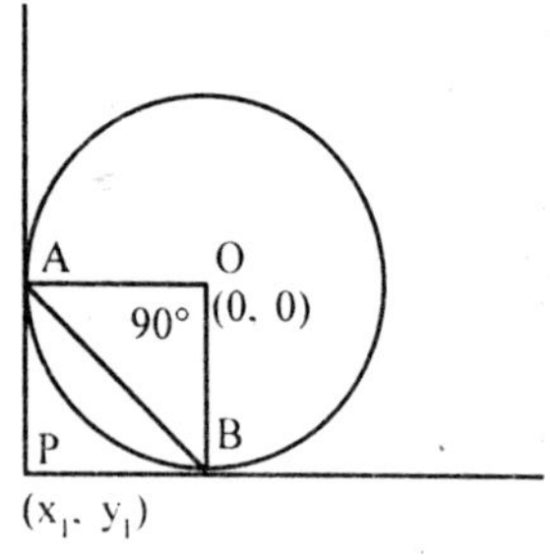

Hence the combined equation of OA and OB will be

$$\Rightarrow x^2+y^2-a^2\left(\frac{x\,x'}{a^2}+\frac{y\,y'}{a^2}\right)^2=0.$$

$$\Rightarrow x^2 + y^2 - a^2\left[\frac{x^2\,x'^2+y^2\,y'^2+2x\,x'\,y\,y'}{a^4}\right]=0$$

$\Rightarrow \quad a^2x^2 + a^2y^2 - x^2x'^2 - y^2y'^2 - 2xyx'y' = 0$

$\Rightarrow \quad x^2(a^2 - x'^2) + y^2(a^2 - y'^2) - 2xyx'y' = 0$ ...(3)

If OQ and OB are perpendicular, the sum of the coeffs. of x^2 and y^2 in (3) must be zero. Therefore the required condition is

$$a^2 - x'^2 + a^2 - y'^2 = 0$$

or $\quad x'^2 + y'^2 = 2a^2.$ **Ans.**

Example 119:

Find the equation to the pair of tangents drawn

(1) from the point (11, 3) to the circle $x^2 + y^2 = 65$,

(2) from the point (4, 5) to the circle

$$2x^2 + 2y^2 - 8x + 12y + 21 = 0.$$

Solution:

(1) The circle is $x^2 + y^2 = 65$; ...(1)

hence $\quad S \equiv x^2 + y^2 = 65$ ...(2)

The point is (11, 3); so $S_1 = (11)^2 + (3)^2 - 65 = 65$ ...(3)

The L.H.S. of the equation of the tangent from (11, 3) on (1) is

$$T \equiv 11x + 3y - 65.$$

Therefore the required equation is

$$SS_1 = T^2$$

$\Rightarrow \quad (x^2 + y^2 - 65)(65) = (11x + 3y - 65)^2$

$\Rightarrow \quad 65x^2 + 65y^2 - 4225 = 121x^2 + 9y^2 + 66xy - 1430x - 390y + 4225$

On simplification, we get

$$28x^2 + 33xy - 28y^2 - 715x - 195y + 4225 = 0.$$

(ii) The equation of the circle is

$$2x^2 + 2y^2 - 8x + 12y + 21 = 0$$

or $\quad S \equiv x^2 + y^2 - 4x + 6y + \dfrac{21}{2}$ ...(1)

The point is (4, 5); hence $S_1 = (4)^2 + (5)^2 - 4.4 + 6.5 + \dfrac{21}{2} = \dfrac{131}{2}$.

L.H.S. of the tangent from (4, 5) on (1) (as if the point is on the circle) is

$$T \equiv x.4 + 5.y - 2(x + 4) + 3(y + 5) + \frac{21}{2}$$

$$= 2x + 8y + \frac{35}{2}$$

As the equation of the pair of tangents is

$$SS_1 = T^2$$

hence $\left(x^2 + y^2 - 4x + 6y + \frac{21}{2}\right)\left(\frac{131}{2}\right) = \left(2x + 8y + \frac{35}{2}\right)^2$

Simplifying, we get

$$1.3x^2 - 64xy + 3y^2 + 664x + 226y + 762 = 0.$$ **Ans.**

Example 120:

Find the length of the tangent drawn to the circle $2x^2 + 2y^2 = 3$ from the point $(-2, 3)$.

Solution:

The circle is $2x^2 + 2y^2 = 3$ or $x^2 + y^2 - 3/2 = 0$

Length of the tangent to this circle from $(-2, 3)$ is

$$= \sqrt{\left\{(-2)^2 + (3)^2 - \frac{3}{2}\right\}} = \sqrt{\frac{23}{2}} = \frac{1}{4}\sqrt{46} \text{ units.}$$ **Ans.**

Example 121:

Find the lengths of the tangents drawn to the circle $3x^2 + 3y^2 - 7x - 6y = 12$ from the point $(6, -7)$.

Solution:

The point is $(6, -7)$ and the circle is

$$3x^2 + 3y^2 - 7x - 6y = 12 \text{ or } x^2 + y^2 - \frac{7}{3}x - 2y - 4 = 0.$$

Length of the tangent of this circle from $(6, -7)$ is

$$= \sqrt{(6)^2 + (-7)^2 - \frac{7}{3}(6) - 2(-7) - 4} = \sqrt{81} = 9 \text{ units.}$$ **Ans.**

Example 122:

Find the lengths of the tangents of the tangents drawn to the circle $x^2 + y^2 + 2bx - 3b^2 = 0$ from the point

$(a + b, a - b)$.

Solution:

The circle is given as $x^2 + y^2 + 2bx - 3b^2 = 0$.

The length of the tangents from the given point (a + b, a − b) on this circle will be

$$\sqrt{[(a+b)^2 + (a-b)^2 + 2b(a+b) - 3b^2]} = \sqrt{(a^2 + 2ab + b^2)}. \textbf{ Ans.}$$

Example 123:

Prove that the polar of a given point with respect to any one of the circles $x^2 + y^2 - 2kx + c^2 = 0$, where k is variable, always passes through a fixed point, whatever be the value of k.

Solution:

The equation of the circle is given as

$$x^2 + y^2 - 2kx + c^2 = 0 \qquad ...(1)$$

where k is variable.

Let the given point be (α, β).

Equation of the polar of (α, β) with respect to (1) is

$$x\alpha + y\beta - k(x + \alpha) + c^2 = 0$$

or $$(x\alpha + y\beta + c^2) - k(x + \alpha) = 0 \qquad ...(2)$$

This equation is of the form $P + \lambda Q = 0$, where

$$P \equiv x\alpha + y\beta + c^2 = 0 \text{ and } Q \equiv x + \alpha = 0.$$

Hence (2) is the equation of the line always passing through the point of intersection of

$$xa + y\beta + c^2 = 0 \text{ and } x + \alpha = 0$$

Solving, we get the fixed point as

$$\left[-\alpha, \frac{1}{\beta}\left(a^2 - c^2\right)\right]. \qquad \textbf{Ans.}$$

Example 124(a):

Prove that the polars of the point (1, − 2) with respect to the circles whose equations are

$$x^2 + y^2 + 6y + 5 = 0 \quad \textit{and} \quad x^2 + y^2 + 2x + 8y + 5 = 0$$

coincide; prove also that their is another point the polars of which with respect to these circles are the same and find its coordinates.

Solution:

The circles are given as $x^2 + y^2 + 2x + 8y + 5 = 0$

and $$x^2 + y^2 + 2x + 8y + 5 = 0. \qquad ...(2)$$

The point is (1, – 2).

The polar of point (1, –2) with respect to (1) is

$$x.1 + y(-2) \text{ with respect to (2) is}$$

$$x.1 + y.(-2) + (x + 1) + 4(y - 2) + 5 = 0$$

or $\quad 2x + 2y - 2 = 0 \quad$ or $x + y - 1 = 0$. ...(4)

As (3) and (4) are the same, hence the two polars coincide.

Again let (α, β) be the other point for which the polars are the same.

Then the polar of (α, β) with respect to (1) is a

$$x\alpha + y\beta + 3(y + \beta) + 5 = 0$$

or $\quad x\alpha + y(\beta + 3) + (3\beta + 5) = 0$ ...(5)

The polar of (x, β) with respect to (2)

$$x\alpha + y\beta + (x + \alpha) + 4(y + \beta) + 5 = 0$$

or $\quad x(\alpha + 1) + y(\beta + 4) + (\alpha + 4\beta + 5) = 0$ ...(6)

As (5) and (6) represent the same straight line, comparing the coefficients, we get

$$\frac{\alpha+1}{\alpha} = \frac{\beta+4}{\beta+3} = \frac{\alpha+4\beta+5}{3\beta+5}$$

$\therefore \quad \dfrac{\alpha+1}{\alpha} = \dfrac{\beta+4}{\beta+3}$, hence $\alpha - \beta - 3 = 0$ ...(7)

Again $\dfrac{\alpha+1}{\alpha} = \dfrac{\alpha+4\beta+5}{3\beta+5}$ or $\alpha^2 + \alpha\beta - 3\beta - 5 = 0$ ...(8)

Putting the values of β from (7) in (8), we have

$$\alpha^2 + \alpha.(\alpha - 3) - 3(\alpha - 3) - 5 = 0$$

or $\quad 2\alpha^2 - 6\alpha + 4 = 3$

or $\quad \alpha^2 - 3\alpha + 2 = 0.$

Solving, we get $\quad \alpha = 1 \quad$ or $\quad \alpha = 2.$

Substituting in (7), we get b = – 2 or – 1.

Hence the two possible points are (1, –2) and (2, –1).

First point is given; hence the second is (2, –1). **Ans.**

Example 124(b):

Find the Pole of the st. line $48x - 54y + 53 = 0$ with respect to the circle

$$3x^2 + 3y^2 + 5x - 7y + 2 = 0.$$

Solution:

The line is given as 48x – 54y + 53 = 0 ...(1)

The equation to the circle is given as

$$3x^2 + 3y^2 + 5x - 7y + 2 = 0 \qquad ...(2)$$

Let the pole of (1) with respect to (2) be (α, β) then the polar of (α, β) with respect to (2) will be

$$3x\,\alpha + 3y\,\beta + \frac{5}{2}(x + \alpha) - \frac{7}{2}(y + \beta) + 2 = 0$$

or $x(6\alpha + 5) + y(6\beta - 7) + (5\alpha - 7\beta + 4) = 0$. ...(3)

As (1) and (3) represent the same straight line, hence comparing the coefficients, we get

$$\frac{6\alpha+5}{48} = \frac{6\beta-7}{-54} = \frac{5\alpha-7\beta+4}{53}.$$

By first two terms, we get $54\alpha + 48\beta - 11 = 0$ ...(4)

By first and third terms, we get $78\alpha + 336\beta + 73 = 0$ ...(5)

Solving (4) and (5), we get

$$\alpha = \frac{1}{2},\ \beta = -\frac{1}{3}.$$

Hence pole is $\left(\frac{1}{2}, \frac{1}{3}\right)$. **Ans.**

Example 125(a):

Find the equation to that chord of the circle $x^2 + y^2 = 81$ which is bisected at the point (–2, 3), and its pole with respect to the circle.

Solution:

Equation of the circle is $x^2 + y^2 = 81$. ...(1)

Let A be the point (–2, 3).

Slope of the line joining A ≡ (–2, 3) to the centre of the circle O ≡ (0, 0) is

$$m = \frac{3-0}{-2-0} = -\frac{3}{2}.$$

If the chord is bisected at (– 2, 3), it must be perpendicular to OA.

Equation of the line passing through A and perpendicular to OA will be

$$(y - 3) = 2/3\,(x + 2)\ 2x - 3y + 13 = 0 \qquad ...(3)$$

As (2) and (3) represent the same straight line, comparing the coeff is, we get

$$\frac{\alpha}{2} = \frac{\beta}{-3} = \frac{-81}{13},$$

whence $\alpha = -\frac{162}{13}$ and $\beta = -\frac{243}{13}$. ...(1)

Hence the pole is $\left(-\frac{162}{13}, \frac{243}{13}\right)$. **Ans.**

Example 125(b) :

Find the pole of the st. line $x + 2y = 1$ with respect to the circle $x^2 + y^2 = 5$.

Solution :

Equation of the polar is given as $x + 2y = 1$. ...(1)

Equation of the circle is given to $x^2 + y^2 = 1$. ...(2)

Let the pole of (1) with respect to (2) be (α, β).

Polar of (α, β) with respect to (2) is $x\alpha + y\beta = 5$ ...(3)

As (1) and (3) represent the same equations, comparing the coeffis., we get

$$\frac{1}{\alpha} = \frac{2}{\beta} = \frac{1}{5} \quad \text{or } \alpha = 5, \alpha = 10$$

Hence the pole is (5, 10). **Ans.**

Example 126

Find the pole of the line $2x - y = 6$ with respect to the circle $5x^2 + 5y^2 = 9$.

Solution :

The line as $2x - y = 6$ ...(1)

The circle is given as $5x^2 + 5y^2 = 9$. ...(2)

Let the pole (1) with respect to (2) be (α, β).

Polar of (α, β) with respect to circle (2) will be

$$5.x\alpha + 5y.\beta = 7 \qquad ...(3)$$

As (3) and (1) represent the same line, comparing the coeffs. of x and y and constant terms, we get

$$\frac{5\alpha}{2} = \frac{5\beta}{-1} = \frac{9}{6}$$

or $\alpha = \frac{3}{5}, \beta = -\frac{3}{10}$ So the pole is $\left(\frac{3}{5}, -\frac{3}{10}\right)$.

Example 127:

The polar of P with respect to the circle $x^2 + y^2 = a^2$ touches the circle $(x - \alpha)^2 + (y - \beta)^2 = b^2$; prove that its locus is the curve given by the equation $(\alpha x + \beta y - a^2)^2 = b^2 (x^2 + y^2)$.

Solution:

The circles are given as

$$x^2 + y^2 = a^2 \quad ...(1)$$

and $$(x - \alpha)^2 + (y - \beta)^2 = b^2. \quad ...(2)$$

Let the co-ordinates of the moving point P be (h, k).

Polar of (h, k) with respect to (1) is

$$xh + yk - a^2 = 0. \quad ...(3)$$

If (3) is a tangent to (2), the length of the perpendicular from the centre of (2), *i.e.* (α, β) to (3) must be equal to be, the radius of (2). The length of perpendicular from (a, b) on (3) is

$$\frac{\alpha h + \beta k - a^2}{\sqrt{(h^2 + k^2)}} = b$$

or $$(ah - bk - a^2) = b \sqrt{(h^2 + k^2)}$$

Squaring and generalising, we get the required locus as

$$(\alpha x + \beta y - a^2) = b^2 (x^2 + y^2).$$

2

SYSTEM OF CIRCLES

ANGLE OF INTERSECTION OF TWO CURVES

The angle of intersection between two curves is defined to be the angle between the tangents to the curve at that point. If the angle of intersection is 90°, the curves are said to intersect each other *Orthogonally*.

If the two circles inter set at P, the radii O, P and O_2P which are perpendicular to the tangent at P must also be at right angles. Hence O, O_2 = 0, $P^2 + O_2P^2$ *i.e.* The square of the distance between the centres must be equal to the sum of the squares of teh radii.

Also the tangent from O_2 to the circles be othogonal the length of the tangent drawn from the centre of one circle to the second circle is equal to the radius of the first.

CONDITION THAT TWO CIRCLES MAY INTERSECT ORTHOGONALLY

Let the Equation of two intersecting circles be given as

$$x^2 + y^2 + 2gx + 2fy + c = 0 \quad \text{...(i)}$$

and

$$x^2 + y^2 + 2gx_1 + 2fy_1 + c_1 = 0. \quad \text{...(ii)}$$

Let PT and PT' be the tangents to the circle (i) and (ii) respectively. Since the circles intersect each other orthogonally *i.e.* the angle between their tangents PT and PT' is 90°, the tangents PT' and PT when produced always pass through the centres O' and O respectively. This gives OPO' a right angled triangle.

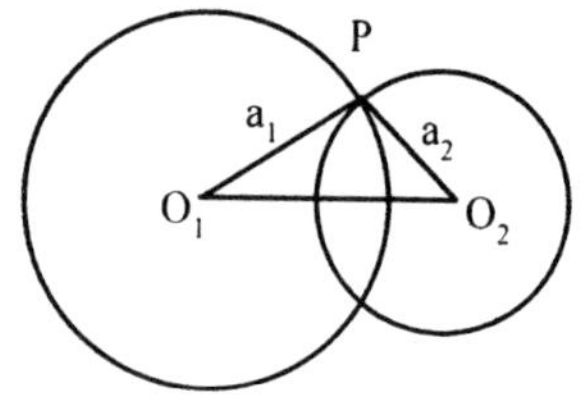

Now $\quad OO'^2 = OP^2 + O'P^2$

$$\Rightarrow \quad (-g + g_1)^2 + (-f + f_1)^2 = g^2 + f^2 - c + g_1^2 + f_1^2 - c_1$$

$\Rightarrow \qquad 2gg_1 + 2ff_1 = c + c_1.$

This is the required condition that two circles may cut each other orthogonally.

CURVE THROUGH THE INTERSECTION OF TWO CURVES

Let S = 0 and S' = 0 be two given curves then the equation $S + \lambda S' = 0$, where λ is a constant or any function of x and y is some curve which passes through the intersection of curves S = 0 and S' = 0. Let (h, k) be any point on the intersection of these two curves. This implies that S = 0 and S' = 0 when x and y are replaced by (h, k). This makes $S + \lambda S_1 = 0$ when x, y are replaced by (h, k). Hence (h, k) lies on the curve $S + \lambda S' = 0$.

RADICAL AXIS

Definition: The radical axis of two circles is the locus of the point which moves such that lengths of the tangents drawn from point to the two circles are same. In the figure MP is the radical axis of two circles with centre O and O'

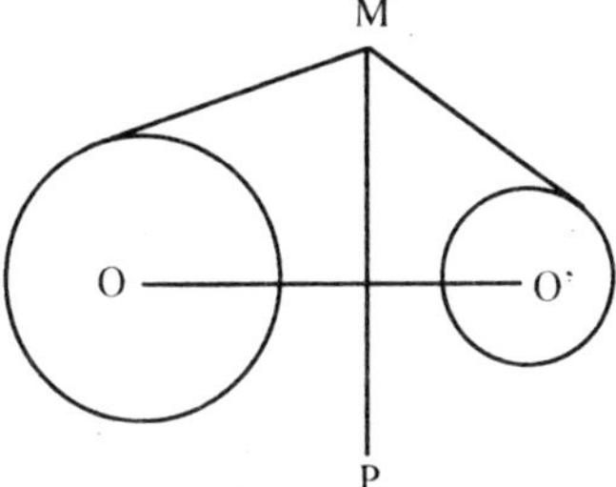

EQUATION OF RADICAL AXIS OF TWO CIRCLES

Let equation of two circles are given as

Let $x^2 + y^2 + 2gx + 2fy + c = 0$

and $\qquad x^2 + y^2 + 2g'x + 2f'y + c' = 0$

be the equations of two circles. Take (h, k) a point on the locus. $\therefore$ the lengths of the tangents from this point on two circles are equal.

This gives $\sqrt{h^2 + k^2 + 2gh + 2fk + c} = \sqrt{h^2 + k^2 + 2g'h + 2f'k + c'}$

$\Rightarrow h^2 + k^2 + 2gh + 2fk + c = h^2 + k^2 + 2g'h + 2f'k + c'$

$\Rightarrow 2(g - g')h + 2(f - f')k + c - c' = 0,$

Hence the locus is given as

$2(g - g')x + 2(f - f')y + c - c' = 0,$

which is the equation of the radical axis.

If we denote S = 0 and S' = 0 as the equation of two circles, then the equation of the radical axis can be written as S – S' = 0. Since S – S' = 0 represents a curve passing through the intersection of S = 0 and S' = 0, therefore, the radical axis of two circles is their common chord.

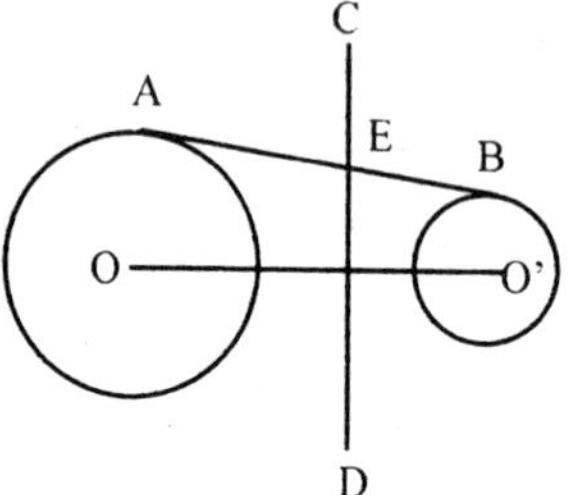

PROPERTIES OF RADICAL AXIS

(i) The radical centre of three circles described on the sides of a triàngle as diameter is the orthocentre of the triangle.

Let ABC be a triangle and AD is perpendicular from A on BC. Since ∠s ADB and ADC are at right angles, the circles with AB and AC as diameter pass through AD.

∴ AB is the radical axis. Similary, the radical axis of the circles on AB and BC as diameter will be the altitude BE and altitude CF for the third pair of circles. This shows that the altitudes of the triangle are the radical axis, *i.e.* The radical centre is the orthocentre of the triangle.

(ii) The radical axis of two circle bisects their common tangents.

Let AB be the common tangent and CD the radical axis of these two circles. E is the point of intersection of the common tangent and the radical axis. Since E lies on the radical axis of these two circles, therefore,

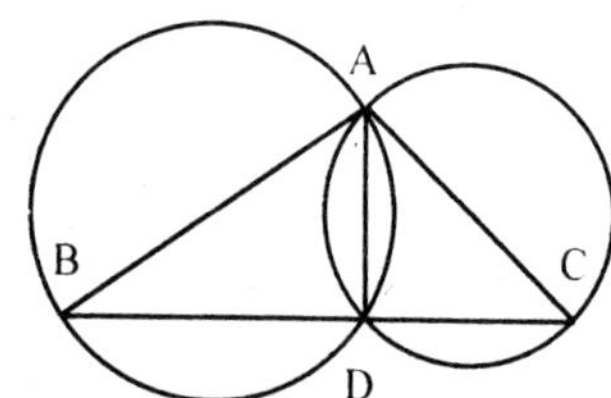

$$EA = EB,$$

which proves the result.

(iii) If two circles cut a third circle orthogonally, the radical axis of the two circles passes through the centre of the third circle.

Let the equations to the two circles be

$$x^2 + y^2 + 2gx + c = 0 \quad ...(i)$$

and $$x^2 + y^2 + 2g'x + c = 0 \quad ...(ii)$$

Let the circle cutting (i) and (ii) orthogonally be

$$x^2 + y^2 + 2Gx + 2Fy + C = 0$$

then $$2Gg = c + C.$$

$$2Gg' = c + C.$$

This gives G = 0 and C = – c

$\therefore$ the equation to the third circle is

$$x^2 + y^2 + 2Fy - c = 0.$$

The centre of this circle is (0, – F), which lies on the radical axis of the two circles whose equation is x = 0.

(iv) The difference of the squares of the tangents to two circles from any point in their plane varies as the distance of the point from their radical axis.

Take the line joining the centres of two circles as the x-axis and the radical axis as y-axis. The equations to the two circles can be taken as

$$x^2 + y^2 + 2gx + c = 0, \quad x^2 + y^2 + 2g'x + c = 0.$$

Let $P(x_1, y_1)$ be any point. If PA, PB be the tangents to the two circles then

$$PA^2 - PB^2 = (x_1^2 + y_1^2 + 2gx_1 + c) - (x_1^2 + y_1^2 + 2g'x_1 + c)$$

$$= 2(g - g')x_1$$

= 2 (g – g') × (distance of the point P from the radical axis, since radical axis is y-axis.)

Hence $PA^2 - PB^2$ varies as the distance of the point from the radical axis.

(v) The radical axis of three circles taken in pairs meet in a point.

Let the equation of three circles be

$$S = 0, \quad S' = 0 \quad \text{and} \quad S'' = 0$$

taking co-efficients of x^2 and y^2 as unity in all the three cases. Equation of radical axis of these circles taken in pairs are

$$S - S' = 0$$

$$S' - S'' = 0$$

$$S'' - S = 0.$$

Since on addition the left hand side vanishes, the lines represented by them must be a point. The common point of the radical axis of three circles taken in pairs is called the *Radical Centre.*

(vi) The radical axis of two circles is perpendicular to the line joining their centers.

Let the equation of two given circle be

$$x^2 + y^2 + 2gx + 2fy + c = 0 \quad \text{...(i)}$$

$$x^2 + y^2 + 2g'x + 2f'y + c' = 0 \quad \text{...(i)}$$

and their radical axis be

$$2(g-g')x+2(f-f')y+c-c'=0 \qquad ...(iii)$$

The centres of circles (i) and (ii) are $(-g, -f)$ and $(-g', -f')$ respectively. The line joining their centres is

$$y+f=\frac{f'-f}{g'-g}(x+g) \qquad ...(iv)$$

The slope of (iii) is $= \dfrac{-(g'-g)}{f'-f}$.

The slope of (iv) is $\dfrac{f'-f}{g'-g}$.

Since the product of their slopes is

Given by $-\dfrac{(g'-g)}{f'-f}\times\dfrac{(f'-f)}{(g'-g)}=-1,$

therefore, the lines are perpendicular. Hence the radical axis of two circles is perpendicular to the line joining their centres.

Worleing Rule : To get the radical axis of two circles first make the co-efficient of x^2 and y^2 in both the equations and subtract.

CO-AXAL CIRCLES

A system of circles is said to be co-axal if when they have a common radical axis *i.e.*, when the radical axis of each pair of circles of the system is same.

EQUATION OF SYSTEM OR CO-AXAL CIRCLES IN THE SIMPLEST FORM

Take the line of centres as the x-axis and the common radical axis as y-axis.

Let $$x^2+y^2+2gx+2fy+c=0. \qquad ...(i)$$

represent the members of the co-axal system for different values of g, f and c.

Since the centre of all circles of the system lie on x-axis, the y-co-ordinate of the system must vanish. This gives $f = 0$.

When the above equation becomes

$$x^2+y^2+2gx+c=0. \qquad ...(ii)$$

The equation of any two circles of the system can be put in the following form

$$x^2 + y^2 + 2g_1x + c_1 = 0. \quad ...(iii)$$

$$x^2 + y^2 + 2g_2x + c_2 = 0. \quad ...(iv)$$

Equation of the radical axis becomes as

$$2(g_1 - g_2)x + c_1 - c_2 = 0.$$

Since y-axis is the radical axis, comparing the equation with $x = 0$, then we have

$$c_1 - c_2 = 0$$

$$\Rightarrow \quad c_1 = c_2.$$

Equations (iii) and (iv) becomes

$$x^2 + y^2 + 2g_1x + c = 0$$

and $$x^2 + y^2 + 2g_2x + c = 0.$$

Hence $x^2 + y^2 + 2gx + c = 0$ is the equation of co-axal system of circles in the simplest form.

VARIOUS TYPES OF CO-AXAL CIRCLES

Let the system of co-axal circles be represented by

$$x^2 + y^2 + 2gx + c = 0.$$

The points of intersection of any circle of the system and the radical axis (y-axis) *i.e.*, $x = 0$ are given by $y^2 + c = 0$.

Case 1 if $c > 0$

In this case y^2 becomes negative and therefore the values of y become imaginary *i.e.*, the system of circles do not cut the y-axis

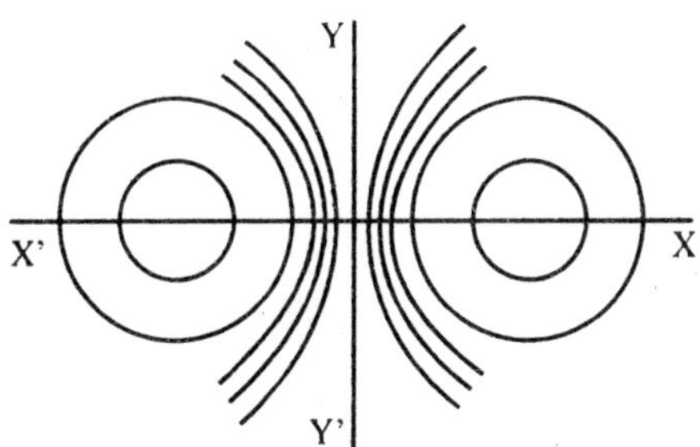

Case 2 if $c = 0$

This gives both the values of y as zero *i.e.*, y-axis becomes the tangent to the family of co-axal system at the origin.

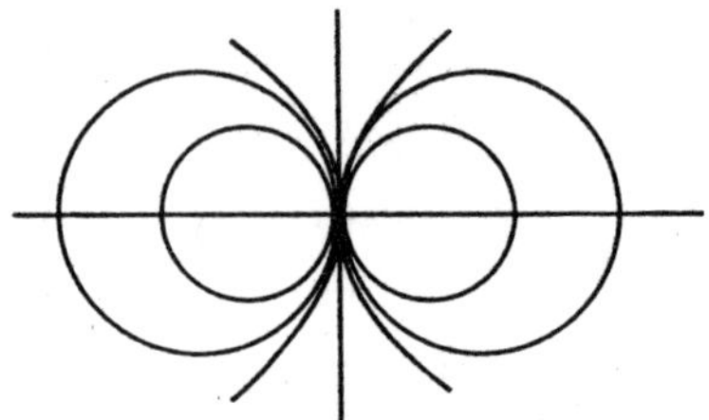

Case 3 if $c < 0$

This gives the values of y as real and distinct.

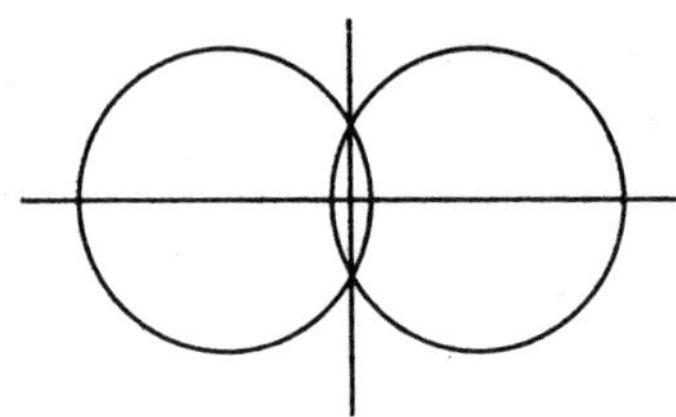

EQUATION OF THE SYSTEM OF CO-AXAL CIRCLES HAVING GIVEN EQUATION OF TWO CIRCLES OF THE SYSTEM

Let $S = x^2 + y^2 + 2gx + 2fy + c$

and $S' = x^2 + y^2 + 2g'x + 2f'y + c'$

be the equations of the two circles of the system. The family of circles passing through the point of intersection of these two circles is

$$S + \lambda_1 S' = 0. \quad \text{...(iii)}$$

We take any two circles of the family (iii)

$$S + \lambda_1 S' = 0 \quad \text{...(iv)}$$

and $$S + \lambda_2 S' = 0. \quad \text{...(v)}$$

The equation of the radical axis is given by

$$\frac{S+\lambda_1 S'}{1+\lambda_1} - \frac{S+\lambda_2 S'}{1+\lambda_2} = 0$$

$$\Rightarrow \quad (1 + \lambda_1)(S + \lambda_1 S') - (1 + \lambda_2)(S + \lambda_2 S') = 0$$

$$\Rightarrow \quad (\lambda_1 + \lambda_1)(S - S') = 0$$

$$\Rightarrow \quad S = S' = 0.$$

This proves that every pair of system $S + \lambda S' = 0$ has the same radical axis. Hence the required equation of the system of co-axal circles is $S + \lambda S' = 0$.

LIMITING POINTS OF THE SYSTEM OF CO-AXAL CIRCLES

The equation of the system of co-axal circles is

$$x^2 + y^2 = 2gx + c = 0. \qquad ...(i)$$

Radius of (i) is $\sqrt{(g^2 - c)}$. This becomes zero when $g^2 = c$ *i.e.*, $g = \pm\sqrt{c}$. The point $(+\sqrt{c}, 0)$ and $(-\sqrt{c}, 0)$ are called limiting points of the system.

Hence at a paticular point $(\pm \sqrt{c}, 0)$ we have point circles which belongs to the system. These points circles are called the limiting point of the system. It c is negative these points are imaginary.

It c is positive the limiting point L_1 and L_2 are real and in this case circles intersect in imaginary points.

EQUATION OF A SYSTEM OF CO-AXAL CIRCLES WHEN THE EQUATIONS OF THE RADICAL AXIS AND ONE CIRCLE OF THE SYSTEM ARE GIVEN

Let the equation of the given circle be $S = 0$ and the equation of the radical axis be $L = 0$, where

$$S = x^2 + y^2 + 2g + 2fy + c$$

and $\quad L = Ax = By + C.$

The equation $S + \lambda L = 0$ represents a system of circles passing through the intersection of $S = 0$ and $L = 0$. Consider the following two circles from the system

$$S + \lambda_1 L = 0$$

$$S + \lambda_2 L - 0.$$

The equation of radical axis becomes as

$$(\lambda_1 - \lambda_2) L = 0$$

or $$L = 0.$$

This proves that any two circles of the system have the same radical axis. Hence the system of circles $S + \lambda L = 0$ represents a system of co-axal circles.

MISCELLANEOUS EXAMPLES

Example 1:

Prove that a common tangent to two circles is bisected by the radical axis. (Hence, by joining the middle points of any two of the common tangents, we have a construction for the radical axis).

Solution:

Let us take the common tangent of the two circles as x-axis and a line perpendicular to the tangent as y-axis and say the equations or the circles are

$$(x - \alpha)^2 + (y - \beta)^2 = \beta^2 \quad ...(1)$$

and $$(x - a)^2 + (y - b)^2 = b^2 \quad ...(2)$$

as abscissa of each will be equal to its radius.

Let PN be the perpendicular from P to the centre of (1), whose coordinates are (α, β) on x-axis. As the perpendicular is parallel to y-axis, its equation will be $x = \alpha$. So the coordinates of N, *i.e.*, the point of contact of (1) will be $(\alpha, 0)$. Similarly coordinates of the point of contact of (2) are $(a, 0)$.

The coordinates of the mid-point of MN is

$$\left\{\frac{1}{2}(\alpha + a), 0\right\} \quad ...(3)$$

Again, the equation of the radical axis of (1) and (2) is

$$\{(x - \alpha)^2 + (y - \beta)^2 - \beta^2\} - \{(x - a)^2 + (y - b)^2 = b^2\} = 0$$

$$\Rightarrow \quad 2x\,(x - \alpha) + 2y\,(b - \beta) + \alpha^2 - a^2 = 0$$

$$\Rightarrow \quad 2x\,(x - \alpha) + 2y\,(b - \beta) + \alpha^2 - a^2.$$

Solving with a-axis, *i.e.*, $y = 0$, we get $\frac{a + \alpha}{2}$.

It is same as the mid-point of MN by (3). **Hence proved.**

Example 2:

Find the general equation of all circles any pair of which have the same radical axis as the circles

$$x^2 + y^2 = 4$$

and $$x^2 + y^2 + 2x + 4y = 6$$

Solution:

The equations of the circles are given as

$$x^2 + y^2 = 4 \quad ...(1)$$

and $$x^2 + y^2 + 2x + 4y = 6. \quad ...(2)$$

The general equation of all the circles having the same radical axis is given by $S + \lambda S' = 0$, where $S = 0$ and $S' = 0$ are the two circles and λ is a constant. Hence equation is

$$(x^2 + y^2 - 4) + \lambda (x^2 + y^2 + 2x + 4y - 6) = 0$$

or $$(x^2 + y^2)(\lambda + 1) + 2\lambda (x + 2y) = 4 = 6\lambda.$$

Example 3:

If two circles cut orthogonally, prove that the polar of any point P on the first circle with respect to the second passes through the other end of the diameter of the first circle which goes through P.

Hence. (by considering the orthogonal circle of three circle as the locus of a point such that its polars with respect to the circles, meet in a point) prove that the orthogonal circle of three circles, given by the general equation is

$$\begin{vmatrix} x+g_1, & y+f_1 & g_1x+f_1y+c_1 \\ x+g_2, & y+f_2 & g_2x+f_2y+c_2 \\ x+g_3, & y+f_3 & g_3x+f_3y+c_3 \end{vmatrix} = 0.$$

Solution:

(a) Let the equation of the first circle be

$$x^2 + y^2 = a^2 \qquad ...(1)$$

and let the other circle be

$$x^2 + y^2 + 2gx + 2fy + c = 0. \qquad ...(2)$$

If these two circles cut orthogonally, then

$$2g.0 + 2f.0 = a^2 - c$$

$$\Rightarrow \qquad c = a^2$$

Hence (2) becomes

$$x^2 + y^2 + 2gx + 2fy + a^2 = 0 \qquad ...(3)$$

Let (a cos α, a sin α) be any point on (1). The polar of (a cos α, a sin α) with respect to (3) will be

$$(x.a \cos\alpha + y.a \sin\alpha + g (x + a \cos\alpha)$$
$$+ f (y + a \sin\alpha) + a^2 = 0 \qquad ...(4)$$

The co-ordinates of the other end of the diameter of the circle (1) of which one end is (a cos α, a sin α) is clearly

$$(-a \cos\alpha, -a \sin\alpha).$$

Substituting these co-ordinates in the L. H. S. of (4),

$$-a\cos\alpha \,.\, a\cos\alpha - a\sin\alpha \,.\, a\sin\alpha + g(-a\cos\alpha + a\cos\alpha)$$
$$+ f(-a\sin\alpha + a\sin\alpha) + a^2 = 0$$

$\Rightarrow - a(\cos^2 \alpha + a \sin^2 \alpha) + a^2 = 0 =$ R. H. S.

$\therefore (-a \cos \alpha, -a \sin \alpha)$ satisfies (4).

Hence proved

(b) Let the three circles be

$$x^2 + y^2 + 2g_1x + 2f_1y + c_1 = 0 \quad ...(1)$$

$$x^2 + y^2 + 2g_2x + 2f_2y + c_2 = 0 \quad ...(2)$$

and $$x^2 + y^2 + 2g_3x + 2f_3y + c_3 = 0 \quad ...(1)$$

Let the co-ordinates of the moving point be (h, k).

The equation of the polar of (h, k) with respect to (1) is given as

$$xh = yk + g_1(x + h) + f_1(y + k) = c_1 = 0$$

$$\Rightarrow \quad x(h + g_1) + y(k + f_1) + (g_1h + f_1k + c_1) = 0 \quad ...(4)$$

Similarly the polars of (h, k) with respect to (2) and (3) are respectively

$$x(h + g_2) + y(k + f_2) + (g_2h + f_2k + c_2) = 0 \quad ...(5)$$

$$\Rightarrow \quad x(h + g_3) + y(k + f_3) + (g_3h + f_3k + c_3) = 0 \quad ...(6)$$

If the circles are orthogonal, then the three polars should pass through one point. So if (4), (5) and (6) are concurrent the reqd. condition will be

$$\begin{vmatrix} h+g_1 & k+f_1 & g_1h+f_1k+c_1 \\ h+g_1 & k+f_2 & g_2h+f_2k+c_2 \\ h+g_3 & h+f_3 & g_3h+f_3k+c3 \end{vmatrix} = 0.$$

Generalising for (h, k) the required locus is

$$\begin{vmatrix} x+g_1 & y+f_1 & g_1x+f_1y+c_1 \\ x+g_1 & y+f_2 & g_2x+f_2y+c_2 \\ x+g_3 & y+f_3 & g_3x+f_3y+c3 \end{vmatrix} = 0$$

Proved.

Example 4:

The polars of a point P with respect to two fixed circles meet in the point Q. Prove that the circle on PQ as diameter passes through two fixed points, and cuts both the given circles at right angles.

Solution:

Taking the line joining the centre of the two given circles as the axis of x and the radical axis as axis of y (as radical axis is perpendicular to the line of centres), the equations of the circles can be written as

$$x^2 + y^2 - 2gx + c = 0 \quad ...(1)$$

and $$x^2 + y^2 - 2gx + c = 0 \qquad ...(2)$$

Let the point P be (h, k)

The polars of (h, k) with respect to (1) and (2) are resp.

$$xh + yk - g(x + h) + c = 0$$

$$\Rightarrow \quad x(h - g) + yk + c - gh = 0. \qquad ...(3)$$

$$xh + yk - g_1(x + h) + c = 0$$

$$\Rightarrow \quad x(h - g_1) + yk + c - g_1h = 0. \qquad ...(4)$$

Solving (3) and (4), we get the co-ordinates of Q as

$$\left(-h, \frac{h^2 - c}{k}\right)$$

The equation of the circle having PQ as diameter will be

$$(x - h)(x + h) + (y - k)\left(-y, \frac{h^2 - c}{k}\right) = 0$$

$$\Rightarrow \quad kx^2 + ky^2 - y(k^2 + h^2 - c) - ck = 0. \qquad ...(5)$$

The equation is always satisfied by $x = \pm\sqrt{(c)}$, $y = 0$. So this circle always passes through $(\pm\sqrt{c}, 0)$ which are fixed points. **Proved.**

Again (5) is same as $x^2 + y^2 - y.\ \dfrac{k^2 + h^2 - c}{k} - c = 0 \qquad ...(6)$

Two circles cut orthogonally if $2gg' + 2ff - c - c' = 0$.

Putting the values of g, g' etc., from (1) and (6), we get

$$\text{L. H. S.} = 2g.\ 0 + 2.0\left(-\frac{k^2 + h^2 - c}{k}\right) = c - c = 0$$

So the y cut orthogonally. Similarly for the (2) **Proved.**

Example 5:

Prove that a common tangent to two circles of a co-axal system subtends a right angle at either limiting point of the system.

Solution:

Let the equations to any coaxial circles having y-axis as radical axis be

$$x^2 + y^2 + 2g_1x + c = 0 \qquad ...(1)$$

and $$x^2 + y^2 + 2g_2x + c = 0 \qquad ...(2)$$

where the line of centres is the x-axis.

limiting points will be

($\sqrt{c}$, 0) and ($-\sqrt{c}$, 0) say L_1 and M

(Cf. Art. 189)

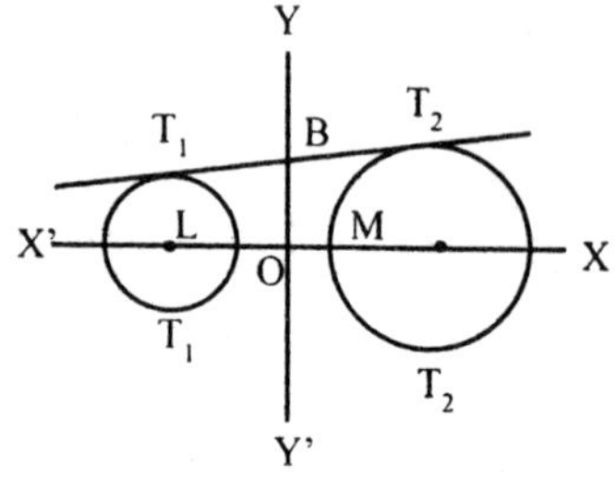

Let the common tangent $T_1 T_2$ meet the radical axis at $P \equiv (0, k)$, then $PT_1 = 1/2\ T_1 T_2$.

Again, if PT_1 is the tangent on (1) and PT_2 on (2),

$$PT_2^{\,2} = PT_1^{\,2} = k^2 + c \qquad ...(3)$$

Now $PT^2 = (0 - \sqrt{c})^2 + (k - 0)^2 = k^2 + c$ ...(4)

So $PT_1 = PL = PT_2$ [by (3) and (4)].

Hence $\angle T_1LT_2$ must be 90°

Similarly $\angle T_1MT_2$ will be 90°. **Proved.**

Example 6:

Find the equation to the circle cutting orthogonally the three circles

$$x^2 + y^2 = a^2,\ (x - c)^2 + y^2 = a^2 \quad \text{and} \quad x^2, (y - b)^2 = a^2.$$

Solution:

The circles are given as

$$x^2 + y^2 = a^2 \qquad ...(1)$$

$$(x^2 + c)^2 + y^2 = a^2$$

or $x^2 + y^2 - 2cx = a^2 - c^2$ ...(2)

and $x^2 + (y - b)^2 = a^2$.

or $x^2 + y^2 - 2by = a^2 - b^2$ ...(3)

Subtracting (2) from (1) the radical axis of (1) and (2) is

$$2xc = c^2 \quad \text{or}\, x = c/2 \qquad ...(4)$$

Subtracting (3) from (1), the radical axis of (1) and (3) is

$$2by = b^2 \quad \text{or}\, y = b/2 \qquad ...(4)$$

By (4) and (5), the radical centre is (c/2, b/2)

The length of the tangent from (c/2, b/2) upto (1)

$$= \sqrt{\{(c/2)^2 + (b/2)^2 - a^2\}}.$$

Hence the required circle is

$$(x - c/2)^2 + (y - b/2)^2 = (c/2)^2 = (b/2)^2 - a^2$$

$\Rightarrow \qquad x^2 + y^2 - cx - by + a^2 = 0.$ **Ans.**

Example 7:

From the preceding question show that the centres similitude (i.e., the points in which the common tangents to two circles meet the line of centres) divide the line joining the centres internally and externally in the ratio of the radii.

Solution:

The centres and radii of the circles given by (1) and (2) are respectively

$(k, 0), (k', 0), \sqrt{(k^2 - \delta)}$ and $\sqrt{(k^2 - \delta)}$.

Hence the abscissa of the points which divide the line of the centres in the ratio of the radii internally and externally, respectively, are

$$\frac{k\sqrt{\left(k^2-\delta\right)}+k\sqrt{\left(k'^2-\delta\right)}}{\sqrt{\left(k^2-\delta\right)}+\sqrt{\left(k^2-\delta\right)}} \qquad ...(1)$$

$$\frac{k'\sqrt{\left(k^2-\delta\right)}-k\sqrt{\left(k'^2-\delta\right)}}{\sqrt{\left(k^2-\delta\right)}-\sqrt{\left(k'^2-\delta\right)}} \qquad ...(2)$$

Solving equation with x-axis, *i.e.*, y = 0, we get

$$x^2-2\frac{kk'+\delta}{k+k'}x+\delta=0$$

$\Rightarrow \qquad (k + k')\, x^2 - 2\,(kk' + \delta)\, x + \delta\,(k + k') = 0.$

The abscissae of the points of intersection are given by

$$x=\frac{2(kk'+\delta)\pm\sqrt{\left[4(kk'+\delta)^2-4(k+k').\delta(k+k')\right]}}{2(k=k')}$$

$$=\frac{(kk'+\delta)\pm\left[(kk'+\delta)^2-\delta(k+k')^2\right]}{k+k'}$$

$$=\frac{(kk'+\delta)\pm\sqrt{\left[\left(k^2+\delta\right)\left(k'^2-\delta\right)\right]}}{k+k'}$$

which are the same as we get on rationalising the deno. of (1) and (2).

Hence proved.

Example 8:

Prove that the two circles, which pass through the two points (0, a) and (0, – a) and touch the straight line $y = mx + c$, will cut orthogonally if $c^2 = a^2 (2 + m^2)$.

Solution:

Let the equation of the circle be

$$x^2 + y^2 + gx + 2fy + \lambda = 0. \qquad ...(1)$$

If the circle passes through (0, a), we have

$$a^2 + 2fa + \lambda = 0. \qquad ...(2)$$

and if the circle passes through (0, – a), we have

$$a^2 - 2fa + \lambda = 0. \qquad ...(3)$$

Solving (2) and (3), we get $\lambda = -a^2$ and $f = 0$

Putting the value in (1), the equation of the circle becomes

$$x^2 + y^2 + 2gx - a^2 = 0. \qquad ...(4)$$

Its centre will be $(-g, 0)$ and radius $\sqrt{(g^2 + a^2)}$.

If $y = mx + c$ touches (4), then the length of perpendicular from the centre upto the line must be equal to the radius

So $$\frac{-gm + c}{\sqrt{(1 + m^2)}} = \sqrt{(g^2 + a^2)}$$

$$\Rightarrow (-gm = c)^2 = \{\sqrt{(1 + m^2)}.\ \sqrt{(g^2 + a^2)}\}^2$$

$$\Rightarrow g^2m^2 + c^2 - 2gmc = (1 + m^2) = (g^2 + a^2) = g^2m^2 + g^2 + a^2 + a^2 m^2$$

$$\Rightarrow g^2 + 2gmc + a^2 - c^2 + a^2m^2 = 0 \qquad ...(5)$$

Now (5) will give us two values of g; so two circles satisfying the given conditions are possible. Let these values be g_1 and g_2, then the circles will be

$$x^2 + y^2 + 2g_1x - a^2 = 0$$

and $$x^2 + y^2 + 2g_2x - a^2 = 0$$

If these two cut orthogonally, then we must have

$$2g_1g_2 = -a^2 - a^2. \qquad ...(6)$$

As g_1 and g_2 are the roots of (5), so

$$g_1g_2 = a^2 - c^2 + a^2m^2$$

Putting in (6), we get

$$2\,(a^2 - c^2 + a^2m^2) = -\,2a^2$$

$\Rightarrow \quad c^2 = 2a^2 + a^2m^2$

$\Rightarrow \quad c^2 = a^2\,(2 + m^2).$ **Hence proved.**

Example 9:

Find the equation to the circle which cuts orthogonally each of the circles

$$x^2 + y^2 + 2gx + c = 0$$

$$x^2 + y^2 + 2g'x + c = 0$$

and $\quad x^2 + y^2 + 2hx + 2ky + a = 0$

Solution:

Equation of the circles are given as

$$x^2 + y^2 + 2gx + c = 0 \qquad ...(1)$$

$$x^2 + y^2 + 2g'x + c = 0 \qquad ...(2)$$

and $\quad x^2 + y^2 + 2hx + 2ky + a = 0 \qquad ...(3)$

The radical axis of (1) and (2) is

$$(x^2 + y^2 + 2g'x + c) - (x^2 + y^2 + 2gx + c) = 0$$

$\Rightarrow \quad x = 0 \qquad ...(4)$

The radical axis of (1) and (3) is

$$(x^2 + y^2 + 2hx + 2ky + a) - (x^2 + y^2 + 2gx + c) = 0$$

$\Rightarrow \quad 2x\,(h - g) + 2ky + (a - c) = 0 \qquad ...(5)$

Solving (5) and (4), $x = 0$, $y = (c - a)/2k$.

Hence the radical centre is $[0, (c - a)/2k]$

Again the length of the tangent from $[0, (c - a)/2k]$ upto (1)

$$\sqrt{\left\{\left(\frac{c-a}{2k}\right)^2 + c\right\}}$$

So the required circle is

$$(x-0)^2 + \left(y - \frac{c-a}{2k}\right)^2 = \left\{\left(\frac{c-a}{2k}\right)^2 + c\right\}$$

or $\quad x^2 + y^2 - 2y\dfrac{c-a}{2k} + \left(\dfrac{c-a}{2k}\right)^2 = \left(\dfrac{c-a}{2k}\right)^2 + c$

or $\quad k(x^2 + y^2) + y(a - c) - ck = 0$ **Ans.**

Example 10:

It $x + y\sqrt{-1} = \tan(u + v\sqrt{-1})$, where x, y, u and v are all real, prove that the curves u = constant give a family of coaxial circles passing through the points (0, ± 1), and that the curves v = constant give a system of circles cutting the first system orthogonally.

Solution:

We are given

$$\tan(u + iv) = x + iy. \quad ...(1)$$

Therefore $\quad \tan(u - iv) = x - iy. \quad ...(2)$

and $\quad \tan[(u - iv) + (u - iv)] = \dfrac{\tan(u+iv)+\tan(u-iv)}{1-\tan(u+iv)+\tan(u-iv)}$

$\Rightarrow \quad \tan 2u = \dfrac{x+iv+x-iy}{1-(x+iy)+(x-iy)}$

$\Rightarrow \quad \tan 2u = \dfrac{2x}{1-x^2-y^2}$

$\Rightarrow \quad \tan 2u\,(1 - x^2 - y^2) = 2x$

$\Rightarrow \quad x^2 + y^2 + 2x\cot 2u - 1 = 0. \quad ...(3)$

Similarly, $\tan\{(u - iv) - (u - iv)\} = \dfrac{(x+iy)-(x-iy)}{1+(x+iy)\,(x-iy)}$

$\Rightarrow \quad i\tan 2v = \dfrac{2iy}{1+x^2+y^2}$

$\Rightarrow \quad x^2 + y^2 + 2y\cot 2v + 1 = 0. \quad ...(4)$

Solving (3) with y-axis, *i.e.*, x = 0, we get

$$y^2 - 1 = 0$$

or $\quad y = \pm 1$

Hence (3) passes through (0, ± 1) for all values of u.

Clearly (3) pass (4) represent different systems of coaxial circles.

Again for (3) and (4) respectively, g = cot 2u, g' = 0, f = 0,

f = both 2v, c = 1 and c' = – 1.

Hence $\quad 2gg' + 2ff - c - c' = 0 = 1 + 1 = 0$

Therefore the two systems cut orthogonally. **Proved.**

Example 11:

Show that the equation to the circle cutting orthogonally the circles

$$(x - a)^2 + (y - b)^2 = b^2$$

$$(x - b)^2 + (y - a)^2 = a^2$$

and $$(x - a - b - c)^2 + y^2 = ab + c^2$$

is $$x^2 + y^2 - 2x(a + b) - y(a + b) + a^2 + 3ab + b^2 = 0.$$

Solution:

The equations of the circles are given as

$$(x - a)^2 + (y - b)^2 = b^2$$

$$\Rightarrow \quad x^2 + y^2 - 2ax - 2by + a^2 = 0, \qquad ...(1)$$

$$(x - a)^2 + (y - a)^2 = a^2$$

$$\Rightarrow \quad x^2 + y^2 - 2bx - 2ay + b^2 = 0, \qquad ...(2)$$

and $$(x - a - b - c)^2 + y^2 = ab + c^2$$

is $$x^2 + y^2 - 2x(a + b + c) + (a + b + c^2) - ab = c^2 = 0 \qquad ....(3)$$

Subtracting (1) from (2), the equation to the radical axis or (1) and (2) is

$$2x(a - b) - 2y(a - b) - (a^2 - b^2) = 0$$

$$\Rightarrow \quad 2x - 2y - (a + b) = 0 \qquad ...(4)$$

Subtracting (3) from (1), the radical axis of (1) and (3) is

$$2x(b + a) - 2by + a^2 + ab + c^2 - (a + b + c)^2 = 0$$

$$\Rightarrow \quad 2x(b + c) - 2by + b^2 - ab - 2ac - 2bc = 0 \qquad ...(5)$$

Solving (4) and (5), we get

$$x = (a + b)$$

and $$y = \frac{a+b}{2}$$

So the radical centre is $\{(a + b).\ 1/2\ (a + b)\}$.

The length of the tangent from $\{(a + b).\ 1/2\ (a + b)\}$ upto (1)

$$= \sqrt{\left\{(a+b)^2 + \frac{1}{4}(a+b)^2 - 2a(a+b) - 2b\frac{1}{2}(a+b) + a^2\right\}}$$

Hence the equation to the circle cutting (1), (2) and (3) orthogonally is

$$\{x - (a + b)\}^2 + \left\{y - \frac{1}{2}(a+b)^2\right\}$$

$$= (a + b)^2 + \frac{1}{4}(a + b)^2 - 2a(a + b) - b(a + b) + a^2$$

$$\Rightarrow \quad x^2 + (a + b^2) - 2x(a + b) + y^2 + \frac{1}{4}(a + b)^2 - 2y.\frac{1}{2}(a + b)$$

$$= (a + b)^2 + \frac{1}{4}(a + b)^2 - 2a(a + b) - b(a + b) + a^2$$

$$\Rightarrow \quad x^2 + y^2 - 2x(a + b) - y(a + b) + a^2 + 3ab + b^2 = 0$$

which is the same as given equation. **Hence proved.**

Example 12:

Prove that the polars of any point with respect to a system of coaxial circles all pass through a fixed point, and that the two points are equidistant from the radical axis and subtend a right angle at a limiting point of the system. If the first point be one limiting point of the system prove that the second point is the other limiting point.

Solution:

If L is one of the limiting points, its co-ordinates will be $(\sqrt{c}, 0)$; then

We have slope of PL $= m_1 = \dfrac{k-0}{h-\sqrt{c}}$,

Slope of QL $= m_2 = \dfrac{\dfrac{h^2-c}{k}-0}{h-\sqrt{c}} = -\dfrac{h^2-c}{k\left(h+\sqrt{c}\right)}$

$$\therefore m_1 m_2 = \frac{k}{h-\sqrt{c}} \times -\frac{h^2-c}{k\left(h+\sqrt{c}\right)} = -1$$

Hence PL and QL are at right angles.

Similarly we can prove for the other limiting point $(\sqrt{c}, 0)$. **Proved.**

Example 13:

Prove that the polar of a limiting point of a coaxial system with respect to any circle of the system is the same for all circles of the system.

Solution:

The system of the coaxial circles is given by

$$x^2 + y^2 + 2gx + c = 0$$

where the radical axis is y-axis and line of centres is the x-axis.

The limiting points are $(\sqrt{c}, 0)$ and $(\sqrt{c}, 0)$

The polar of $(\sqrt{c}, 0)$ with respect to the circle is

$$x.\sqrt{c} + y.0 + g(x + \sqrt{c}) + c = 0$$

$$\Rightarrow \quad x(\sqrt{c} + g) + \sqrt{c}(g + \sqrt{c}) = 0$$

$$\Rightarrow \quad x + \sqrt{c} = 0,$$

which is independent of the variable g.

Hence the polar is same for all values of g. Proved

Let the system of the coaxial circles be represented by

$$x^2 + y^2 + 2\lambda x + c = 0 \quad ...(1)$$

and the co-ordinates of any point be (h, k).

The polar of P ≡ (h, k) with respect to (1) is

$$xh + yk + \lambda(x + h) + c = 0$$

$$\Rightarrow \quad (xh + yk + c) + \lambda(x + h) = 0 \quad ...(2)$$

As (2) is of the form $P + \lambda Q = 0$, it always passes through the point of intersection of the curves (st, lines)

$$xh + yk = c = 0$$

and $$x + h = 0$$

Hence it passes through a fixed point. **Proved.**

Again, solving these two equations, the co-ordinates of the point of intersection, Q *i.e.*, fixed point, are

$$\left(-h, \frac{h^2 - c}{k}\right)$$

The radical axes of (1) is y-axis or x = 0

Distance of P ≡ (h, k) from the radical axis = h.

Distance of the fixed point $Q = \left(-h, \frac{h^2 - c}{k}\right)$ from radical axis = h (numerically). **Hence proved.**

Example 14:

A fixed circle is cut by a series of circles all of which pass through two given points; prove that the straight line joining the intersections of the fixed circle with any circle of the system always passes through a fixed point.

Solution:

Taking the two fixed points as (0, a) and (0, – a), the equation of the system of circles can be written as

$$x^2 + y^2 + 2gx - a^2 = 0. \qquad ...(1)$$

let the fixed circle be $x^2 + y^2 + 2ax + 2by + c = 0.$ ...(2)

The straight line passing through the point of intersection of (1) and (2) is the common chord whose equation will be

$$(x^2 + y^2 + 2ax + 2by + c) - (x^2 + y^2 + 2gx - a^2) = 0$$

$$\Rightarrow \quad (2ax + 2by + c + a^2) - 2gx = 0. \qquad ...(3)$$

(3) being of the form P + lQ = 0, passes through a fixed point which is the point of intersection $2gx = 0$ or $x = 0$ and $2ax + 2by + c + a^2 = 0.$

Solving the two, the co-ordinates of the fixed point are $\left(0, \frac{c+a^2}{2b}\right)$

Proved

Example 15:

Find the equation to the circle cutting orthogonally the three circles

$$x^2 + y^2 - 2x + 3y - 7 = 0,$$

$$x^2 + y^2 + 5x - 5y + 9 = 0,$$

and $$x^2 + y^2 + 7x - 9y + 29 = 0$$

Solution:

The three circles are given as

$$x^2 + y^2 - 2x + 3y - 7 = 0 \qquad ...(1)$$

$$x^2 + y^2 + 5x + 5y + 9 = 0 \qquad ...(2)$$

and $$x^2 + y^2 + 7x - 9y + 29 = 0 \qquad ...(3)$$

Subtracting (1) from (2), the radical axis of (1) and (2) is

$$7x - 8y + 16 = 0 \qquad ...(4)$$

Subtracting (1) from (3), the radical axis of (1) and (3) is

$$9x - 12y + 36 = 0 \text{ or } 3x - 4y + 12 = 0 \qquad ...(5)$$

Solving (4) and (5), we get $x = 8$ and $y = 9$.

So the radical centre is (8, 9).

The length of the tangent from (8, 9) upto (1)

$$= \sqrt{[(8)^2 + (9)^2 - 2 \cdot 8 + 3 \cdot 9 - 7]} = \sqrt{(149)}$$

Hence the equation to the required circle is

$$(x - 8)^2 + (y - 9)^2 = 149$$

$$\Rightarrow \quad x^2 + y^2 - 16x - 18y - 4 = 0$$

Ans.

Example 16:

Prove that a system of coaxial circles inverts with respect to either limiting point into a system of concentric circles and find the position of the common centre.

Solution:

Let the equation representing the coaxial system be

$$x^2 + y^2 + 2gx + c = 0 \qquad ...(1)$$

If P be limiting point as $(\sqrt{c}, 0)$, then changing the origin to P, we get

$$(x + \sqrt{c})^2 + (y + 0)^2 + 2g\ (x + \sqrt{c})^2 + c = 0$$

$$\Rightarrow \qquad x^2 + y^2 + 2x\ (\sqrt{c} + g) + 2c + 2g\ \sqrt{c} = 0$$

Changing into polar co-ordinates, we get

(by putting $x = r \cos\theta$ and $y = r \sin\theta$)

$$r^2 + 2r \cos\theta\ (\sqrt{c} + g) + 2\ \sqrt{c}\ (\sqrt{c} + g) = 0$$

$$\Rightarrow \qquad r^2 + r\ 2(\sqrt{c} + g)\ (r \cos\theta + \sqrt{c}) = 0.$$

To get the inverted equation, put $r = a^2/r$, so we get the inverse as

$$\left(\frac{a^2}{r}\right)^2 + 2(\sqrt{c} + g)\left(\frac{a^2}{r}.\cos\theta + \sqrt{c}\right) = 0.$$

$$\Rightarrow \qquad a^4 + 2\ (\sqrt{c} + g)\ (a^2 r \cos\theta + \sqrt{c}r^2) = 0$$

The equation represents a concentric system of circles for different values of g, having the centre as

$$\left(\frac{a^2}{r}, 0\right)$$

Ans.

Example 17:

Prove that tangent drawn from any point of a fixed circle of a coaxial system to two other fixed circles of the system are in a constant ratio.

Solution:

Let the 3 coaxial circles be given as

$$x^2 + y^2 + 2g_1x + c = 0 \qquad ...(1)$$

$$x^2 + y^2 + 2g_2x + c = 0 \qquad ...(2)$$

and $\qquad x^2 + y^2 + 2g_3x + c = 0 \qquad ...(3)$

Let P be any point (h, k) on the circle (1); then the co-ordinates will satisfy (1); hence

$$h^2 + k^2 + 2g_1h + c = 0 \quad \text{or} \quad h^2 + k^2 + c = -2g_1h \qquad ...(4)$$

If PT_1 and PT_2 be the tangents from P upto (2) and (3) respectively, then

$$PT_1 = h^2 + k^2 + 2g_2h + c = 2g_2h - 2g_1h$$

$$\{\because h^2 + k^2 + c = -2g_1h \text{ by } (4)\}$$

Similarly $PT_2^2 = h^2 + k^2 + 2g_3h + c = 2g_3h - 2g_1h.$

So $$\frac{PT_1^2}{PT_2^2} = \frac{2h(g_2 - g_1)}{2h(g_3 - g_2)} \text{ or } \frac{PT_1}{PT_2} = \sqrt{\frac{(g_1 - g_2)}{(g_3 - g_2)}}$$

which is constant as g_1, g_2 and g_3 are constant. **Hence proved.**

Example 18:

Find the co-ordinate of the limiting points of the co-axal system to which the circles $x^2 = y^2 + 4x + 2y + 5 = 0$ and $x^2 = y^2 + 2x + 4y + 7 = 0$ belong.

Solution:

The equation of the co-axal system is

$$x^2 = y^2 + 4x + 2y + 5 + \lambda\,(x^2 = y^2 + 2x + 4y + 7) = 0$$

$$x^2 = y^2 + \frac{2(2+\lambda)}{1+\lambda}x + \frac{2(1+2\lambda)}{1+\lambda}y + \frac{5+7\lambda}{1+\lambda} = 0.$$

This gives radius of circle as

$$\sqrt{\frac{2(2+\lambda)^2 + (1+2\lambda)^2 - (5+7\lambda)(1+\lambda)}{(1+\lambda)^2}}$$

and the coordinates of centre as

$$\left[-\left(\frac{2+\lambda}{1+\lambda}\right), -\left(\frac{1+2\lambda}{1+\lambda}\right)\right].$$

For limiting point the radius becomes zero. This means

$$2\lambda^2 + 4\lambda = 0$$

$$\lambda = 0 \qquad \text{or} \qquad \lambda = -2.$$

Hence the limiting point are (– 2, – 1) and 0, – 3)

Example 19:

A straight line is drawn touching one of a system of coaxial circles in P and cutting another in Q and R. Show that PQ and PR subtend equal or supplementary angles at one of the limiting points of the system.

Solution:

Let the system of coaxial circles $x^2 + y^2 + 2gx + c = 0$ having any two circles

$$x^2 + y^2 + 2g_1x + c = 0 \qquad ...(1)$$

$$x^2 + y^2 + 2g_2x + c = 0. \qquad ...(2)$$

Let a line AB touch (1) at P and cut (2) at Q and R. If the co-ordinates of Q be (h, k) and L be one of the limiting points say $(\sqrt{c}, 0)$, then length of tangent QP from (h, k) on (1) $= \sqrt{(h^2 + k^2 + 2g_1h + c)}$

and $$QL = \sqrt{\{(h - \sqrt{c})^2 + (k - 0)^2\}} = \sqrt{\{(h^2 + c - 2h\sqrt{c} + k^2)}$$

Hence $$\frac{QP^2}{QL^2} = \frac{h^2 + k^2 + c + 2g_1h}{h^2 + k^2 + c + 2h\sqrt{c}} \qquad ...(3)$$

As $Q \equiv (h, k)$ lies on (2), so

$$h^2 + k^2 + 2g_2h + c = 0$$

$$\Rightarrow \quad k^2 + k^2 + c = -2g_2h.$$

Putting in (3), $$\frac{QP^2}{QL^2} = \frac{-2g_2h + 2g_1h}{-2g_2h - 2h\sqrt{c}} = \frac{g_2 - g_1}{g_2 + \sqrt{c}} \qquad ...(4)$$

Similarly if R be (h', k'), $$\frac{RP^2}{RL^2} = \frac{g_2 - g_1}{g_2 - \sqrt{c}} \qquad ...(5)$$

By (4) and (5), $$\frac{QP^2}{QL^2} = \frac{RP^2}{RL^2} \text{ or } \frac{QP}{QL} = \pm\frac{PR}{RL}$$

$$\Rightarrow \frac{QP}{RP} = \pm\frac{QL}{RL}$$

Hence the required result. **Proved**

Example 20:

Prove that the circle of similitude of the two circles

$$x^2 + y^2 - 2kx + \delta = 0 \text{ and } x^2 + y^2 - 2k'x + \delta = 0$$

(*i.e.*, the locus of the points at which the two circles subtend the same angle) is the co-axal circle

$$x^2 + y^2 - 2\,\frac{kk' + \delta}{k + k}x + \delta = 0.$$

Solution:

Let A and B be the centres of the given circles

$$x^2 + y^2 - 2kx + \delta = 0 \qquad ...(1)$$

and
$$x^2 + y^2 - 2k'x + \delta = 0 \qquad ...(2)$$

If P is any point on which the circles subtend equal angles, it means $\angle T_1PT_2 = \angle S_1PS_2$ where PT_1 and PT_2 are the tangents on (1) and PS_1 and PS_2 are the tangents on (2). By plane geometry clearly PA and PB are the bisectors of $\angle T_1PT_2$ and $\angle S_1PS_2$ respectively.

So $\angle T_1PA = \angle S_1PB$ Also $\angle AT_1P = \angle BS_1P = 90°$ each.

So triangles PAT_1 and PBT_2 are similar.

Hence $$\frac{PT_1}{PS_1} = \frac{PA}{PB} = \frac{r_1}{r_2},$$

where r_1 and r_2 the radii of (1) and (2)

or $$\frac{PT_1^2}{PS_1^2} = \frac{r_1^2}{r_2^2}$$ **Ans.**

If the co-ordinates of P be (x, y) then

$$\frac{x^2+y^2-2kx+\delta}{x^2+y^2-2k'x+\delta} = \frac{k^2-\delta}{k'^2-\delta}$$

Simplifying, the required locus is

$$x^2 + y^2 - \frac{(2kk'+\delta)}{k+k'}x + \delta = 0.$$

Example 21:

Find the locus of the point of contact of parallel tangents which are drawn to each of a series of coaxial circles.

Solution:

Let the system be represented by

$$x^2 + y^2 - 2gx + c = 0 \qquad ...(1)$$

and the slope of each parallel tangent be m.

If (h, k) is any point of contact, the equation of the tangent at (h, k) to (1) will be

$$xh + yk - g(x + h) + c = 0. \qquad ...(2)$$

Slope of (2) $\dfrac{g-h}{k} = m$ (by hypothesis). ...(3)

Again, as (h, k) lies on (1),

$$h^2 + k^2 - 2gh + c = 0$$

The required locus will be found by eliminating g from (3) and (4); so putting the values of g from (3) in (4), we get

$$h^2 + k^2 - 2h(km + h) + c = 0.$$

Simplifying and generalising, we get the required locus as

$$y^2 - x^2 + 2mxy = c.$$ **Ans.**

Example 22:

Prove that the two circles which pass through the points (0, a) and (0, – a) and touch the line $y = mx + c$ will cut orthogonally if $c^2 = a^2(2 + m^2)$.

Solution:

Let the circle be given as

$$x^2 + y^2 + 2gx + 2fy + d = 0 \qquad ...(i)$$

It passes through (0, a) and (0, – a) then we have

$$\therefore \quad a^2 + 2fa + d = 0 \qquad ...(ii)$$

$$a^2 - 2fa + d = 0 \qquad ...(iii)$$

Solving (ii) and (iii), we have $d = -a^2$ and $f = 0$.

Equation (i) becomes

$$x^2 + y^2 + 2gx - a^2 = 0$$

This circle touches the line $y = mx + c$ if

$$\frac{-mg + c}{\sqrt{m^2 + 1}} = \sqrt{g^2 + a^2}$$

$$\Rightarrow \quad g^2 + 2mcg + \{a^2(m^2 + 1) - c^2\} = 0.$$

This equation gives two values of g (say g_1 and g_2) such that

$$g_1 + g_2 = -2mc \qquad ...(iv)$$

$$g_1 g_2 = a^2(m^2 + 1) - c^2 \qquad ...(v)$$

and the two circles touching $y = mx + c$ are

$$x^2 + y^2 + 2g_1x - a^2 = 0$$

$$x^2 + y^2 + 2g_2x - a^2 = 0.$$

These circles cut each other orthogonally if

$$2g_1g_2 = -2a^2$$

$\Rightarrow \quad g_1g_2 = -a^2$

$\Rightarrow \quad a^2(m^2 + 1) - c^2 = -a^2$ (from iv)

$\Rightarrow \quad c^2 = a^2(m^2 + 2).$

Example 23:

Find the locus of the centre of the circle which cuts two given circles orthogonally

Solution:

Let the two given circles be

$$x^2 + y^2 + 2g_1x + 2f_1y + c_1 = 0 \quad ...(1)$$

and $$x^2 + y^2 + 2g_2x + 2f_2y + c_2 = 0. \quad ...(2)$$

Say the centre of the required circle be (h, k). The equation of any circle with (h, k) as centre will be

$$x^2 + y^2 - 2hx - 2ky + c = 0. \quad ...(3)$$

If (1) and (3) cut orthogonally, then we have

$$-2hg_1 - 2kf_1 = c + c_1. \quad ...(4)$$

If (2) and (3) cut orthogonally, then

$$-2hg_2 - 2kf_2 = c + c_2. \quad ...(5)$$

To get the reqd. locus, we have to eliminate the variable c from (4) and (5), so subtracting (4) from (5), we get

$$2h(g_1 + g_2) + 2k(f_1 - f_2) = c_2 - c_1.$$

Generalising, we get the reqd., locus as

$$2x(g_1 + g_2) + 2y(f_1 - f_2) + (c_1 - c_2) = 0.$$ **Ans.**

Example 24:

Prove that the following pairs of circles intersect orthogonally:

$$x^2 + y^2 - 2ax + c = 0$$

and $$x^2 + y^2 + 2by - c = 0.$$

Solution:

The circles are given as

$$x^2 + y^2 - 2ax + c = 0 \quad ...(1)$$

and $$x^2 + y^2 + 2by - c = 0 \quad ...(2)$$

Here we have $g = -a,\ g' = 0,\ f = 0,\ f = b,\ c = c,\ c' = -c.$

The circles cut orthogonally if

$$2gg' + 2ff = c + c'. \quad ...(3)$$

Putting the values in (3),

$$\text{L. H. S.} = 0 + 0 + 0$$

and $\text{L. H. S.} = c - c = 0.$ **Hence Proved**

Example 25:

Prove that the following pairs of circles intersect orthogonally

$$x^2 + y^2 - 2ax + 2by + c = 0$$

and $$x^2 + y^2 + 2by + 2ay - c = 0$$

Solution:

The circles are given as

$$x^2 + y^2 - 2ax + 2by + c = 0 \quad \text{...(1)}$$

and $$x^2 + y^2 + 2bx + 2ay - c = 0 \quad \text{...(2)}$$

Here $g = -a,\ f = b,\ g' = b,\ f = a,\ c = c,\ c' = -c.$

Putting in the conditions for orthogonal circles

$$\text{L. H. S.} = 2(-a)(b) + 2(b)(a) = 0$$

and $\text{R. H. S.} = c - c = 0.$ **Hence Proved.**

Example 26:

Find the equations to the straight lines joining the origin to the points of intersection of

$$x^2 + y^2 - 4x - 2y = 4$$

and $$x^2 + y^2 - 2x - 4y - 4 = 0.$$

Solution:

The equations are given as

$$x^2 + y^2 - 4x - 2y = 4 \quad \text{...(1)}$$

and $$x^2 + y^2 - 2x - 4y - 4 = 0. \quad \text{...(2)}$$

The required equation will be obtained by making (1) homogeneous with the help of the common chord of (1) and (2). From (1) (2), subtracting, we get

$$(x^2 + y^2 - 2x - 4y - 4) - (x^2 + y^2 - 4x - 2y - 4) = 0$$

$$\Rightarrow \quad 2x - 2y = 0 \text{ or } \quad y - x = 0$$

Hence the equation of the common chord is $y - x = 0$. As it passes through the origin, the equation of the lines joining points of intersection of (1) and (2) with origin is

$$(y + x)(y - x) = 0 \text{ or } \quad (y - x)^2 = 0$$ **Ans.**

Example 27:

Prove that the general equation of all circles cutting orthogonally the circles

$$x^2 + y^2 + 2a_1x - 2b_1y + c_1 = 0$$

$$x^2 + y^2 + 2a_2x - 2b_2y + c_2 = 0$$

is
$$\begin{vmatrix} x^2+y^2 & x & y \\ c_1 & a_1 & b_1 \\ c_2 & a_2 & b_2 \end{vmatrix} + \lambda \begin{vmatrix} x & y & 1 \\ a_1 & b_1 & 1 \\ a_2 & b_2 & 1 \end{vmatrix}$$

Solution:

Let the equation to the circle passing through the origin be

$$x^2 + y^2 + 2gx - 2fy = 0.$$

It cuts the given circles orthogonally, therefore we have

$$c_1 + 2ga_1 + 2fb_1 = 0$$

$$c_2 + 2ga_2 + 2fb_2 = 0$$

Eliminating g, f we get

$$\begin{vmatrix} x^2+y^2 & x & y \\ c_1 & a_1 & b_1 \\ c_2 & a_2 & b_2 \end{vmatrix} = C$$

The radical axis of the orthogonal system is the line of centres of the given system. Therefore, the equation to the radical axis is

$$\begin{vmatrix} x & y & 1 \\ a_1 & b_1 & 1 \\ a_2 & b_2 & 1 \end{vmatrix} = 0.$$

Hence the equation to the orthogonal system is

$$\begin{vmatrix} x^2+y^2 & x & y \\ c_1 & a_1 & b_1 \\ c_2 & a_2 & b_2 \end{vmatrix} + \lambda \begin{vmatrix} x & y & 1 \\ a_1 & b_1 & 1 \\ a_2 & b_2 & 1 \end{vmatrix} = 0.$$

3
PARABOLA

Definition : *Parabola is a locus of a point, which moves so that its distance from a fixed point bears a constant ratio e (= 1) to its distance from a fixed straight line. The fixed point is called focus and the fixed and usually denoted b S.*

The constant ratio is called the *ecentricity* and is denoted by e.

The fixed Straight line is called *directrix.*

The straight line passing through the focus and perpendicular to the directrix is called the *axis.*

When the econitricity e is equal to unity. The conic section is called a parabola.

EQUATION OF PARABOLA

Let S be the focus and ZM be the directrix. From S draw a line SAX' ⊥ to the directrix. Take A as the middle point of SB. From A draw a line AY ⊥ to BS. AS and AY are taken as the x-axis and y-axis respectively. Let AS = a. This gives AB = a and A lies on the locus. Take any point P(h, k) on the locus. Let PC be perpendicular to ZM.

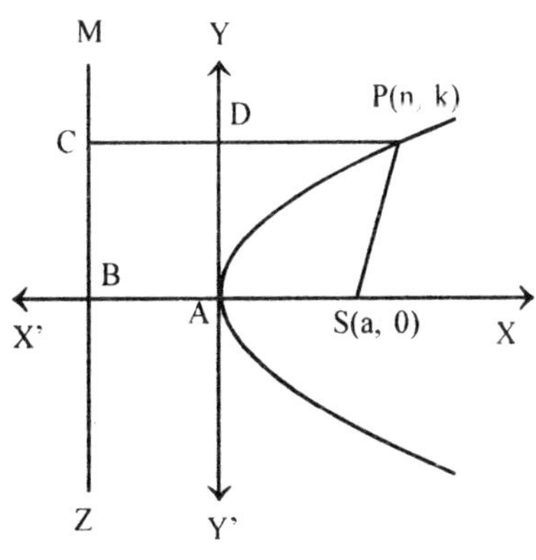

Then we have SP = PC. ...(i)

∴ (i) becomes

$$\sqrt{(h-a)^2+k^2}=(h+a).$$

Squaring, both side we get

$$h^2 - 2ah + a^2 + k^2 = h^2 + 2ah + a^2$$

or $\qquad k^2 = 4ah.$

Hence the locus is given as

$$y^2 = 4ax,$$

which is required equation of the parabola with origin as the vertex, x -axis as the axis of the parabola and y-axis as the tangent at the vertex.

TRACING OF THE PARABOLA

The parabola can be traced according to the following steps

(i) Changing of y to –y does not change the equation of the parabola. This implies that the curve is symmetrical about x-axis.

(ii) Since the negative value of x makes the value of y (obtained from the equation of the parabola) imaginary, no part of the curve can lie to the left of the y-axis

(iii) The curve passes through origin.

(iv) The value of y increases as x increases. Using the above steps, we plot the curve and shape of parabola is shown in the figure.

OTHER FORMS OF THE PARABOLA

The other forms of the parabola are represented as follows:

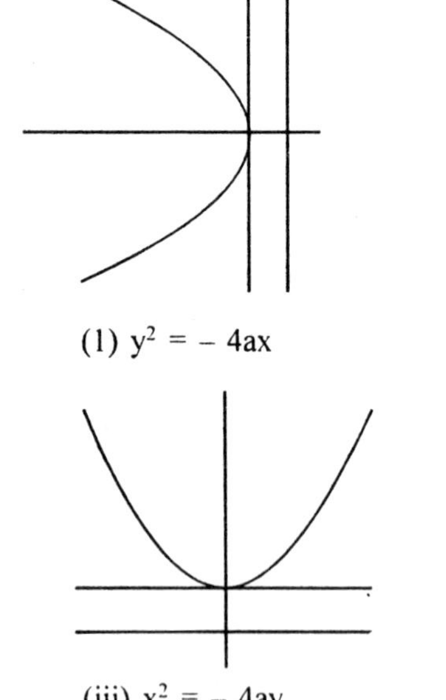

(1) $y^2 = -4ax$

(iii) $x^2 = -4ay$

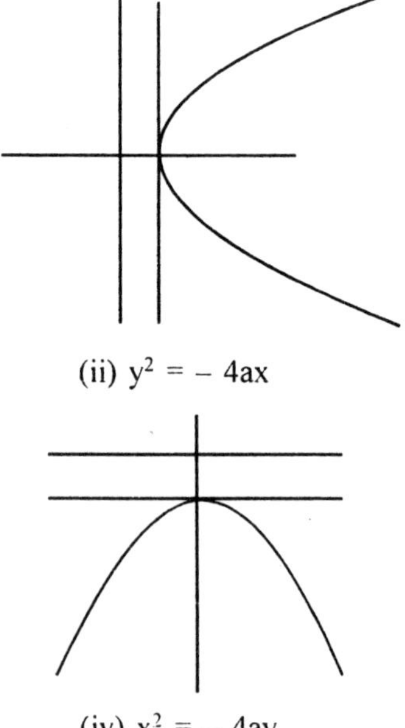

(ii) $y^2 = -4ax$

(iv) $x^2 = -4ay$

POSITION OF A POINT RELATIVE TO A PARABOLA

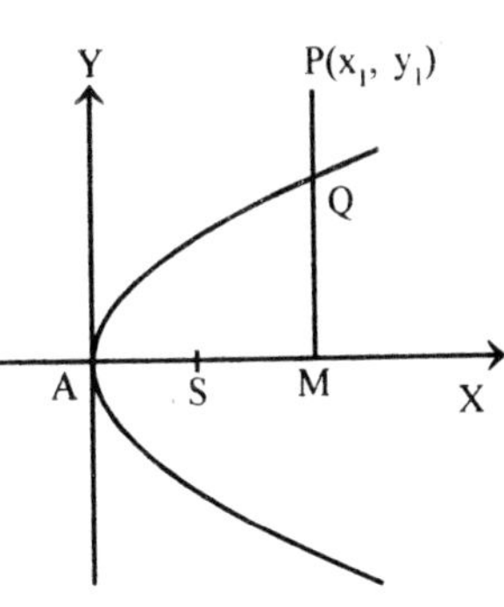

The point (x_1, y_2) lies outside, on or inside the parabola $y^2 = 4ax$ according as the expression $y_1^2 - 4ax_1$ is positive, zero or negative.

If (x_1, y_1) is a given point and $y_1^2 - 4ax_1 = 0$, then by definition (x_1, y_2) lies on the parabola. In case $y_1^2 - 4ax_1$ is non zero, we proceed as follows :

Let P(x_1, y_1) be a point lying outside the parabola and PM be perpendicular to x-axis. This gives

$$PM > QM$$

$$\Rightarrow PM^2 - QM^2 > 0$$

Now $PM^2 = y_1$, and $QM^2 = 4ax_1$, as the co-ordinate of Q is (x_1, QM) and Q lies on the parabola.

Substituting the values of PM and QM, we get

$$y_1^2 - 4ax_1 > 0$$

which is the required condition for the point P(x_1, y_1) to be outside the parabola. Similarly, the condition for P(x_1, y_1) to lie inside the parabola can be obtained.

LATUS RECTUM

A chord of the parabola passing through the focus and perpendicular to the axis is called *Latus Rectum.*

In the figure BSB' is the latus rectum.

Now $\quad BSB' = 2BS.$

The co-ordinates of B is (a, BS). Since B lies on the parabola, we must have

$$BS^2 = 4a.a = 4a^2$$

$$\Rightarrow \quad BS = 2a.$$

$$\therefore \quad BSB' = 4a,$$

which is the length of latus rectum.

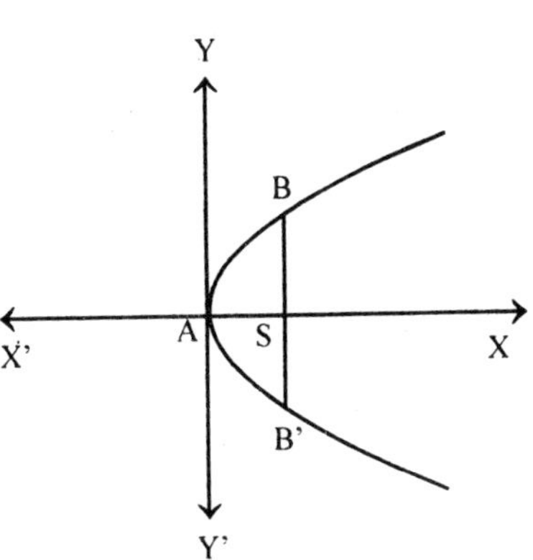

INTERSECTION OF A STRAIGHT LINE AND A PARABOLA

Let the equation of the straight line be given as

$$y = mx + c \qquad ...(i)$$

and the eqn. of parabola is

$$y^2 = 4ax. \qquad ...(ii)$$

The point of intersection of the curves (i) and (ii) can be obtained as

$$(mx + c)^2 = 4ax$$

$$\Rightarrow \quad m^2x^2 + 2x(cm - 2a) + c^2 = 0. \qquad ...(iii)$$

since the equation (iii) is quadratic in x, it gives two points of intersection between a straight line and a parabola. These points can be real and different, real and coincident or imaginary according as the roots of the equation (iii) are real and different, real and equal or imaginary.

Example 1(a):

Prove that the equation to the parabola, whose vertex and focus are on the axis of x at distances a and a' from the origin respectively is $y^2 = 4(a' - a)(x - a)$.

Solution:

Let A(a, 0) be the vertex, S(a', 0) be the focus and ZM be the directrix of a parabola. Now AZ = AS = a' − a ($\because$ directrix and focus are equidistant from the vertex)

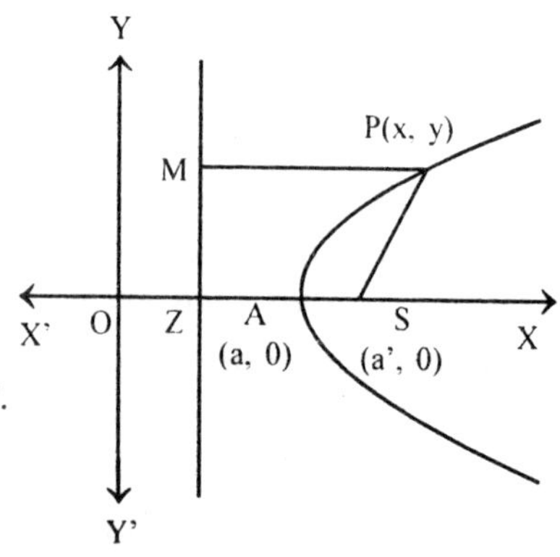

$\therefore$ OZ = OA − AZ

$= a - a' + a = 2a - a'$

$\therefore$ Equation of ZM is $x = 2a - a'$.

Let P(x, y) be a point on the parabola, then

$$PS^2 = (x - a')^2 + y^2.$$

Also $PM = \dfrac{x - 2a + a'}{1}$.

But PS = PM. or $PS^2 = PM^2$

$$\Rightarrow \quad (x - a')^2 + y^2 = (x - 2a + a')^2$$

$$\Rightarrow \quad y^2 = (x + a')^2 - (x + a')^2 + 4a^2 - 4a(x + a')$$

$$\Rightarrow \quad y^2 = 4(a' - a)(x - a).$$

Example 1(b):

Show that the area of the triangle inscribed in the parabola $y^2 = 4ax$ is $1/8a\ (y_1 - y_2)\ (y_2 - y_3)\ (y_3 - y_1)$, where y_1, y_2, y_3 are the ordinates of its vertices.

Solution:

The equation of the parabola is given as

$$y^2 = 4ax.$$

Let the vertices of the triangle inscribed in the parabola (i) be $A(x_1, y_1)$, $B(x_2, y_2)$, $C(x_3, y_3)$

Since these points lie on the parabola (i) then we have

$$\therefore \quad y_1^2 = 4ax_1$$

$$\text{or} \quad x_1 = \frac{y_1^2}{4a}$$

$$\text{Similarly } x_2 = \frac{y_1^2}{4a}, \quad x_3 = \frac{y_3^2}{4a}$$

$$\therefore \quad \text{The vertices are } \left(\frac{y_1^2}{4a}, y_1\right), \left(\frac{y_2^2}{4a}, y_2\right), \left(\frac{y_3^2}{4a}, y_3\right)$$

$\therefore$ Area of the inscribed triangle is given as

$$= \frac{1}{2}\left[\frac{y_1^2}{4a}(y_2 - y_3) + \frac{y_2^2}{4a}(y_3 - y_1) + \frac{y_3^2}{4a}(y_1 - y_2)\right]$$

$$\left[\because \ \Delta = \frac{1}{2} x_1 (y_2 - y_3) + x_2 (y_3 - y_1) x_3 (y_1 - y_2)\right]$$

$$= \frac{1}{8a}\ [y_1\ y_2\ (y_1 - y_2) - y_3\ (y_1^2 - y_2^2) + y_3^2\ (y_1 - y_2)]$$

$$= \frac{1}{8a}\ (y_1 - y_2)\ (y_2 - y_3)\ (y_3 - y_1).$$

PARAMETRIC CO-ORDINATES

The point given by

$$x = at^2,$$

$$y = 2at \qquad \text{...(i)}$$

lies on the parabola for all values of t. The equations (i) are known as the parametric equations to the parabola and the point $(at^2, 2at)$ on the parabola is known as the point t.

TANGENT AND NORMAL AT THE POINT (at^2, 2at)

We know that the equation of tangent and normal at (x_1, y_1) to the parabola is given by the following equation

Tangent $yy_1 = 2a(x + x_1)$

Normal $y - y_1 = \dfrac{-y_1}{2a}(x - x_1)$.

Substituting $x_1 = at^2$, $y_1 = 2at$, the corresponding equations of tangent and normal becomes

$$2aty = 2a(x + at^2)$$

$$y - 2at = -t(x - at^2)$$

$$\Rightarrow \quad ty = x + at^2 \qquad ...(i)$$

$$y + tx = 2at + at^3 \qquad ...(ii)$$

Equations (i) and (ii) are the required equations of tangent and normal to the parabola at the point (at^2, 3at)

CONDITION THE LINE y = mx + c IS TANGENT TO THE PARABOLA $y^2 = 4ax$

The points of intersection of the line

$$y = mx + c \qquad ...(i)$$

and parabola $y^2 = 4ax$ ...(ii)

are given by the equation

$$(mx + c)^2 = 4ax$$

$$\Rightarrow \quad m^2x^2 + 3x\,(mc - 2a) + c^2 = 0. \qquad ...(iii)$$

The equation (iii) being a quadratic in x gives two values of x. In case the line y = mx + c is to be tangent to the parabola, the equation (iii) should give only one point of intersection, *i.e.*, the discriminant of equation must be zero. Thus we have

$$4\,(mc - 2a)^2 - 4m^2c^2 = 0$$

$$\Rightarrow \quad 4m^2c^2 - 4amc + 4a^2 - 4m^2c^2 = 0$$

$$\Rightarrow \quad mc = c \quad \text{or} \quad c = a/m.$$

Which is the required condition. Thus y = mx + a/m is tangent to the parabola for all values of m. This is the equation of the tangent to the parabola in the slope form. Substituting c = a/m in (iii) we obtain

$$m^2c^2 + 2\,(a - 2a)\,x + a^2/m^2 = 0$$

$$(mx - a/m^2 = 0.$$

Which gives $x = a/m^2$. Equation (iii) gives $y = 2a/m$. Thus $y = mx + a/m$ is tangent to the parabola $y^2 = 4ax$ at the point $(a/m^2, 2a/m)$.

EQUATION OF TANGENT TO THE PARABOLA AT (x_1, y_1)

Equation of the parabola is

$$y^2 = 4ax. \qquad \text{...(i)}$$

Let as take any two point $P(x_1, y_1)$ and $Q(x_2, y_2)$ on parabola equation of chord PQ is given by

$$y - y_1 = \frac{y_1 - y_2}{x_1 - x_2}(x - x_1) \qquad \text{...(ii)}$$

Since $P(x_1, y_1)$ and $Q(x_2, y_2)$ lie on parabola

$$\therefore \quad y_1^2 = 4ax_1 \qquad \text{...(iii)}$$

and $$y_2^2 = 4ax_2 \qquad \text{...(iv)}$$

Subtracting (iv) from (iii) and simplifying, we get

$$(y_1 - y_2)(y_1 + y_2) = 4a(x_1 - y_2)$$

or $$\frac{y_1 - y_2}{x_1 - x_2} = \frac{4a}{y_1 + y_2}.$$

Subtracting the values the values of $\frac{y_1 - y_2}{x_1 - x_2}$ in (ii), we obtain

$$\Rightarrow \quad y - y_1 = \frac{4a}{y_1 + y_2}(x - x_1).$$

The chord PQ will be tangent to the parabola if the point Q coincides with P *i.e.* if $x_2 = x_1$ and $y_2 = y_1$. this gives

$$y - y_1 = \frac{2a}{y_1}(x - x_1)$$

$$\Rightarrow \quad yy_1 - y_1^2 = 2ax - 2ax_1$$

$$\Rightarrow \quad yy_1 = y_1^2 + 2ax - 2ax_1 - 2ax_1 + 2ax_1$$

$$\Rightarrow \quad yy_1 = (y_1^2 - 4ax_1) + 2a(x + x_1)$$

$$\Rightarrow \quad yy_1 = 2a(x + x_1). \qquad [\because (y_1^2 - 4ax_1) = 0]$$

which is the required equation of the tangent to the parabola $y^2 = 4ax$ at the point (x_1, y_1).

EQUATION OF NORMAL

Let $y^2 = 4ax$ be the equation of parabola and (x_1, y_1) be any point on it. We have to find the equation of normal at the point (x_1, y_1) is given by

Equation of a line passing through (x_1, y_1) is

$$y - y_1 = m(x - x_1) \qquad ...(i)$$

Equation of a tangent to the parabola at (x_1, y_1) is given by

$$yy_1 = 2a(x + x_1) \qquad ...(ii)$$

The slope of tangent is $\frac{2a}{y_1}$. Since normal is a line perpendicular to tangent, therefore the slope of normal is $\frac{y_1}{2a}$ then. Equation (i) becomes

$$y - y_1 = \frac{-y_1}{2a}(x - x_1), \qquad ...(iii)$$

This is the required equation of normal at the point (x_1, y_1) to the parabola $y^2 = 4ax$.

Let $\frac{-y_1}{2a} = m$, so that $y_1 = -3am$.] Substituting y_1 in the equation $y_1^2 = 4ax_1$, we get $x_1 = am^2$. Equation (iii) can now be re-written as

$$y + 2am = m(x - am^2)$$

$$\Rightarrow \quad y = mx - 2am - am^3. \qquad ...(iv)$$

The equation (iv) is the equation of the normal to the parabola in the slope form.

CO-NORMAL POINTS

Show that three normals can be drawn from a given point to a parabola and that the algebraic sum of the ordinates of their feet is zero. We know that

The equation of normal at the point $(at^2, 2at)$ to the parabola $y^2 = 4ax$ is

$$y + tx = 2at + at^3$$

Let it passes through (h, k). Which gives

$$at^3 + 2at - th - k = 0 \qquad ...(i)$$

This equation, being cubic in t, will give three values of t and to each value of t there corresponds a normal. Hence from a given point three normals can be drawn to the parabola. If t_1, t_2, t_3 are the roots of (i), then ordinates of the feet of these normals are $2at_1$, $2at_2$, $2at_3$. These points are called *Co-normal Points: The sum of the ordinates of these normals is*

$$2at_1 + 2at_2 + 2at_3 = 2a(t_1 + t_2 + t_3)$$

$= 0$ [as $\Sigma t = 0$ from equation (i)]

which is the require result

Example 1(a):

Show that the distance between a tangent to the parabola $y^2 = 4ax$ and a parallel normal is a cosec $\theta \sec^2 \theta$, where θ is the angle which either makes with the axis.

Solution:

The general equation of a tangent and normal are given as

$$y = mx + \frac{a}{m} \quad \text{...(i)}$$

$$y = mx - 2am - am^3. \quad \text{...(ii)}$$

Now tangent and normal meets the x-axis ($y = 0$) at the point $\left(\frac{-a}{m^2}, 0\right)$ and $(2a + am^2, 0)$ respectively

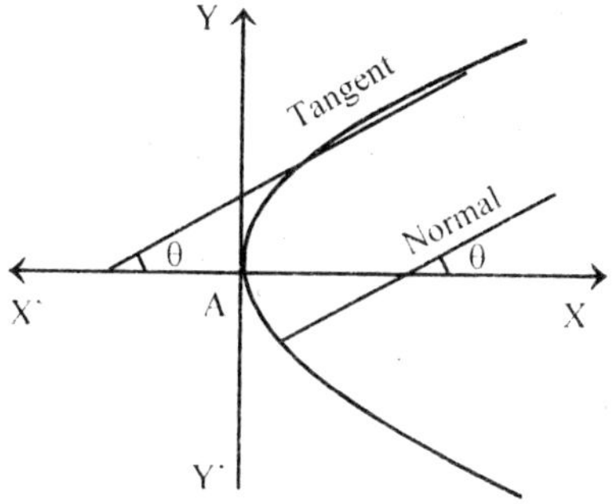

$\therefore$ The portion they intercept on the x-axis is given as

$$= \left[(2a + am)\left(\frac{a}{m^2}\right)\right]$$

$$= 2a + a\tan^2\theta + a\cot^2\theta \qquad (\because\ m = \tan\theta)$$

$$= a(\text{cosec}^2\theta + \sec^2\theta) = a\,\text{cosec}^2\theta\sec^2\theta.$$

Hence the required distance

$$= a\,\text{cosec}^2\theta + \sec^2\theta\sin\theta.$$

$$= a\,\text{cosec}^2\theta + \sec^2\theta.$$

Example 1(b):

Prove that the straight line $lx + my + n = 0$ touches the parabola $y^2 = 4ax$ if $ln = am^2$.

Solution:

The given parabola is $y^2 = 4ax$ ...(i)

and the line is $lx + my + n = 0$. ...(ii)

Substituting the value of $y = -\frac{(lx+n)}{m}$ obtained from (i) is (ii), we have

$$\left(\frac{(lx+n)}{m}\right)^2 = 4ax$$

$\Rightarrow \qquad l^2x^2 + 2(ln - 2am^2)\,x + n^2 = 0.$ ...(iii)

The line (ii) touches the parabola $y^2 = 4ax$, if the roots of (iii) are equal *i.e.*

if $\qquad 4(ln - 2am^2)^2 = 4l^2n^2$

$\Rightarrow \qquad l^2n^2 + 4a^2\,m^4 - 4alnm^2 = l^2n^2$

$\Rightarrow \qquad 4am^2(am^2 - ln) = 0$

$\Rightarrow \qquad am^2 - ln = 0 \qquad (\because\ a \neq 0. m \neq 0 \Rightarrow am^2 \neq 0)$

$\Rightarrow \qquad ln = am^2$. is the required condition.

FOCAL CHORD

Definition: *Any chord of the parabola which passes through focus of a parabola is called a focal chord of the parabola.*

PROPERTIES OF THE PARABOLA

(a) The tangent at any point P on the parabola bisects the angle between the focal chord through P and the perpendicular from P on the directrix.

Proof:

Let P $(at^2, 2at)$ be any point on the parabola, PT is the tangent to the parabola $y^2 = 4ax$, s is the focus, HZ' is the directrix and PM is perpendicular from P on ZZ'.

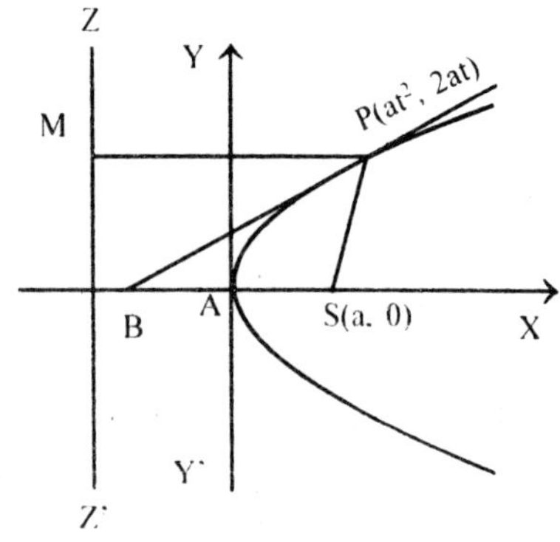

Equation of tangent PT is given by $\quad ty = x + at^2$ given by

$\therefore$ Slope of PT $= \frac{1}{t}$

Also slopes of PM and SP and respectively 0 and

$$\frac{2at - 0}{at^2 - a} \text{ or } 0 \text{ and } \frac{2t}{t^2 - 1}$$

If θ and oϕ be the angles MPT and SPT respectively, then

We have $\tan\theta = \dfrac{\frac{1}{t}-0}{1+\left(\frac{1}{t}\right).0} = \dfrac{1}{t}$

and $\tan\phi = \dfrac{\frac{2t}{t^2-1}-\frac{1}{t}}{1+\frac{2t}{(t^2-1)}.\frac{1}{t}} = \dfrac{1}{t}.$

$\therefore \tan\theta = \tan\phi$ or $\theta = \phi$

which is the result

(b) Show that locus of the foot of perpendicular from the focus on any tangent to the parabola is the tangent at the vertex.

Proof:

Equation of any tangent to the parabola $y^2 = 4ax$ is

$$y = mx + a/m. \qquad ...(i)$$

The equation of any line through the focus (a, 0) and perpendicular to equation is (i) is given as

$$y = -\frac{1}{m}x + a/m. \qquad ...(ii)$$

Subtracting (i) from (ii), we have

$$\left(m + \frac{1}{m}\right)x = 0$$

$\Rightarrow \qquad x = 0,$

which is the tangent at the vertex. Hence the result

(c) The tangents at the extremities of a focal chord intersect at right angles on the directrix.

Proof:

Let $P(at_1^2, 2at_1)$ and $Q(at_2^2, 2at_2)$ be the extremities of the focal chord PSQ. Equation of PQ is given by

$$y - 2at_1 = \frac{2at_1 - 2at_2}{at_1^2 - at_2^2}(x - at_1^2).$$

$$y - 2at_1 = \frac{2}{t_1 + t_2}(x - at_1^2).$$

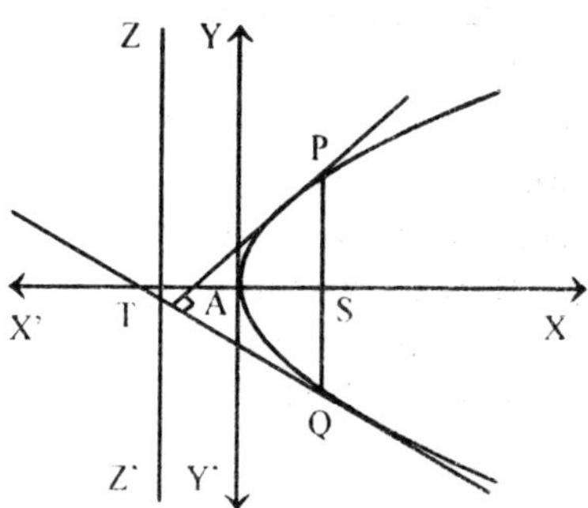

Since the chord passes through the focus (a, 0) then we have

$$0 - 2at_1 = \frac{2}{t_1 + t_2}(a - at_1^2)$$

$$-2at_1 (t_1 + t_2) = 2a - 2at_1^2$$

or we have $\quad t_1 t_2 = -1.$ ...(i)

Equation of tangents at P and Q are

Given by $t_1 y = x + at_1^2$ ...(ii)

and $\quad t_2 y = x + at_2^2$ ...(iii)

The point of intersection of these tangents is obtained by solving equation (ii) and (iii) simultaneously. This gives

$$x = at_1 t_2.$$ ...(iv)

The slope of (ii) is $\frac{1}{t_1}$ and of (iii) is $\frac{1}{t_2}\cdot$ The product of these two slopes is $\frac{1}{t_1 t_2}\cdot$ ...(v)

Using the result $t_1 t_2 = -1$ (iv) gives $x = -a$, *i.e.*, the tangent meet at the directrix. Thus (v) gives the product of the slopes of these tangents as -1, *i.e.*, the tangents are at right angles.

(d) The portion of a tangent to a parabola cut off between the directrix and the curve subtends a right angle at the focus.

Proof:

Let P $(at^2, 2at)$ be any point on the parabola. The equation of tangent at P is given by

$$ty = x + at^2. \quad ...(i)$$

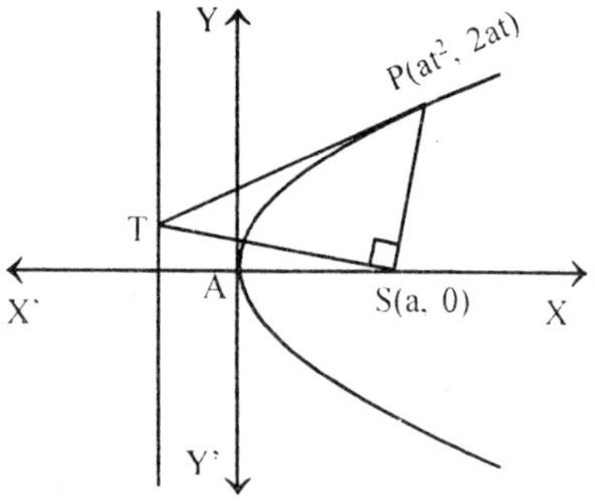

The co-ordinates of T are obtained by solving equation (i) and $x = -a$, the equation of the directrix. Which gives

$$ty = -a + at^2$$

$$\Rightarrow \quad y = \frac{(t^2 - 1)}{t}a.$$

$\therefore$ The co-ordinates of T are as follows:

$$\left(-a, \frac{a(t^2 - 1)}{t}\right).$$

Slope of PS $= m_1 = \dfrac{2at}{at^2 - a} = \dfrac{2at}{a(t^2-1)}$. ...(iii)

Slope of TS $m_2 = \dfrac{a(t^2-1)}{\frac{t}{-2a}}$. ...(iv)

Now $m_1 \times m_2 = -1$ hence the result

(e) If SY be perpendicular to the tangent at a point P of a parabola, prove that $SY^2 = AS.SP$.

Proof:

Take any point P $(at^2, 2at)$.

Equation of tangent at P is given by

$$ty = x + at^2. \quad ...(i)$$

Equation of SY passing through S (a, 0) and perpendicular to PT is given by

$$y = (-t)(x - a)$$

i.e. $tx + y = at.$ ...(ii)

Solving (i) and (ii), we obtain

$$x = 0 \text{ and } y = at.$$

$\therefore$ the co-ordinates of Y are (0, at).

$$\therefore \quad SY^2 = a^2 + a^2 t^2$$

$$= a^2 (1 + t^2)$$

$$= a.(a + at^2) = \angle S.SP.$$

Hence $SY^2 = AS.\ SP$ hence the result

(f) the orthocentre of any triangle formed by three tangent to a parabola lies on the directrix.

Proof:

Let the point be $P(at_1^2, 2at_1)$, $Q(at_2^2, 2at_2)$ and $R(at_3^2, 2at_3)$. The equation of the tangents at P, Q and R are

$$t_1 y = x + at_1^2 \quad ...(i)$$

$$t_2 y = x + at_2^2 \quad ...(ii)$$

and $t_3 y = x + at_3^2$. ...(iii)

The lines (i) and (ii) intersect at $\{at_1t_2, a(t_1 + t_2)\}$.

The equation of the perpendicular from this point on (iii) is given by

$$t_3x + y = a(t_1 + t_2 + t_1t_2t_3). \quad ...(iv)$$

Similarly, the equation of the perpendicular on (i) from the point of intersection of (ii) and (iii) is given by

$$t_1x + y = a(t_2 + t_3 + t_1t_2t_3). \quad ...(v)$$

The x-co-ordinate of the orthocentre which is the point of intersection of (iv) and (v) is–a. The orthocentre, therefore, lies on the directrix. Hence the result

PROPERTIES OF SUBTANGENT AND SUB-NORMALS

(a) The subtangent of any point on the parabola is bisected at the vertex.

Proof:

Take any point $P(at^2, 2at)$. Equation of PT is

$$ty = x + at^2.$$

It meets the axis of X when $y = 0$. This gives the co-ordinate of T as $(-\ at^2, 0)$. Also co-ordinate of M are $(at^2, 0)$. This means

$$AM = at^2.$$

and $\quad AT = at^2.$

$\therefore \quad AM = AT.$

Hence A bisects TM.

(b) The subnormal at any point of a parabola is constant and equal to the semi-latus rectum.

Proof:

Equation of normal at P is given by

$$tx + y\ 2at + at^3.$$

It meets the axis of X at $2at + at^2$. Therefore, the co-ordinates of G are $(2a + at^2, 0)$.

Now $\quad MG = AG - AM$

$$= 2a + at^2 - at^2$$

$$= 2a.$$

Thus the subnormal is of constant length which is equal to the semi-rectum of the parabola.

SUBTANGENT AND SUBNORMAL

Let the tangent and normal at a point P of the parabola meet the axis in T and G respectively. From P draw PM perpendicular to X-axis. TM is called subtangent and MG is called subnormal of the point P.

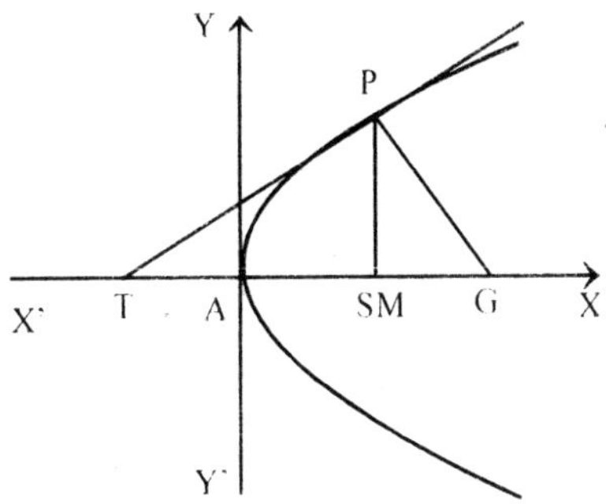

Example 1:

If a chord of the parabola joining the points t_1, t_2 passes through the focus of the parabola, prove that $t_1t_2 = -1$.

Solution:

Let the equation of parabola be $y^2 = 4ax$. ...(i)

The equation of the chord joining t_1, t_2 is

$$(t_1 + t_2)\,y = 2x + 2at_1t_2. \quad \text{(ii)}$$

Since it passes through the focus (a, 0) of parabola (i).

$\therefore \quad 0 = 2a + 2a\,t_1t_2$

or $\quad t_1t_2 = -1$.

Example 2:

A focal chord of the parabola $y^2 = 4ax$, meets it in point P and Q. If S be the focus, prove that

$$\frac{1}{SP} + \frac{1}{SQ} = \frac{1}{a}.$$

Solution:

The equation of parabola is $y^2 = 4ax$

Let $P(at^2, 2at)$ be a point and PSQ be the focal chord so that the co-ordinates of Q are $\left(\frac{a}{t^2}, \frac{-2a}{t}\right)$.

$\therefore$ SP = focal distance of point P $= a + x = a + at^2 = a(1 + t^2)$

Similarly, $SQ = a + \frac{a}{t^2} = \frac{a(1+t^2)}{t^2}$.

$$\therefore \qquad \frac{1}{SP} = \frac{1}{SQ} = \frac{1}{a(1+t^2)} + \frac{t^2}{a(1+t^2)}$$

$$= \frac{(1+t^2)}{a(1+t^2)} = \frac{1}{a}.$$

Hence $\frac{1}{SP} = \frac{1}{SQ} = \frac{1}{a}$.

Example 3:

Prove that the length of the intercept on the normal at the point $P(at^2, 2at)$ of the parabola $y^2 = 4ax$ made by the circle described on the line joining the focus and P as diameter is a $\sqrt{(1 = t^2)}$.

Solution:

The equation of the normal at $P(at^2, 2at)$ is given as

$$y + xt = 2at + at^3.$$

Let the circle with PS as diameter cuts the normal PG at B. This gives

$$PB = \sqrt{(PS^2 - SB^2)}$$

where $PS = \sqrt{(at^2 - a)^2 + (2at - 0)^2}$

$$= \sqrt{a^2 (t^4 - 2t^2 + 1 + 4t^2)}$$

$$= a\,(t^2 + 1),$$

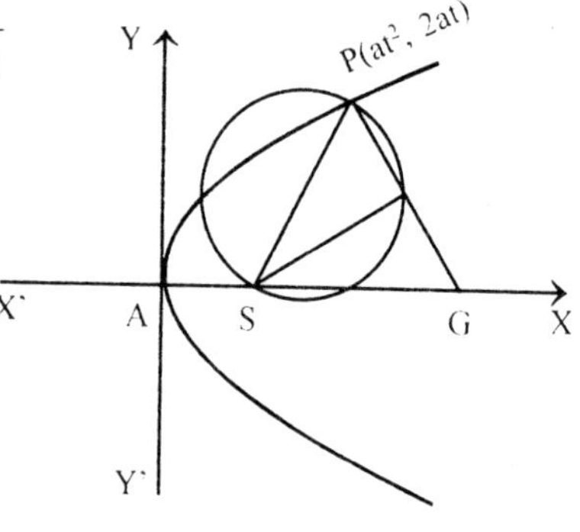

SB = Length of perpendicular from S(a, 0) on the line $y + xt = 2at + at^3$

$$= \frac{-at + 2at + at^3}{\sqrt{(t^2 + 1)}}$$

$$= at\sqrt{(t^2 + 1)}.$$

Substituting the values of PS and SA in (i), we get

$$PB = a\sqrt{a^2(t^2+1)^2 - a^2t^2(t^2+1)}$$

$$= a\sqrt{(t^2+1)}\left(\sqrt{t^2+1-t^2}\right) = a\sqrt{t^2+1}$$ (Hence the result follows).

TANGENT THROUGH A POINT

Let $y^2 = 4ax$ be the equation of the parabola and (h, k) be the co-ordinate of the given point

We know that the line

$$y = mx + a/m \qquad ...(i)$$

Is always tangent to the parabola $y^2 = 4ax$ for all values of m. If this tangent passes through (h, k), then we have

$$k = mh + a/m$$

or $$m^2h - mk + a = 0. \qquad ...(ii)$$

Equation (ii) being quadratic in m gives two values of m, say m_1 and m_2. Now m_1 and m_2 lies on (ii), therefore,

We have $k = m_1h + a/m_1$

and $k = m_2h + a/m_2$.

This implies that there exists two tangents

$$y = m_1x + a/m_1$$

and $$y = m_2x + a/m_2$$

To the parabola $y^2 = 4ax$ passing through (h, k). This proves that through any point two tangents can be drawn to the parabola. The tangents can be real, coincident or imaginary depending upon the roots of the equation (ii) are real, equal or imaginary respectively. The roots of (ii) are real, coincident or imaginary according as

$$k^2 - 4ah = > = \text{ or } < 0$$

i.e., according as the point (h, k) lies outside, on or inside the parabola.

EQUATION OF A PAIR OR TANGENTS FROM AN EXTERNAL POINT

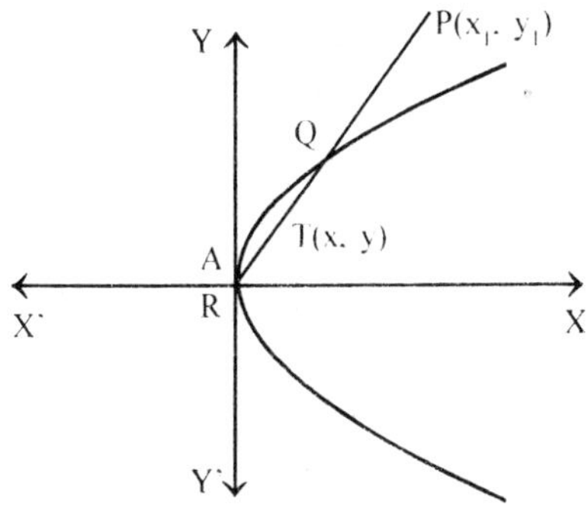

Let $P(x_1, y_1)$ be a point and let T(x, y) be any point on the line through P cutting the parabola in two points Q and R. If Q divides the line PT in some ratio $\lambda : 1$, then the co-ordinates of Q are given by

$$\left(\frac{\lambda x + x_1}{\lambda + 1}, \frac{\lambda y + y_1}{\lambda + 1}\right).$$

The point Q less on the parabola., therefore we have

$$\left(\frac{\lambda y+y_1}{\lambda+1}\right)^2 = 4a\left(\frac{\lambda x+x_1}{\lambda+1}\right)$$

$\Rightarrow \quad (\lambda y + y_1)^2 - 4a(\lambda x + x_1)(\lambda + 1) = 0$

$\Rightarrow \quad \lambda^2(y^2 - 4ax) + 2\lambda[yy_1 - 2a(x + x_1)] + y_1^2 - 4ax_1 = 0.$

This equation being quadratic in λ, gives the values of λ which corresponds to points Q and R. For the line to be tangents both the values of λ must be same. Which gives

$$[yy_1 - 2a(x + x_1]^2 = (y^2 - 4ax)(y_1^2 - 4ax_1)$$

this is the required equation of the pair of tangents drawn from the point (x_1, y_1) to the parabola $y^2 = 4ax$.

CHORD OF CONTACT

To find the equation of the chord of contact of tangents drawn to a parabola from a given point outside it.

Let O(h, k) be any point lying out side the parabola $y^2 = 4ax$. From O two tangents OP and OQ are drawn touching the parabola at the point P and Q.

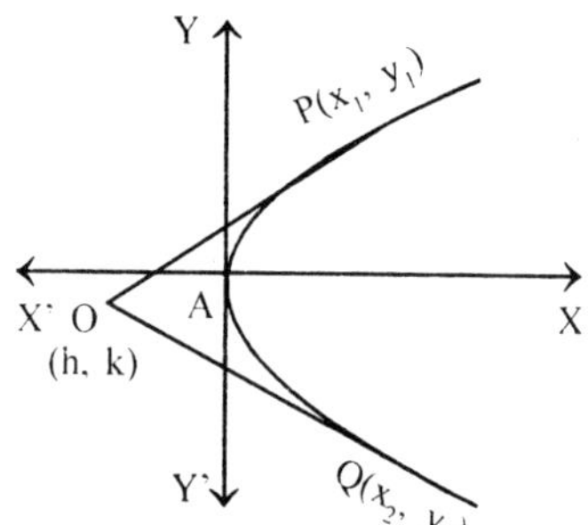

The equation of the tangents at P and Q are given as

$yy_1 = 2a(x + x_1)$ ...(i)

$yy_2 = 2a(x + x_2)$ ...(ii)

Since these tangents pass through O(h, k), so we have

$ky_1 = 2a(h + x_1)$ ...(iii)

$ky_2 = 2a(h + x_2).$ ...(iv)

Equations (iii) and (iv) show that both the points $P(x_1, y_1)$ and $Q(x_2, y_2)$ satisfy the equation

$ky = 2a(x + h).$ ...(v)

Hence (v) represents the required equation of the chord of contact.

PROPERTIES OF POLE AND POLAR

(1) The chord of a parabola having a given point as its middle point is parallel to the polar of that point.

Let the equation of the parabola be $y^2 = 4ax$ and (h, k) the co-ordinates of the given middle point. Equation of a line through (h, k) is given by

$$y - k = m(x - h) \quad \text{...(i)}$$

and equation of the polar of (h, k) is given by

$$yk = 2a(x + h). \quad \text{...(ii)}$$

The points of intersection of line (i) and the parabola is given by

$$y - k = m\left(\frac{y^2}{4a} - h\right)$$

$$\Rightarrow \quad my^2 - 4ay + 4ak - 4ahm = 0$$

It shows that $y_1 + y_2 = 4a/m$. Since (h, k) is the middle point of the chord, therefore, $\frac{y^2 + y_2}{2} = 2a/m$ or $m = 2a/k$, which is also the slope of the line (ii). Hence the lines (i) and (ii) are parallel.

(2) If the polar of P with respect to a parabola passes through Q, then the polar of Q passes through P.

Let the parabola be $y^2 = 4ax$ and the co-ordinates of P and Q be (x_1, y_1) and (x_2, y_2) respectively.

The polar of P is given by

$$yy_1 = 2a(x + x_1). \quad \text{...(i)}$$

The polar of Q is given by

$$yy_2 = 2a(x + x_2). \quad \text{...(ii)}$$

If the polar of P Passes through Q, then we have

$$y_1y_2 = 2a(x_1 + x_2). \quad \text{...(iii)}$$

Again if the polar of Q passes through P, then we have

$$y_1y_2 = 2a(x_1 + x_2). \quad \text{...(iv)}$$

From (iii) and (iv), we conclude that if the polar of P passes through Q, then the polar of Q will pass through P. The points P and Q are called *Conjugate Points.*

If the pole of a line $u = 0$ lies on the line $v = 0$, then the pole of $v = 0$ will lie on $u = 0$.

Let (x_1, y_1) and (x_2, y_2) be the poles of $u = 0$ and $v = 0$ then

we have $u \equiv yy_1 - 2a(x + x_1) = 0$ ...(i)

$$v \equiv yy_2 - 2a(x + x_2) = 0 \quad \text{...(ii)}$$

If the pole of the line u = 0 lies on v = 0, *i.e.*, (x_1, y_1) lies on v = 0, then we have

$$y_1y_2 - 2a(x_1 + x_2) = 0. \qquad ...(iii)$$

This is also the condition that (x_2, y_2), the pole of the line v = 0 lies on u = 0. Hence the result. The lines u =0 and v = 0 are called *conjugate lines.*

Cor. *Obtain the condition that the two straight lines* $l_1x + m_1y + n_1 = 0$ *and* $l_2x + m_2y + m_2 = 0$ *are conjugate with respect to the parabola* $y^2 = 4ax$.

The pole of $l_1x + m_1y + n_1 = 0$ *with respect to parabola* $y^2 = 4ax$ *is*

$$\left(\frac{n_1}{l_1}, \frac{-2am_1}{l_1}\right).$$

This pole will lie on $l_1x + m_1y + n_1 = 0$ if

$$l_1\left(\frac{n_1}{l_1}\right) + m_2\left(\frac{-2am_1}{l_1}\right) + n_2 = 0$$

or $\qquad l_1n_2 + l_2n_1 = 2am_1m_2,$

which is the required condition.

POLE AND POLAR

Definition : *Let P be any point. The polar of P is the locus of the points of intersection of the tangents drawn to the parabola at the ends of chord passing through P.*

TO FIND THE EQUATION OF THE POLAR OF A GIVEN POINT WITH RESPECT TO A PARABOLA

Let $P(x_1, y_1)$ be the given point. Through P a chord is drawn meeting the parabola is Q and R. The tangents at Q and R intersect at T. The locus of T is called the polar of P and P is called the pole. Let the co-ordinates of T be (h, k). Since QR is the chord of contact of T, therefore its equation is given as

$$yk = 2a(x + h). \qquad ...(i)$$

As QR passes through P, it gives

$$y_1k = 2a(x_1 + h). \qquad ...(ii)$$

$\therefore$ locus of T is $yy_1 = 2a(x + x_1)$.

This is the required equation of the polar.

POLE OF A STRAIGHT LINE

To find the pole of the line $lx + my + n = 0$ with respect to the parabola $y^2 = 4ax$.

Let (x_1, y_1) be the pole of the line $lx + my + n = 0$ with respect to the parabola $y^2 = 4ax$, then we have

$$yy_1 = 2a(x + x_1) = 0$$

must represent the same straight line as $lx + my + n = 0$.

Comparing the co-efficients, we have

$$\frac{y_1}{m} = \frac{-2a}{l} = \frac{-2ax_1}{n}$$

$$\Rightarrow \qquad x_1 = \frac{n}{l}, \text{ and } y_1 = \frac{-2am}{l}$$

Hence the pole of $lx + my + n = 0$ is

$$\left(\frac{n}{l}, \frac{-2am}{l}\right).$$

EQUATION OF A CHORD WHOSE MIDDLE POINT IS GIVEN

Let $P(x_1, y_1)$, $Q(x_2, y_2)$ be the end points of a chord PQ whose middle point is (h, k). Equation of the chord passing through (h, k) is given by

$$y - k\ m(x - h). \qquad \text{...(i)}$$

Slope of the line $PQ = \frac{y_2 - y_1}{x_2 - x_1}$. Since P and Q lie on the parabola $y^2 = 4ax$, so we have

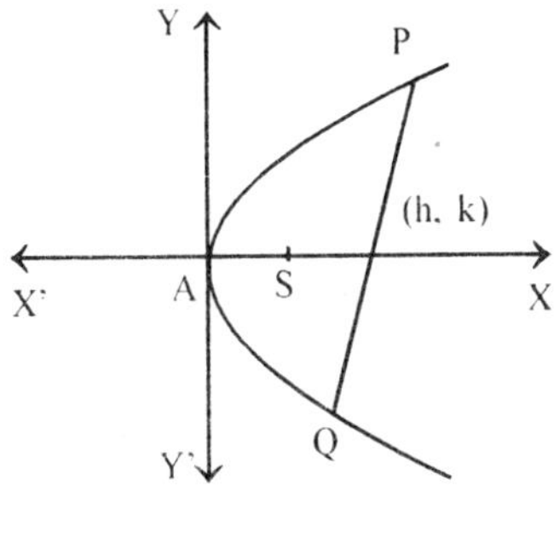

$$y_1^2 = 4ax_1.$$

$$y_2^2 = 4ax_2.$$

Subtraction and simplification gives

$$\frac{y_2 - y_1}{x_2 - x_1} = \frac{4a}{y_1 + y_2}.$$

$\therefore$ Slope of (i) is given by

$$m = \frac{4a}{y_1 + y_2}$$

Since (h, k) is the middle points of PQ, so we get

$$k = \frac{y_1 + y_2}{2}$$

This gives $\qquad m = \frac{4a}{k} = \frac{2a}{k}$

and (i) becomes

$$y - k = \frac{2a}{k}(x - h),$$

which is the required equation of the chord whose middle point is (h, k).

LOCUS OF THE MIDDLE POINTS OF A SYSTEM OF PARALLEL CHORDS OF A PARABOLA

Let the equation of the parabola be

$$y^2 = 4ax \qquad \text{...(i)}$$

Let the slope of the system of parallel chords be m. Let us take any chord whose middle point is (h, k). Then the equation of the chord is

$$y - k = \frac{2a}{k}(x - h).$$

But the slope of the chord is m, therefore, we have

$$m = \frac{2a}{k}$$

or $$k = \frac{2a}{m}$$

Hence the locus is $y = \frac{2a}{m}$, which is a straight line parallel to the axis of the parabola. This locus is called diameter.

TANGENTS AT THE EXTREMITY OF A DIAMETER

The equation of a diameter of the parabola bisects system of parallel chords having slope m is given by

$$y = \frac{2a}{m}.$$

The intersection of this diameter and the parabola $y^2 = 4ax$ is

$$\frac{4a^2}{m^2} = 4ax$$

$$\Rightarrow \qquad x = \frac{a}{m^2}$$

Therefore the co-ordinates of the extremity is $(a/m^2. 2a/m)$.

Equation of the tangent at the point is given by

$$y - \frac{2a}{m} = 2a(x + a/m^2)$$

$$\Rightarrow \qquad y = mx + a/m,$$

which is parallel to the given system of chords. This proves that the tangents at the extremity of a diameter of parabola is parallel to the chords which that diameter bisects.

TO SHOW THAT TANGENTS AT THE ENDS OF A CHORD OF A PARABOLA MEET ON THE DIAMETER WHICH BISECTS THE CHORD

Let $P(at_1^2, 2at_1)$ and $Q(at_2^2, 2at_2)$ be the ends of a chord of the given parabola $y^2 = 4ax$. Then

The slope of PQ is $\frac{2}{t_1 + t_2}$.

$\therefore$ The equation of the diameter bisecting the chord is given by

$$y = a(t_1 + t_2).$$

The points of intersection of the tangents at P and Q is $[a\, t_1t_2, a(t_1 + t_2)]$, which is obtained by solving the equation

$$t_1y = x + at_1^2$$

$$t_2y = x + at_2^2.$$

Since y-co-ordinate is $a(t_1 = t_2)$, the result follows.

MISCELLANEOUS EXAMPLES

Example 1:

Find the equation of a common tangent of the parabola

$$y^2 = 4ax \text{ and } x^2 = 4by.$$

Solution:

The equation to the parabolas are given as

$$y^2 = 4ax \quad ...(1)$$

and $$x^2 = 4by \quad ...(2)$$

Equation to any tangent to (1) is $y = mx + a/m$. ...(3)

If this line is a tangent to (2) also, we must get only one point of intersection of (2) and (3).

Putting the value of y from (3) in (2), we get

$$x^2 = 4b\,(mx + a/m)$$

$$\Rightarrow \quad mx^2 - 4bm^2x - 4ab = 0. \quad ...(4)$$

If (3) is a tangent to (2), (4) must be a perfect square.

So discriminant must be zero; hence

$$16b^2m^4 - 4m(-4ab) = 0$$

$\Rightarrow$ $bm^4 + am = 0.$

whence either m = 0

$\Rightarrow$ $m = (-a/b)^{1/3}.$

If m = 0, we do not get any finite value of a/m, so putting $m = (-1/b)^{1/3}$ in (3), the equation to the tangent becomes

$$y = (-1/b)^{1/3}\, x + a \div (-1/b)^{1/3}$$

$$b^{1/3}\, y + a^{1/3}\, x - a^{2/3}\, b^{2/3}$$ **Ans.**

Example 2:

Prove that two straight lines, one a tangent to the parabola $y^2 = 4a(x + a)$ and the other to the parabola $y^2 = 4a'(x + a')$, which are at right angles to one another, meet on the straight line $x + a + a' = 0$.

Show also that this straight line is the common chord of the two parabolas.

Solution:

The equations to the parabolas are given as

$$y^2 = 4a(x + a) \qquad ...(1)$$

and $$y = 4a'(x + a') \qquad ...(2)$$

Changing the origin to the (– a, 0), (1) becomes

$$y^2 = 4ax. \qquad ...(3)$$

Equation to any tangent to (3) is

$$y = mx + \frac{a}{m}.$$

Changing back to the first origin, this equation becomes

$$y = m(x + a) + \frac{a}{m}. \qquad ...(4)$$

Example 3:

Find the focus and lotus rectum of the parabola

$$x^2 - 2ax + 2ay = 0.$$

Solution:

The equation is given as $x^2 - 2ax + 2ay = 0$

$\Rightarrow$ $$(x - a)^2 = -2ay + a^2 = -4.a/2\,(y - a/2) \qquad ...(1)$$

Changing the origin to (a.a/2) the equation (1) becomes

$$X^2 = 4(-a/2)Y \quad ...(2)$$

According to the new axis, the vertex of (2) is (0, 0), axis is Y-axis *i.e.* X = 0, latus rectum is 4a/2 = 2a, and focus (0, – a/2).

Changing back to the original axis, the co-ordinates of vertex will become (a,a/2), axis is x – a = 0 or x = a, the latus rectum is 2a and the focus is

$$(0 + a, -a/2 + a/2) \quad \text{or} \quad (a, 0)$$ **Ans.**

Example 4:

Find the latus rectum and focus of the parabola

$$y^2 = 4y - 4x.$$

Solution:

The given curve is $y^2 = 4y - 4x$

or $\quad y^2 - 4y = -4x \quad$ or $\quad (y - 2)^2 = -4.1.(x - 1) \quad ...(1)$

Changing the origin to (2, 1), equation No. (1) becomes

$$Y^2 = -4.1X$$

where X and Y are new variables.

According to the new axes, the vertex to the parabola given by (2) is (0, 0), axis is axis of X, *i.e.* Y = 0, latus rectum is 4 and focus is (– 1, 0).

Changing back to the original axes the vertex will be (1, 2), axis is y – 2 = 0 or y = 2 latus rectum is 4 and the focus is (– 1 + 1, 0 + 2) or (0, 2). **Ans.**

Example 5:

Find the vertex, axis, latus rectum and focus of the parabolas:

$$y^2 = 4x + 4y$$

Solution:

The equation is

$$y^2 = 4x + 4y$$

or $\quad y^2 - 4y = 4x$

or $\quad y^2 - 4y + 4x = 4x + 4$

or $\quad (y - 2)^2 = 4(x + 1), \quad ...(1)$

Changing the origin to (– 1, 2) we get (1) is

$$Y^2 = 4X \quad ...(2)$$

where X and Y are new variables.

Comparing with the standard equation $y^2 = 4ax$, the vertex of (2) X = 0, Y = 0, focus is (1, 0) as a = 1, axis is Y = 0 and latus rectum is 4.

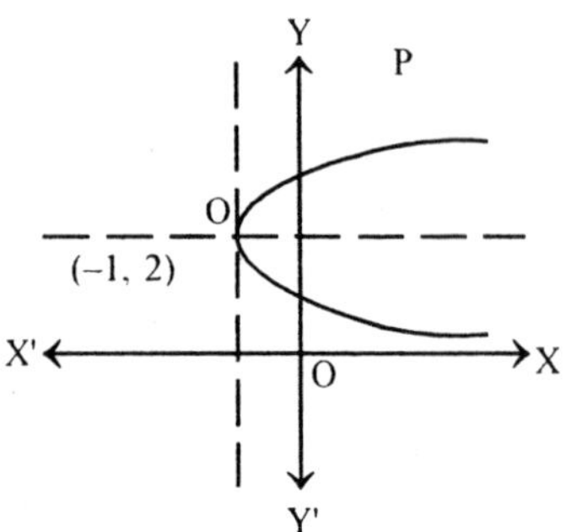

Hence according to the original axis. Changing again origin to (0, 0) the vertex is (− 1, 2), latus rectus is 4 (no change), focus is [(1 − 1), (0 + 2)] or (0, 2) and the axis is y = 2. **Ans.**

Example 6:

For what point of the parabola $y^2 = 18x$ is the ordinate equal to three times the abscissa?

Solution:

Parabola is $y^2 = 18x$. ...(1)

Let (h, k) be any point on the parabola; then

$$k^2 = 18h. \quad ...(2)$$

If ordinate is 3 times the abscissa, k = 3h.

Putting the values in (2)

$$9h^2 = 18h \quad \text{or} \quad 9h^2 = 18h = 0 \quad \text{or} \quad 9h\,(h - 2) = 0.$$

Hence either h = 0 or h = 2 the corresponding values of k are 0 and 6.

Hence either h = 0 or h = 2 the corresponding values of k are 0 and 6.

Hence the required co-ordinates are (0, 0) or (2, 6). **Ans.**

Example 7:

In the parabola $y^2 = 6x$, find (1) the equation to the chord through the vertex and the negative end of the latus rectum, and (2) the equation to any chord through the point on the curve whose abscissa is 24.

Solution:

The given parabola is $y^2 = 6x$

or $$y^2 = 4.\frac{3}{2}\ x.$$

(i) Hence the value of a = 3/2; the focus will be (3/2, 0) and the vertex is clearly (0, 0). As the ends of the latus rectum are (a, ± 2a), the negative and will be (3/2, ± 3).

Equation of the chord joining origin (0, 0) and (3/2, − 3) is

$$y-0=\frac{-3-0}{\frac{3}{2}-3}(x-0) \qquad \text{or} \quad 2x+y=0.$$ **Ans.**

(ii) Let the ordinate of the point whose abscissa is 24 be k so (24, k) will satisfy (1); so $k^2 = 4.(3/2)\ 24 = 144$ or $k = 12$, so point is (24, 12).

Equation of any line passing through (24, 12) is given as

$$y - 12 = m\,(x - 24).$$ **Ans.**

Example 8:

Prove that the locus of the centre of a circle, which intercepts a chord of given length 2a on the axis of x and passes through a given point on the axis of y distant b from the origin, is the curve

$$x^2 - 2yb + b^2 = a^2.$$

Trace this parabola.

Solution:

Let the centre of the required circle be (h, k); so its equation can be written as $x^2 + y^2 - 2hx - 2ky + c = 0$. ...(1)

The circle passes through a point on y-axis which is at a distance of be from the origin The co-ordinates of the point will be (0, b). These co-ordinates will satisfy (1); hence

$$0 + b^2 - 0 - 2kb + c = 0$$

$$\Rightarrow \qquad b^2 - 2kb + c = 0. \qquad ...(2)$$

Again, solving (1) with x-axis *i.e.* $y = 0$, we get

$$x^2 - 2hx + c = 0 \qquad ...(3)$$

If the circle cuts x-axis at A and B, the abscissae of A and B say x_1 and x_2 are given by (3). The distance AB will be $(x_1 - x_2) = \sqrt{\{(x_1 - x_2)^2 - 4x_1 x_2\}} = 2a$ (by hypothesis).

By (3), $x_1 + x_2 = 2h$ and $x_1x_2 = c$. So $2a = \sqrt{\{2h)^2 - 4c\}}$

or $4a^2 = 4h^2 - 4c$ or $h^2 - c - a^2 = 0$ ...(4)

The required locus will be obtained by eliminating c from (2) and (4).

So adding these two, we get $b^2 - 2kb - h^2 - a^2 = 0$.

Generalising, the required locus is

$$x^2 - 2yb + b^2 - a^2 = 0.$$ **Proved.**

This can be written as $x^2 = 2yb + a^2 - b^2$

$$\text{or } x^2 = 2b\left(y + \frac{a^2 - b^2}{2b}\right) \qquad \ldots(5)$$

Changing the origin to $\left(0, -\frac{a^2 - b^2}{2a}\right)$ the curve becomes

$$X^2 = 2bY \qquad \ldots(6)$$

Where X and Y are new variables. The vertex of the parabola (6) is (0, 0) focus (0, – b/2), axis is X = 0 *i.e.* Y-axis and latus rectum is 2b. Transferring to the original axis, the co-ordinates of the vertex will be $\left(0, -\frac{a^2 - b^2}{2a}\right)$ axis of the parabola will be the same x-axis and latus rectum will be 2b; hence the curve may be drawn.

Example 9:

Find the vertex, axis, latus rectum and focus of the parabola

$$x^2 + 2y = 8x - 7$$

Solution:

The given equation is

$$x^2 + 2y = 8x - 7 \quad \text{or} \quad x^2 - 8x = -2y - 7$$

$$\Rightarrow x^2 + 8x + 16 = -2y - 7 + 16 \quad \text{or} \quad (x - 4)^2 = -2y + 9$$

$$\Rightarrow (x - 4)^2 = -2\left(y - \frac{9}{2}\right) \quad \text{or} \quad (x - 4)^2 = -4\,\frac{1}{2}\cdot\left(y - \frac{9}{2}\right) \qquad \ldots(1)$$

Changing the origin to (4, 9/2), the equation (1) becomes

$$X^2 = -4.\ \frac{1}{2}.\ Y. \qquad \ldots(2)$$

where X and Y are new variables.

According to t he new axis, the vertex of (2) is (0, 0), its axis Y axis or X = 0; latus rectum is 4.(1/2) = 2, and the focus is (0, – 1/2) (as a = – 1/2).

Changing back to the original axes the vertex will be (4, 9/2), axis will be X = 0 *i.e.* x – 4 = 0; latus rectum is 2 and the focus is (0 + 4, – 1/2 = 9/2) *i.e.* (4, 4) **Ans.**

Example 10:

Find the value of p when the parabola $y^2 = 4px$ goes through the point (i) (3, – 2) and (ii) (9, – 12).

Solution:

The curve is $y^2 = 2px$. ...(1)

(i) If the curve passes through (3, – 2), the co-ordinates will satisfy (1); so $4 = 4p.\ 3$ or $p = 1/3$ **Ans.**

(ii) If the curve passes through (9, – 12), the co-ordinates will satisfy (1); r $144 = 4p.\ 9$ or $p = 4$. **Ans.**

Example 11:

Show that the locus of the middle point of chords of the parabola $y^2 = 4ax$, which are of constant length 2l is

$$(4ax - y^2)(y^2 + 4a^2) = ,4a^2l^2.$$

Solution:

Let $R(x_1, y_1)$ be the middle point of the chords PQ of the parabola $y^2 = 4ax$. We also assume that the co-ordinates of P and Q are $(at_1^2, 2at_1)$ and $(at_2^2, 2at_2)$.

Since the length of the chord is $2l$, it gives

$$\sqrt{[(at_1^2 - at_2^2)^2 + (2at_1 - 2at_2)^2]} = 2l$$

$$\Rightarrow \quad a^2(t_1^2 - t_2^2)^2 + 4a^2(t_1 - t_2)^2 = 4al^2 \qquad ...(i)$$

Also $$x_1 = \frac{at_1^2 + at_2^2}{2} \Rightarrow t_1^2 + t_2^2 = 2x_1/a \qquad ...(ii)$$

$$y_1 = \frac{2a(t_1 + t_2)}{2} \Rightarrow t_1 + t_2 = y_1/a. \qquad ...(iii)$$

Eliminating t_1, t_2 from equations (i), (ii) and (iii), we have

$$(4x_1a - y_1^2)(y_1^2 + 4a^2) = 4l^2a^2.$$

$\therefore$ Locus is

$$(xa - y^2)(y^2 + 4a^2) = 4a^2l^2.$$

Example 12:

A tangent to the parabola $y^2 + 4bx = 0$ meets the parabola $y^2 = 4ax$ at P and Q. Prove that the locus of the middle point of PQ is $y^2(2a + b) = 4a^2x$.

Solution:

Let the mid-point of PQ be (x_1, y_1). PQ being a chord of $y^2 = 4ax$ having (x_1, y_1) as its mid-point, its equation as

$$yy_1 - 2a(x + x_1) = y_1^2 - 4ax_1 \qquad (S_1 = T)$$

$$\Rightarrow \qquad y = \frac{2ax}{y_1} + \frac{y_1^2 - 2ax_1}{y_1} \qquad \text{...(i)}$$

The equation (i) will be tangent to the parabola $y^2 = -4bx$

$$\text{if} \qquad \frac{y_1^2 - 2ax_1}{y_1} = \frac{-b}{2a/y_1} \qquad \left(c = \frac{a}{m}\right)$$

$$\Rightarrow \qquad y_1^2 - 2ax_1 + -(b/2a)\, y_1^2$$

$$\Rightarrow \qquad y_1^2\,(2a + b) = 4a^2x_1.$$

Hence the locus of (x_1, y_1) is

$$y^2\,(2a + b) = 4a^2x.$$

Example 13:

Show that the locus of the poles of the normal chords of the parabola $y^2 = 4ax$ is

$$(x + 2a)\, y^2 + 4a^3 = 0.$$

Solution:

We know that the equation of normal in the slope form is

$$y = mx - 2am - am^3 \qquad \text{...(i)}$$

Let its pole be (h, k).

The equation of the polar of (h, k) with respect to the parabola $y^2 = 4ax$ is

$$yk = 2a(x + h) \qquad \text{...(ii)}$$

$\therefore$ (i) and (ii) are identical, so we get after comparing coefficients

$$\frac{k}{1} = \frac{2a}{m} = \frac{2ah}{-2am - am^3}.$$

This gives $m = \dfrac{2a}{k}$

$$\text{and} \qquad 2am + am^3 = \frac{-2ah}{k}$$

$$\Rightarrow \qquad 2a \cdot \frac{2a}{k} + a\left(\frac{2a}{k}\right)^3 = \frac{-2ah}{k}$$

$$\Rightarrow \qquad 4a^2k^2 + 8a^2 = -2ahk^2.$$

$$\Rightarrow \qquad (h + 2a)\, k^2 + 4a^3 = 0.$$

$\therefore$ locus of (h, k) is

$$(x + 2a)\, y^2 + 4a^3 = 0.$$

Example 14:

Prove that the equation $y^2 + 2Ax + 2By + C = 0$ represents a parabola, whose axis is parallel to the axis of a, and find its vertex and the equation to its latus rectum.

Solution:

The given equation is $y^2 + 2Ax + 2By + C = 0$

$$\Rightarrow \quad y^2 + 2By + 2Ax - C$$

$$\Rightarrow \quad (y + B)^2 = -2Ax - C + B^2 = -2A\left(x - \frac{B^2 - C}{2A}\right) \quad ...(1)$$

Transferring the origin at $\left(\frac{B^2 - C}{2A}, -B\right)$ the equation (1) becomes

$$Y^2 = -2AX. \quad ...(2)$$

where X and Y are new variables.

For equation No. (2), the vertex is (0, 0) axis of parabola is X-axis or Y = 0, and the focus is (– A/2, 0); hence the equation to latus rectum is X = – A/2.

Changing the origin back to the initial one, the co-ordinates of the vertex according to the original axes will be

$$\left(\frac{B^2 - C}{2A}, -B\right) \quad \textbf{Ans.}$$

The axis of the parabola will be y = – B or y + B = 0 which is parallel to axis of X.

Also, the equation to latus rectum will be

$$x = \frac{B^2 - C}{2A} - \frac{A}{2}$$

or

$$x = \frac{B^2 - A^2 - C}{2A} \quad \textbf{Ans.}$$

Example 15:

Prove that the equation to the parabola whose vertex and focus are on the axis of x at distances a and a' from the origin respectively, is

$$y^2 = 4(a' - a)(x - a).$$

Solution:

By hypothesis vertex is (a, 0) and focus is (a', 0). The directrix is at a distance equal to the distance of focus and on the opposite side of the vertex. The distance between the vertex and focus is (a' – a) = (2a – a'). As the directrix is perpendicular to the axis of parabola and here the axis is same as axis of x (as the line joining the vertex and focus is x-axis), the directrix will be parallel to y-axis. So its equation will be $x = 2a - a'$.

Let there be any point $P \equiv (h, k)$ on the parabola.

Distance of P from the directrix $= h - (2a - a')$. ...(1)

Distance between $P \equiv (h, k)$ and the focus (a, 0)

$$= \sqrt{\{(h - a)^2 + (k - 0)^2\}}.$$

As the curve is a parabola, by the definition,

$$\{h - (2a - a')\} = \sqrt{\{(h - a)^2 + k^2\}}.$$

Squaring and simplifying, we get

$$k^2 = 4a'h + 4a^2 - 2ha - 4aa'$$

$$\Rightarrow \quad k^2 = 4\ [(a' - a)(h - a)].$$

Generalising, the required curve is

$$y^2 = 4(a' - a)(x - a).$$

Proved.

Example 14:

Draw the curves (i) $y^2 = -4ax$,

(2) $x^2 = 4ay$, and

(3) $x^2 = -4ay$.

Solution:

(i) $y^2 = -4ax$.

Vertex is (0, 0), axis i.e, y = 0, focus is (– a, 0) and latus rectum is 4a. For positive values of x, y is imaginary; so the curve is in (2nd) and (3rd) quadrants and is symmetrical about x-axis. Hence the curve is as shown in the adjoining figure.

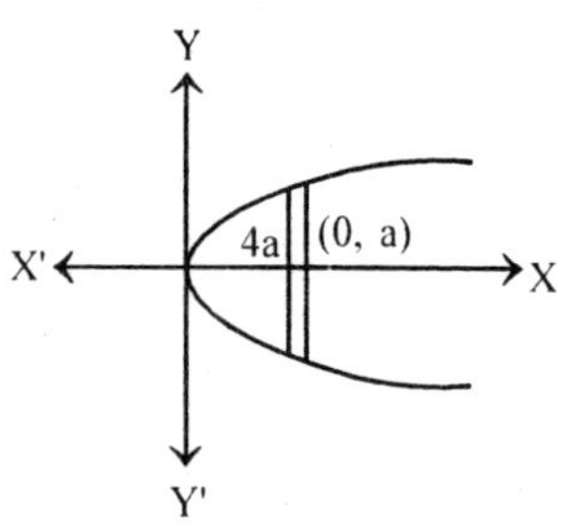

(ii) $x^2 = 4ay$.

The curve lies entirely in (Ist) and (2nd) quadrants because for – ve values of y, x is imaginary and is symmetrical about y-axis. Now vertex is (0, 0), axis is y-axis, focus is (0, a), and the latus rectum is 4a. Hence the curve is as shown in the diagram.

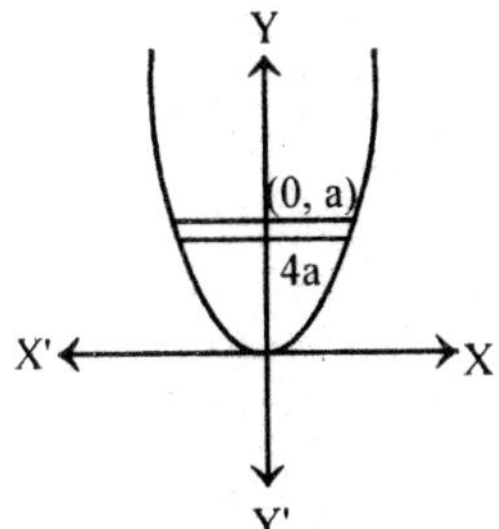

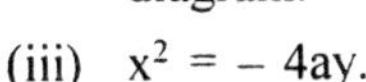

(iii) $x^2 = -4ay$.

The curve lies entirely in (3nd) and (4th) quadrants because for all the + ve values of y, the values of x are imaginary. The curve is symmetrical about y-axis.

Its vertex is (0, 0), axis is y-axis, focus is (0, – a) and the latus rectum is 4a. hence the curve is as shown in the diagram.

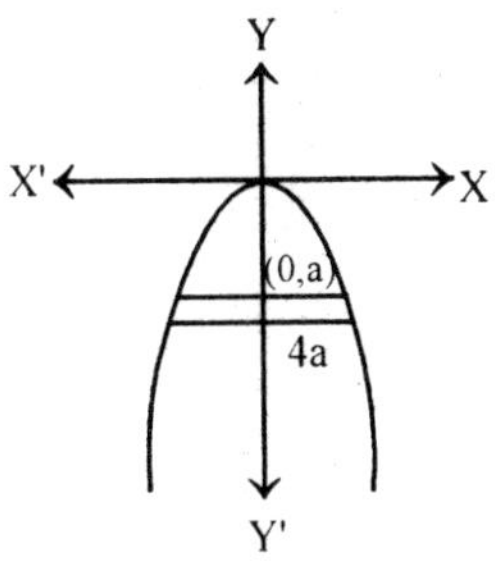

Example 16:

PQ is a double ordinate of a parabola. Find the locus of its points of trisection.

Solution:

Let the equation to the parabola be $y^2 = 4ax$.

If x_1 is the abscissa of any point, the value of the ordinate will be $\pm 2\sqrt{(ax_1)}$. Hence the end points of the double ordinate are $\{a_1, 2\sqrt{(ax_1)}\}$ and $\{a_1 - 2\sqrt{(ax_1)}\}$.

Say (h, k) are the co-ordinates of the point dividing the double ordinate in the ration of 1:2, then

$$h = \frac{x_1 + 2x_1}{3} = x_1 \qquad ...(1)$$

and $$k = \frac{-2\sqrt{(ax_1)} + 2.2\sqrt{(ax_1)}}{3} = \frac{2\sqrt{(ax_1)}}{3}. \qquad ...(2)$$

The required curve is the eliminant of x_1 from (1) and (2).

Hence squaring (2) and putting the value of x_1 from (1), we get

$$k^2 = \frac{4}{9} a.h$$

Generalising, we get the required locus as

$$y^2 = \frac{4}{9}ax \qquad \text{or} \qquad 9y^2 = 4\ ax.$$ **Ans.**

Example 17:

At the ends of the latus rectum of the parabola $y^2 = 12x$.

Solution:

Parabola is $y^2 = 12x = 4.\ 3x$. Hence $a = 3$.

As the ends of the latus rectum are (a, 2a) and (a, – 2a), the points here will be (3, 6) and (3, – 6).

Equation of the tangent at (3, – 6) to parabola $y^2 = 12x$ is

$$y\ .\ 6 = 6\ (x + 3) \qquad \Rightarrow \qquad y - x = 3.$$ **Ans.**

Normal at (3, 6) will be

$$y - 6 = -\frac{6}{6}\ (x - 3) \qquad \Rightarrow \qquad y + x = 9.$$ **Ans.**

Again, equation of the tangent at (3, – 6) will be

$$y\ (-6) = 6\ (x + 3) \qquad \Rightarrow \qquad x + y + 3 = 0$$ **Ans.**

and the normal at (3, – 6) will be

$$y - (-6) = -\frac{-6}{6}\ (x - 3) \qquad \Rightarrow \qquad x - y = 9.$$ **Ans.**

Example 18:

A parabola is drawn to pass through A and B, the ands of a diameter of a given circle of radius a, and to have as directrix a tangent to a concentric circle of radius b; the axes being AB and a perpendicular diameter, prove that the locus of the focus of the parabola is

$$\frac{x^2}{b^2} + \frac{y^2}{b^2 - a^2} = 1.$$

Solution:

Take origin at the centre of the circle, x-axis along the diameter AB and y-axis along the perpendicular to AD through the centre.

As the radius of the first circle is given as a, the co-ordinates of A and B may be taken as (– a, 0) and (a, 0) respectively.

Centre to any concentric circle to this will be again the origin. If the radius of the circle be b, its equation will be

$$x^2 + y^2 = b^2. \qquad ...(1)$$

Equation of any tangent to (1) is given as

$$y = mx + b\{\sqrt{(1 + m^2)}\} \quad ...(2)$$

Hence the directrix given by (2).

Let (h, k) be the co-ordinates of the focus S.

If A lies on the parabola. Its distance from the focus and the directrix must be the same; so we have

$$\sqrt{\left\{(h+a)^2+k^2\right\}} = \frac{-ma+b\sqrt{\left(1+m^2\right)}}{\sqrt{\left(1+m^2\right)}}$$

$$\Rightarrow \sqrt{(1 + m^2)}\ \sqrt{(h + a)^2 + k^2\}} = -ma + b\sqrt{(1 + m^2)} \quad ...(3)$$

Similarly as B lies on the parabola, we will get

$$\sqrt{(1 + m^2)}\ \sqrt{(H + a)^2 + k^2\}} = ma + b\sqrt{(1 + m^2)} \quad ...(4)$$

The reqd. locus will be obtained by eliminating m from (3) and (4). Adding (3) and (4), we get

$$\sqrt{(1 + m^2)}\ [\sqrt{\{(h - a)^2 + k^2\}} + \sqrt{\{(h + a)^2 + k^2\}}] = 2b\sqrt{(1 + m^2)}$$

$$\Rightarrow \sqrt{\{(h - a)^2 + k^2\}} = 2b - \sqrt{\{(h + a)^2 + k^2\}}$$

Squaring, we get

$$h^2 + a^2 - 2ah + k^2 = 4b^2 + h^2 + a^2 + 2ah\ k^2 - 4b\ \sqrt{\{(h + a)^2 + k^2\}}$$

$$\Rightarrow - 4b^2 - 4ah + - 4b\ \sqrt{\{(h + a)^2 + k^2)}$$

Dividing by 4 as it is a common factor and again squaring,

$$b^4 + a^2h^2 + 2ab^2h = b^2\ (h^2 + a^2 + 2ah + k^2).$$

$$\Rightarrow h^2\ (b^2 + a^2) + k^2b^2 = (b^4 - b^2a^2)$$

$$\Rightarrow h^2\ (b^2 - a^2) + k^2b^2 = b^2\ (b^2 - a^2).$$

Dividing by R. H. S. and generalising, the reqd, locus is

$$\frac{x^2}{b^2} + \frac{y^2}{b^2 - a^2} = 1 \quad \textbf{Ans.}$$

Example 19:

If a circle be drawn so as always to touch a given straight line and also a given circle, prove that the locus of its centre is a parabola.

Solution:

As in the last question, taking the centre of the fixed circle as origin, the axes being perpendicular and parallel to the given line respectively, the equation of the given circle will be

$$x^2 + y^2 = r^2 \qquad ...(1)$$

and of the line $x = a$ ...(2)

where r is the radius of the circle and a is the distance of the centre of the given circle from the line.

Let the centre of the other circle and b (h, k). So its equation may be written as $x^2 = y^2 - 2hx - 2ky + c = 0$ ...(3)

Radius of (3) is $\sqrt{(h^2 + k^2 - c)}$.

If (1) touches (3), the distance between the centres must be equal to the sum of the radii. So

$$\sqrt{(h^2 + k^2)} = \sqrt{(h^2 + k^2 - c)} = r. \qquad ...(4)$$

If (3) touches (2), length of the perpendicular from (h, k) on (2) must be equal to the radius

i.e. $$h - a = \sqrt{(h^2 + k^2 - c)} = r. \qquad ...(5)$$

The required locus will be obtained by eliminating c from (4) and (5). So putting the value of $\sqrt{(h^2 + k^2 - c)}$ from (5) in (4), we get

$$\sqrt{(h^2 + k^2)} = (h - a) + r$$

Squaring both sides, $h^2 + k^2 = h^2 = a^2 + r^2 - 2ah + 2rh - 2ar$.

Generalising and simplifying, we get the required locus as

$$y^2 = 2x\,(r - a) + (a - r)^2$$

which is clearly a parabola. **Hence proved.**

Example 20(a):

Prove that the locus of a point, which moves so that its distance from a fixed line is equal to the length of the tangent drawn from it to a given circle, is a parabola. Find the position of the focus and directrix.

Solution:

Let AB be the fixed line, and O be the centre of the fixed circle. Takıng O as the origin, a line through O and parallel to AB as axis of y and a line through O perpendicular to AB as axis, the equation of the circle may be written as

$$x^2 + y^2 = r^2 \qquad ...(1)$$

and the equation of the line as $x = a$. ...(2)

(where the radius of the circle is r and the distance of the centre from the line AB is a).

Let P be any point (h, k) on the locus, PT be the tangent from P to (1) and PN be perpendicular from P upon AB. Then

$$PT = \sqrt{(h^2 + k^2 - r^2)} \text{ and } PN = h - a$$

By hypotesis $PT = PT$ or $PT^2 = PN^2$.

Simplifying and generalising, we get

$$y^2 = -2ax + a^2 + r^2, \quad ...(3)$$

which is clearly a parabola. **Proved.**

Again,

(3) may be written as $y^2 = -2a\left(x \frac{a^2 + r^2}{2a}\right)$.

Clearly the focus is $\left\{\left(\frac{a^2 + r^2}{2a} - \frac{a}{2}\right), 0\right\}$ or $\left(\frac{r^2}{2a}, 0\right)$

and the directrix will be $x = \frac{a^2 + r^2}{2a} + \frac{a}{2}$

$\Rightarrow \quad 2ax = 2a^2 + r^2.$ **Ans.**

Example 20(b):

A double ordinate of the curve $y^2 = 4px$ is of length 8p; prove that the lines from the vertex to its two ends are at right angles.

Solution:

The parabola is given as $y^2 = 4px$. ...(1)

The double ordinate is 8p; so ordinates are $\pm$ 4p.

Putting the value in (1), the corresponding abscissa is 4p.

Hence the co-ordinates of the ends of the given double ordinate are (4p, 4p) and (4p, – 4p). Let these points be A and B respectively and the vertex be $O \equiv (0, 0)$.

Slope of OA $= \frac{4p - 0}{4p - 0} = 1 \; m_1$ (say).

Slope of OB $= \frac{-4p}{4p} = -1 \; m_2$ (say).

Clearly $m_1 \times m_2 = -1$; so OA and OB are at right angles. **Proved.**

Example 20(c):

Two parabolas have a common axis and concavities in opposite directions; if any line parallel to the common axis meet the parabolas in P and P', prove that the locus of the middle point of PP' is another parabola, provided that the later recta of the given parabolas are unequal.

Solution:

Let the equation of a parabola by $y^2 = 4ax$. ...(1)

As the other parabola has an opposite concavity, it will be on negative side, and as its latus rectum is different, say b, its equation may be written as $y^2 = -4bx$. ...(2)

The common axis is x-axis. Let any line parallel to x-axis be

$$y = \lambda. \qquad ...(3)$$

Solving (1) with (3), the point of intersection, say P, is $\left(\frac{\lambda^2}{4a}, \lambda\right)$.

Solving (2) and (3), the point of intersection, say Q, is $\left(-\frac{\lambda^2}{4b}, \lambda\right)$.

If (h, k) be the mid point of PQ then $h = \frac{1}{2}\left(\frac{\lambda^2}{4a} - \frac{\lambda^2}{4b}\right)$ ...(4)

and $k = \frac{1}{2}(\lambda + \lambda) = \lambda$. ...(5)

Putting the value of λ from (5), in (4)

$$h = \frac{1}{2}\left(\frac{k^2}{4a} - \frac{k^2}{4b}\right) = \frac{k^2}{8}\left(\frac{b-a}{ab}\right)$$

Generalising and simplifying, the reqd. locus is

$$y^2 = \frac{8ab}{b-a}x$$ which is a parabola. **Proved.**

If the latera recta are the same clearly h = 0 (by 4). So k = l i.e, y = l will be the reqd locus

Example 20(d):

A tangent to the parabola $y^2 = 8x$ makes an angle of 45° with the straight line $y = 3x + 5$. Find its equation and its point of contact.

Solution:

The parabola is given as $y^2 = 8x$. ...(1)

here $4a = 8$ or $a = 2$

Equation to any tangent to (1) is $y = mx + \frac{2}{m}$ ...(2)

The given line is $y = 3x + 5$. ...(3)

If the angle between (2) and (3) is 45°, clearly

$$\tan 45^\circ = \frac{m-3}{1+m.3} = 1 \qquad \text{as } \tan 45^\circ = 1.$$

or $\quad m - 3 = 1 + 3m$, where $m = -2$.

Putting in (2), the equation to the tangent becomes

$$y = -2x + \frac{2}{-2} \quad \text{or} \quad y + 2x + 1 = 0. \qquad \textbf{Ans.}$$

As the point of contact is $\left(\frac{a}{m^2}, \frac{2a}{m}\right)$, it will be

$$\left(\frac{2}{4}, \frac{2.2}{-2}\right) \quad \text{or} \quad \left(\frac{1}{2}, -2\right) \qquad \textbf{Ans.}$$

Again, if the angle between (2) and (3) is – 45°, we have

$$\frac{m-3}{1+m.3} = \tan(-45^\circ) = -1 \text{ whence } m = \frac{1}{2}.$$

Putting in (2), the equation to the tangent becomes

$$y = \frac{1}{2}x + \frac{2}{1/2} \quad \text{or} \quad 2y = x + 8 \qquad \textbf{Ans.}$$

and the corresponding point of contact will be

$$\left(\frac{2}{(1/2)^2}, \frac{2.2}{1/2}\right) \quad \text{or} \quad (8, 8). \qquad \textbf{Ans.}$$

Example 21(a):

If on a given base triangles be described such that the sum of the tangents of the base angles is constant, prove that the locus of the vertices is a parabola.

Solution:

Taking the x-axis along the base and the y-axis as the right bisector of the base, the co-ordinates of the end points of the base, BC of a triangle ABC may be written as (– a, 0) and (a, 0) where BC = 2a. Let the co-ordinates of the vertex A be (a, k).

Slope of the line $AB = \frac{k-0}{h+a}$ = tan B where A, B and C are the angles of the triangle (as base coincide with x-axis).

Again slope of $AC = \frac{k-0}{h-a} = \tan(\pi - C), \tan C = \frac{k}{a-h}$

By hypothesis, tan B + tan C = λ (constt)

$$\Rightarrow \qquad \frac{k}{h+a}+\frac{k}{a-h}=\lambda$$

$$\Rightarrow \qquad ka - kh + kh + ka + \lambda\ (a^2 - h^2).$$

Generalising and re-arranging, the reqd, locus is $\lambda x^2 = \lambda a^2 - 2ay$, which is parabola.

Example 21(b):

Write down the equations to the tangent and normal, At the point (4, 6) of the parabola $y^2 = 9x$.

Solution:

The parabola is $y^2 = 9x$ and the point is (4, 6). So the equation to the tangent will be

$$y.\ 6 = \frac{9}{2}\ (x + 4) \qquad \text{or} \qquad 4y = 3x + 12. \qquad \textbf{Ans.}$$

Equation to the normal will be

$$y - 6 = -\frac{6}{9/2}\ (x - 4) \qquad \text{or} \qquad 3y = 4x + 34. \qquad \textbf{Ans.}$$

Example 21(c):

At the point of the parabola $y^2 = 6x$ whose ordinate is 12.

Solution:

The parabola is $y^2 = 6x$. If the ordinate of the same point on the parabola is 12, corresponding abscissa will be given by $(12)^2 = 6x$ or x = 24. Hence the point is (24, 12)

$$y.12 = 3\ (x + 24) \qquad \text{or} \qquad 4y - x = 24. \qquad \textbf{Ans.}$$

Equation of the normal at (24, 12) is

$$y - 12 = -\frac{12}{3}\ (x - 24) \qquad \text{or} \qquad y + 4x = 108. \qquad \textbf{Ans.}$$

Example 21(d):

The vertex A of a parabola is joined to any point P on the curve and PQ is drawn at right angles to AP to meet the axis in Q. Prove that the projection of PQ on the axis the always equal to the latus rectum.

Solution:

Let the equation to the parabola be

$$y^2 = 4ax. \quad ...(1)$$

If the co-ordinates of any point P on the parabola be (h, k) then

$$k^2 = 4ah. \quad ...(2)$$

Slope of AP is k/h where A is the vertex (0, 0).

The equation of a line PQ which is perpendicular to AP will be

$$y - k = -\frac{k^2}{k}(x-h)$$

or $$yk + hx = k^2 + h^2. \quad ...(3)$$

If Q lies on X-axis, the its ordinate is 0. So putting y = 0 is (3),

$$x = \frac{k^2}{h} + h.$$

Putting the value of k^2 from (2),

$$x = \frac{4ah}{h} + h \qquad h = (4a + h).$$

The projection of PQ on axis of x is the difference between the abscissae of the points P and Q. Hence the projection

$$= (4a + h) - h = 4a$$

which is equal to the latus rectum. **Proved.**

Example 22:

Find the equation of a parabola. Whose focus (a, b) and directrix $\frac{x}{a} + \frac{y}{b} = 1.$

Solution:

The focus S is (a, b) and the directrix AB is

$$\frac{x}{a} + \frac{y}{b} = 1 \quad \text{or} \quad bx + ay - ab = 0 \quad ...(1)$$

If the moving point P be (h, k), PN be the perpendicular from P on AB, then by the property of parabola,

$$PS = PN \quad \text{or} \quad PS^2 = PN^2.$$

So $$(h - a)^2 + (k = b)^2 = \frac{(bh + ak - ab)^2}{b^2 + a^2}$$

$$\Rightarrow \quad (a^2 + b^2)(h^2 + a^2 - 2ah + k^2 - 2bk + b^2)$$

$= b^2h^2 + a^2k^2 + a^2b^2 + 2abhk - 2ab^2h - 2a^2bk$

$= h^2 (a^2 + b^2 + a^2) k^2 (a^2 + b^2 - a^2) - 2abhk - 2ah (a^2 + b^2 - b^2)$

$- 2bk (a^2 + b^2 - a^2) + a^2 + a^2b^2 + b^4 - a^2b^2 = 0$

Simplifying and generalising, we get

$$a^2x^2 + b^2y^2 + 2abxy - 2a^3x - 2b^3y + a^4 + a^2b^2 + b^4 = 0$$

$\Rightarrow \quad (ax - by)^2 - 2a^3x - 2b^3y + a^4 + a^2b^2 + b^4 = 0.$ **Ans.**

Example 23:

Find the equation to the parabola with

Focus (3, – 4) and directrix $6x - 7y + 5 = 0$.

Solution:

The focus, say S, is given as (3, – 4) and the directrix say AB is

$$6x - 7y + 5 = 0.$$

Let the co-ordinates of the moving point P be (h, k).

If PN be the perpendicular from P to AB, then

$$m\,PN = \frac{6h-7k+5}{\sqrt{(6^2+7^2)}} \quad \text{and} \quad PS = \sqrt{\{(h-3)^2+(k+4)^2\}}$$

If the locus is a parabola, PS = PN or $PS^2 = PN^2$

$\Rightarrow \quad (h - 3)^2 + (k + 4)^2 = \dfrac{(6h-7k+5)^2}{36+49}$

$\Rightarrow \quad 85 [h^2 - 6h + 9 + k^2 + 8k + 16]$

$= [36h^2 + 49k^2 + 25 - 84hk + 60h = 70k]$

$\Rightarrow \quad 36h^2 + 36k^2 + 84hk - 570h + 750k + 2100 = 0$

$\Rightarrow \quad (7h + 6k)^2 - 570h + 750k + 2100 = 0.$

Generalising, we get

$(7x + 6y)^2 - 570h + 750k + 2100 = 0.$ **Ans.**

Example 24(a):

Prove that the locus of the middle points of all chords the parabola $y^2 = 4ax$ which are drawn through the vertex is the parabola y^2 $2ax$.

Solution:

The equation to the parabola is given as

$$y^2 = 4ax. \quad ...(1)$$

Any line passing through the origin may be given by

$$y = mx. \quad ...(2)$$

Solving (1) and (2) $m^2x^2 = 4ax$

or $m^2x^2 - 4ax = 0$

or $x(m^2x - 4a) = 0$

whence either $x = 0$ or $x = (4a/m^2)$. The corresponding values of y are 0 and 4a/m. Hence (1) and (2) meet in O and A at (0, 0) and $(4a/m^2, 4a/m)$ resp. If (h, k) be the mid-point of this chord OA,

$$h = \frac{0+4a/m^2}{2} = \frac{2a}{2}. \quad ...(3)$$

and $$k = \frac{0+4a/m}{2} = \frac{2a}{m}. \quad ...(4)$$

The required locus will be the eliminant of m in (3) and (4).

Putting the value $m = (2a/k)$ from (4) in (3), we get

$$h = \frac{2a}{(2a/k)^2} \Rightarrow h = \frac{2ak^2}{4a^2} \text{ or } k^2 = 2ah$$

Generalising the required locus is $y^2 = 2ax$. **Proved.**

Example 24(b):

Prove that the straight line $x + y = 1$ touches the parabola $y = x - x^2$.

Solution:

The equation of straight line is given as

$$x + y = 1. \quad ...(1)$$

and the parabola as $y = x - x^2$. ...(2)

If (1) touches (2), then solving the two, we must get only one point of intersection. So putting the value of y from (1) in (2), we get

$$1 - x = x = x^2$$

or $x^2 - 2x + 1 = 0$

or $(x - 1)^2 = 0$ where $x = 1$.

As we get the two values of x coinciding, there is only one point of intersection. **Hence proved.**

Example 25:

A parabola touches two given straight lines; if its axis passes through the point (h, k) the given lines being the axes of co-ordinates, prove that the locus of the focus is the curve $x^2 - y^2 - hx + ky = 0$.

Solution:

The focii are given by

$$ax = by = x^2 + y^2 + 2xy \cos w \qquad ...(1)$$

and abscissa of the focus is

$$x = \frac{ab^2}{a^2 + 2ab\cos\omega + b^2}.$$

Since the axis goes through (h, k), we have

$$ak - bh = \frac{ab\left(a^2 - b^2\right)}{a^2 + 2ab\cos\omega + b^2} = \frac{x\left(a^2 - b^2\right)}{b}$$

Putting the values of a and b from (1), we get

$$\frac{k}{x} - \frac{h}{y} = xy\left(\frac{1}{x^2} - \frac{1}{y^2}\right); \text{ or } x^2 - y^2 + ky - bx = 0.$$

Example 26:

Prove that the straight line $y = mx + c$ touches the parabola $y^2 = 4a(x + a)$ if $c = ma + (a/m)$.

Solution:

The line is given as $y = mx + c$ ...(1)

and the parabola is $y^2 = 4a(x + a)$. ...(2)

To solve (1) and (2); we put the values of y from (1) in (2), so we get

$$(mx + c)^2 = 4a(x + a). \qquad ...(3)$$

If (1) touches (2) the two values of m given by (3) must coincide, or it must be a perfect square, therefore its discriminant must be zero. Hence

$$\{2(mc - 2a)\}^2 - 4m^2(c^2 - 4a^2) = 0$$

$$\Rightarrow \quad m^2c^2 + 4a^2 - 4amc - m^2c^2 + 4a^2m^2 = 0$$

$$\Rightarrow \quad 4amc + 4a^2m^2 + 4a^2$$

where $\quad mc = am^2 + a,$

$$\Rightarrow \quad c = am + a/m.$$ **Hence proved.**

Example 27:

A parabola touches two given straight lines at given points; prove that the locus of the middle point of the portion of any tangent which is intercepted between the given straight lines is a straight line.

Solution:

We know that (x/f) + (y/g) = 1 touches the parabola $\sqrt{(x/a)}+\sqrt{(y/b)}$ = 1 if (f/a) + (g/b) = 1. (Ref. Article 242). The middle point will be (f/2, g/2). Substituting the condition, the required locus is (2x/a) + (2y/b) = 1..

It is straight line. **Hence proved.**

Example 28:

TP and TQ are any two tangents to a parabola and the tangent at a third point R cuts them in P' and Q'; prove that

$$\frac{TP'}{TP}+\frac{TQ'}{TQ}=1 \quad and \quad \frac{QQ'}{QT'}=\frac{TP'}{P'P}=\frac{Q'R}{RP'}.$$

Solution:

We have know that, (x/f) + (y/g) =1 ...(1)

always touches the curve $\sqrt{(x/a)}+\sqrt{(y/b)}=1$ if (f/a) + (g/b) =1...(2)

If TP, TQ be the axes and (1) the equation of the third tangent, then by (2),

$$\frac{TP'}{TP}+\frac{TQ'}{TQ}=1$$ **Proved.**

From (2), f/a = 1 – g/b = (b – g)/b ...(3)

and g/b = 1 – f/a = (a – f)/a ...(4)

By (3) and (4), on division $\frac{b-g}{g}=\frac{f}{a-f}$.

As $\frac{QQ}{QT}=\frac{b-g}{g}=\frac{f}{a-f}=\frac{TP'}{PP'}$ if we draw RA parallel to TQ, then TH = f^2/a.

Hence $\frac{Q'R}{RP}=\frac{TH}{HP'}=\frac{f^2/a}{f-f^2/a}=\frac{f}{a-f}=\frac{TP'}{PP'}$. **Proved.**

Example 29:

From an external point P tangents are drawn to the parabola find the equation to the locus of P when these tangents make angles θ_1 and θ_2 with the axis, such that $\theta_1 + \theta_2$ is constant (= 2α.)

Solution:

$\theta_1 + \theta_2 = 2\alpha$, (by hypothesis); so $\tan(\theta_1 + \theta_2) = \tan 2\alpha$

$$\Rightarrow \quad \frac{\tan\theta_1+\tan\theta_2}{1-\tan\theta_1\tan\theta_2}=\tan 2\alpha$$

Putting the values from equations (2), we get

$$\frac{k/h}{1-a/h} = \tan 2\alpha \quad \text{or} \quad \frac{k}{h-a} = \tan 2\alpha$$

Generalising and simplifying, we get the locus of (h, k) as

$$y = (x - a) \tan 2\alpha.$$

Example 30(a):

From an external point P tangents are drawn to the parabola find the equation to the locus of P when these tangents make angles θ_1 and θ_2 with the axis, such that $\tan^2 \theta_1 + \tan^2 \theta_2$ is constant $(= \lambda)$.

Solution:

$$\tan^2 \theta_1 + \tan^2 \theta_2 = \lambda \text{ (by hypothesis)}$$

$$(\tan \theta_1 + \tan \theta_2)^2 - 2 \tan \theta_1 \tan \theta_2 = \lambda.$$

Now we have

$$\frac{k^2}{h^2} - \frac{2a}{h} = \lambda \quad \text{or } k^2 - 2ah = \lambda h^2.$$

Generalising, we get the locus of (h, k) as $y^2 - 2ax = \lambda x^2$. **Ans.**

Example 30(b):

Two tangents to a parabola intercept on a fixed tangent segments whose product is constant; prove that the locus of their point of intersection is a straight line.

Solution:

Let $\left(at_1^2, 2at_1\right)$ and $\left(at_2^2, 2at_2\right)$ be the co-ordinates of the point of contact of the variable tangents respectively of the parabola.

$$y^2 = 4ax. \qquad \text{...(1)}$$

Let $(at^2, 2at)$ be the fixed point A.

The co-ordinates of the point of intersection of variable tangents are given by $x = at_1t_2$ ...(2)

and $y = a(t_1 + t_2)$. ...(3)

Again, let B and C be $(att, a(t + t_1)$ and $(att_2, a(t + t_2)\}$ respectively as the co-ordinates of the points of intersection of fixed tangent with variable tangents.

Now, since BAC is a fixed tangent.

$\therefore$ BA $\times$ AC = constant say l (by hypothesis)

$\Rightarrow \quad (at^2 - att_1)(at^2 - att_2) = l$

$\Rightarrow \quad a^2t^2 (t - t_1)(t - t_2) = \lambda$

$\Rightarrow \quad (t - t_1)(t - t_2) = \lambda/a^2t^2$

$\Rightarrow \quad [t^2 - (t_1 + t_2)] t + t_1t_2] = \lambda/a^2t^2.$

Substituting the values of $(t_1 + t_2)$ and t_1t_2, from (1) and (2), we get on generalisation

$$t^2 - \frac{y}{a} + \frac{x}{a} = \frac{\lambda}{a^2t^2} \text{ or } x - y = - \frac{\lambda}{a^2t^2} - at^2$$

which is a straight line. **Proved.**

Example 30(c):

Find the locus of the middle points of chords of the parabola which are normal to the curve.

Solution:

Let the chord PQ be normal at P $\circ$ $(at_1^2, 2at_1)$. If this intersects the parabola again at Q whose co-ordinates are $(at_2^2, 2at_2)$; then

$$t_2 = - t_1 - 2/t_1$$

or $\quad t_1 + t_2 = - \dfrac{2}{t_1}$ ...(1)

If (h, k) be the co-ordinates of the mid-point of PQ, then

$$h = \frac{1}{2} a \left(t_1^2 + t_2^2\right) \quad ...(2)$$

and $\quad k (t_1 + t_2) a.$ ...(3)

To eliminate t_1 and t_2 between relations (1), (2) and (3), we have from (2) and (3) $-2/t_1 = k/a$ or $t_1 - 2a/k$.

As $t_2 = - t_1 - 2/t_1$, we have $t_2 = \dfrac{k}{a} + \dfrac{2a}{k}$ or $t_2 = \dfrac{(k^2 + 2a^2)}{ak}$

Substituting for t_1 and t_2 in (2), we get $h = \dfrac{a}{2}\left[\dfrac{4a^2}{k^2} + \dfrac{(k^2 + 2a^2)^2}{a^2k^2}\right].$

$4a^4 + k^4 + 4a^4 + 4k^2a^2 = 2hk^2a$ or $k^2 (k^2 - 2ah + 4a^2) + 8a^4 = 0.$

Generalising the locus of (h, k) is $y^2 (y^2 - 2ax + 4a^2) + 8a^4 = 0.$

Proved.

Example 31:

Prove that the locus of the poles of tangents to the parabola $y^2 = 4ax$ with respect to the circle $x^2 + y^2 = 2ax$ is the circle $x^2 + y^2 = ax$.

Solution:

Let (h, k) be the co-ordinates of the pole; then polar of (h, k) with respect to the circle $x^2 + y^2 - 2ax = 0$ is

$$xh + yk - a(x + h) = 0$$

$$\Rightarrow \quad y = x\left(-\frac{h}{k}+\frac{a}{k}\right)+\frac{ah}{k}. \quad ...(1)$$

The equation to any tangent on the given parabola is

$$y = mx + \frac{a}{m}. \quad ...(2)$$

Equations (1) and (2) represent the same straight line; hence equating the coefficients, we get

$$m = -\left(\frac{h}{k}+\frac{a}{k}\right) \quad m = \left(-\frac{h}{k}+\frac{a}{k}\right) \text{ and } \frac{a}{m} = \frac{ah}{k}.$$

To eliminate m between two equations, multiplying the two results, we get

$$\frac{ah}{k}\left(-\frac{h}{k}+\frac{a}{k}\right) = a \text{ or } h^2 + k^2 = ah.$$

Generalising, we get the required locus as $x^2 + y^2 = ax$. **Ans.**

Example 32(a):

A pair of tangents are drawn which are equally inclined to a straight line whose inclination to the axis is α; prove that the locus of their point of intersection is the straight line $y\ (x - a)\tan 2\alpha$.

Solution:

By hypothesis, the tangents through (h, k) are equality inclined with the line whose inclination with x-axis is α.

Then $\quad \theta_2 - \alpha = \alpha - \theta_1$ or $\theta_1 + \theta_2 = 2\alpha$.

$$\Rightarrow \quad \tan(\theta_1 + \theta_2) = \tan 2\alpha$$

$$\Rightarrow \quad \frac{\tan\theta_1 + \tan\theta_2}{1 - \tan\theta_1\ \tan\theta_2} = \tan 2\alpha = \frac{m_1 + m_2}{1 - m_1\ m_2}$$

where m_1 and m_2 are the slopes of the tangents.

Putting the values of m_1 and m_2 we have

$$\frac{\left(\frac{k}{h}\right)}{\left(1-\frac{a}{h}\right)} = \tan 2\alpha \quad \text{or} \quad \frac{k}{h-a} = \tan 2\alpha$$

Generalising and simplifying, the required locus of (h, k) is

$$y = (x - a) \tan 2\alpha.$$

Exumple 32(b):

From an external point P tangents are drawn to the parabola; find the equation to the locus of P when these tangents make angles θ_1 and q_2 with the axis, such that cos θ_1 cos θ_2 is constant (= μ).

Solution:

$$\cos \theta_1 \cos \theta_2 = \mu \text{ (by hypothesis)}$$

$$\Rightarrow \quad \sec \theta_1 \sec \theta_2 = \frac{1}{\mu} \text{ or } \sec^2 \theta_1 \sec^2 \theta_2 = \frac{1}{\mu^2}$$

$$\Rightarrow \quad \left(1+\tan^2 \theta_1\right)\left(1+\tan^2 \theta_2\right) = \frac{1}{\mu^2}.$$

$$\Rightarrow \quad 1 + \tan^2 \theta_1 + \tan^2 \theta_2 + \tan^2 \theta_1 \tan^2 \theta_2 = \frac{1}{\mu^2}$$

$$\Rightarrow \quad 1 + (\tan \theta_1 + \tan \theta_2)^2 - 2 \tan \theta_1 \tan \theta_2 + \tan \theta_2$$

$$+ \tan^2 \theta_1.\tan^2 \theta_2 = \frac{1}{\mu^2}.$$

From equations (2) and (3), Q.No. 1, we get $1+\frac{k^2}{h^2}-\frac{2a}{h}+\left(\frac{a}{h}\right)^2 = \frac{1}{\mu^2}$

or $\quad h^2 = \mu^2 \{(h - a)^2 + k^2\}$

Simplifying and generalising, the locus of (h, k) is

$$x^2 = \mu^2 \{(x - a)^2 + y^2\}.$$

Example 33:

From an external point P tangents are drawn to the parabola; find the equation to the locus of P when these tangents make angles θ_1 and θ_2 with the axis, such than tan θ_1.tan θ_2 is constant (= c).

Solution:

We have tan θ_1.tan θ_2 = c (given).

We know that $\tan\theta_1 \tan\theta_2 = a/h$.

So $a/h = c$ or $hc = a$. Generalising, the locus of (h, k) is $cx = a$. **Ans.**

Example 34(a):

From an external point P tangents are drawn to the parabola; find the equation to the locus of P when these tangents make angles θ_1 and θ_2 with the axis, such that $\cot\theta_1 + \cot\theta_2$ is constant (= d).

Solution:

$$\cot q_1 + \cot q_2 = d \text{ (by hypothesis)}$$

$$\Rightarrow \quad \frac{1}{\tan\theta_1} + \frac{1}{\tan\theta_2} = d \quad \text{or} \quad \frac{\tan\theta_2 + \tan\theta_1}{\tan\theta_1 \tan\theta_2} = d.$$

We get $\dfrac{k/h}{c/h} = d$ or $k = ad$.

Generalising, the locus of (h, k) is $y = ad$. **Ans.**

Example 34(b):

Prove that the locus of the point of intersection of two tangents which intercepts a given distance 4c on the tangent at the vertex is an equal parabola.

Solution:

Let (h, k) be the co-ordinates of the point of intersection of tangents on parabola $y^2 = 4ax$. ...(1)

The equation of any tangent is $y = mx + \dfrac{a}{m}$. ...(2)

We have $m_1 + m_2 = \dfrac{k}{h}$ ...(3)

and $m_1 m_2 = \dfrac{a}{h}$. ...(4)

Tangent at the vertex is y-axis *i.e.*, $x = 0$ ...(5)

Solving (2) and (5), points of intersection of tangents on y-axis are

$$\left(0, \frac{a}{m_1}\right) \text{ and } \left(0, \frac{a}{m_2}\right).$$

Now distance between the points $\left(0, \dfrac{a}{m_1}\right)$ and $\left(0, \dfrac{a}{m_2}\right)$.

$$= \frac{a}{m_2} - \frac{a}{m_1} = \frac{a(m_1 - m_2)}{m_1 m_2} = 4c \text{ (by hypothesis)}$$

$$\Rightarrow \quad \frac{a}{m_1 m_2} \sqrt{\left[(m_1 + m_2)^2 - 4m_1 m_2\right]} = 4c.$$

Putting the value from (3) and (4), we get

$$\frac{a}{a/h} \times \sqrt{\left(\frac{k^2}{h^2} - \frac{4a}{h}\right)} = 4c$$

$$\Rightarrow \quad h\sqrt{\left(\frac{k^2 - 4ah}{h^2}\right)} = 4c.$$

Squaring and simplifying, we get $k^2 - 4ah = 16c^2$.

Generalising, the locus of (h, k) is $y^2 - 4ax = 16c^2$

$$\Rightarrow \quad y^2 = 4a\, x + \left[\frac{4c^2}{a}\right].$$

This also a parabola of latus rectum 4a which is equal to the latus rectum of (1). Hence the two parabolas are equal. **Proved.**

Example 34(c):

If the normals at P and Q meet on the parabola, prove that the point of intersection of the tangents at P and Q lies either in a certain straight line, which is parallel to the tangents at the vertex, or on the curve whose equation is $y^2 (x + 2a) + 4a^3 = 0$.

Solution:

Let the points P and Q be $(at_1^2, 2at_1)$ and $(at_2^2, 2at_2)$ and the equation of the parabola be y = 4ax. The equation of the tangents at P and Q are respectively $t_1 y = x + at_1^2$ and $t_2 y = x + at_2^2$.

Co-ordinates of point of intersection of these tangents are

$\{at_1 t_2, a(t_1 + t_2)\}$.

If the point of intersection is (h, k), then

$$h = at_1 t_2 \quad ...(1)$$

and $$k = a(t_1 + t_2). \quad ...(2)$$

The equations to the normals at P and Q are respectively

$$y = -t_1x + 2at_1 + at_1^3$$

and $$y = -t_2x + 2at_2 + at_2^3.$$

As the point of intersection of these normals lies on the parabola, we have as in Question No. 28, Ex. 28, $t_1t_2 = 2$. ...(3)

From (1) and (3), $x = 2a$. It is a line parallel to $x = 0$, *i.e.*, y-axis.

Again. If the normals at P $\left(at_1^2, 2at_1\right)$ interest the parabola at Q $\left(at_2^2, 2at_2\right)$, then $t_2 = -t_1 - 2/t_1$

$$t_1 + t_2 = -2/t_1. \quad ...(4)$$

Now eliminate t_1 from the equations (1), (2) and (4), to find the locus of point of intersection of tangents. By (2) and (4), $k = -2a/t_1$ or $t_1 = -a/k$ and $t_2 = -hk/2a^2$ {from (1)}. Substituting for 't_1' and 't_2' in (2), we have

$$\frac{k}{a} = \frac{-2a}{k} - \frac{hk}{2a^2}.$$

Generalising and simplifying, we get $y^2(x + 2a) + 4a^3 = 0$.

Example 34(d):

Show the locus of the poles of tangents to the parabola $y^2 = 4ax$ with respect to the parabola $y^2 = 4bx$ is the parabola $y^2 = \frac{4b^2}{a}x$.

Solution:

Let (h, k) be the co-ordinates of the pole, then the polar of (h, k) with respect to parabola $y^2 = 4bx$ is

$$yk = 2b(x + h) \text{ or } y = \frac{2b}{k}x \times \frac{2bh}{k}. \quad ...(1)$$

Equation to any tangent to $y^2 = 4ax$ is $y = mx + \frac{a}{m}$ *i.e.*, constant term is a/m.

So $$\frac{2bh}{k} = \frac{a}{2b/k} = \frac{ak}{2b} \text{ or } ak^2 = 4b^2h$$

Generalising, the locus of (h, k) is $y^2 = \frac{4b^2}{a}x$. **Proved.**

Example 35:

Show that the locus of the poles of chords which subtend a constant angle a at the vertex is the curve $(x + 4a)^2 = 4\cot^2\alpha\,(y^2 - 4ax)$.

Solution:

Let (h, k) be the co-ordinates of the polar of the variable chord.

Th equation of polar of (h, k) w.r.t. the parabola $y^2 = 4ax$ is

$$ky = 2a(x + h) \text{ or } \frac{xy - 2ak}{2ah} = 1. \quad ...(1)$$

Now making the equation of parabola homogeneous with the help of (1), we get the equation of the straight lines joining the vertex (0, 0) to the points of intersection as

$$y^2 = 4ax\frac{ky - 2ax}{2ah} \text{ or } 4ax^2 - 2kxy + hy^2 = 0. \quad ...(2)$$

If α is the angle between the lines given by (2), then

$$\tan\alpha = \frac{2\sqrt{(k^2 - 4ah)}}{4a + h} \quad \left(\text{as } \tan\theta = \frac{2\sqrt{(h^2 - ab)}}{a + b}\right)$$

$$\Rightarrow \quad \tan^2\alpha = \frac{4(k^2 - 4ah)}{(4a + h)^2} \Rightarrow (h + 4a)^2 = 4\cot^2 a\,(k^2 - 4ah).$$

Generalising, the required locus is $(x + 4a)^2 = 4\cot^2 a\,(y^2 - 4ax)$.

Example 36(a):

In the preceding question if the constant angle be a right angle the locus is a straight line perpendicular to the axis.

Solution:

Putting $\alpha = 90^\circ$ in last question, we get the locus of the pole as x =+ 4a = 0, $\therefore \cot 90^\circ = 0$. This is a line perpendicular on x-axis, *i.e.*, y = 0.

Example 36(b):

A point P is such that the straight line drawn through it perpendicular to its polar with respect to the parabola $y^2 = 4ax$ touches the parabola $x^2 = 4by$. Prove that its locus is the straight line $2ax + by + 4a^2 = 0$.

Solution:

Let the equation to parabola be $y^2 = 4ax$ and the co-ordinates of the pole be (h, k); then the polar of the point (h, k) w.r.t. the parabola is $yk = 2a(x + h)$. Then the equation of the line through (h, k) and perpendicular to its polar is

$$y - k = -\frac{k}{2a}(x - h)$$

$\Rightarrow \qquad k(x - h) + 2a(y - k) = 0.$

$\Rightarrow \qquad x = \frac{-2a}{k} y + h + 2a \qquad \text{...(1)}$

If (1) touches the parabola $x^2 = 3by$, it should be of the form

$$x = my + \frac{b}{m},$$

i.e., the constant term should be $\frac{b}{m}$.

So $\qquad h + 2a = \frac{b}{-2a/k} = \frac{-kb}{2a}$

$\Rightarrow \qquad 2ah + kb + 4a^2 = 0$

Generalising, the locus of (h, k) is $2ax + by + 4a^2 = 0$. **Proved.**

Example 36(c):

Find the locus of the middle points of chords of parabola which pass through the focus.

Solution:

Let the ends of chord of parabola $y^2 = 4ax$ be A and B whose co-ordinates are $(at_1^2, 2at_1)$ and $(at_2^2, 2at_2)$ respectively.

Equation of AB is $y(t_1 + t_2) = 2x + 2at_1t_2$. This passes through the focus (a, 0); so the co-ordinates will satisfy it, hence

$$2a + 2at_1t_2 = 0 \text{ or } t_1t_2 = -1. \qquad \text{...(1)}$$

Let (h, k) be the co-ordinates of the middle-point of the chord AB; then

$h = a\left(t_1^2 + t_2^2\right)/2$

or $\qquad t_1^2 + t_2^2 = 2h/a \qquad \text{...(2)}$

and $\qquad k = a(t_1 + t_2) \qquad \text{...(3)}$

The required locus will be obtained by eliminating t_1 and t_2 from (1), (2) and (3).

Identically, we have $(t_1 + t_2)^2 - \left(t_1^2 + t_2^2\right) = 2t_1t_2$

Putting the values from (1), (2) and (3), we get

$$\frac{k^2}{a^2} - \frac{2h}{a} = -2$$

$\Rightarrow \qquad k^2 = 2a(h - a).$ **Ans.**

Example 37:

Through each point of the straight line $x = my + h$ is drawn the chord of the parabola $y^2 = 4ax$ which is bisected at the point; prove that it always touches the parabola $(y + 2am)^2 = 8a(x - h)$.

Solution:

Let the equation of the parabola be $y^2 = 4ax$. If (x_1, y_1) be the co-ordinates of the middle point of any chord, it will be parallel to the polar of (x_1, y_1) w.r.t. the parabola.

Equation of the polar of (x_1, y_1) w.r.t. the parabola is

$$yy_1 = 2a(x + x_1).$$

Its slope is $2a/y_1$; hence the slope of the chord, whose middle point is (x_1, y_1) also $2a/y_1$.

Hence the equation of the chord whose middle point is (x_1, y_1) is

$$(y - y_1) = \frac{2a}{y_1}(x - x_1)$$

$$\Rightarrow \qquad y_1(y - y_1) = 2a(x - x_1). \qquad ...(1)$$

Again as the point (x_1, y_1) lies on the line $x = my + h$ (given), so $x_1 = my_1 = h$.

Substituting the values of x_1 in (1), we have

$$y_1(y - y_1) = 2a(x - my_1 - h)$$

$$\Rightarrow \qquad (y + 2am) = \frac{2a}{y_1}(x - h) + \frac{2a}{2a/y_1}. \qquad ...(2)$$

Changing the origin to $(-2am, h)$ and putting $2a/y_1 = M$, equation (2) becomes $Y = MX + 2a/M$ which is always tangent to the parabola

$$Y^2 = 4.\,2aX \text{ or } Y^2 = 8aX. \qquad ...(3)$$

As referred to the original (3) becomes

$$(y + 2am)^2 = 8a(x - h). \qquad ...(4)$$

Hence (2) always touches (4). **Proved.**

Example 38:

Find the locus of the middle points of chords of the parabola which pass through the fixed point (h, k).

Solution:

Let $P \equiv (x_1, y_1)$ be the co-ordinates of the middle point of the chord AB of the parabola $y^2 = 4ax$.

Then we have, $\frac{2x_1}{a} = t_1^2 + t_2^2$ ...(1)

and $t_1 + t_2 = \frac{y_1}{a}$. ...(2)

Equation of the chord AB is $y(t_1 + t_2) = 2x + 2at_1t_2$. As this passes through (h, k), co-ordinates will satisfy it; hence

$$k(t_1 + t_2) = 2h + 2at_1t_2 = 2h + a\{(t_1 + t_2)^2 - (t_1^2 + t_2^2)\}.$$

Putting the values from (1) and (2), we get

$$k.\frac{y_1}{a} = 2h + a\left\{\frac{y_1^2}{a^2} - \frac{2x_1}{a}\right\} = 3h + a\frac{(y_1^2 - 2ax_1)}{a^2}$$

$$\Rightarrow \quad \frac{ky_1}{a} = \frac{2ah + (y_1^2 - 2ax_1)}{a}.$$

Generalising for (x_1, y_1), we get the locus of (x_1, y_1) as

$$y^2 - ky = 2a(x - h).$$ **Ans.**

Example 39:

Prove that the locus of the feet of the perpendiculars drawn from the vertex of the parabola upon chords, which subtend an angle of 45° at the vertex, is the curve $r^2 - 24ar\cos\theta + 16a^2\cos 2\theta = 0$.

Solution:

Let the equation to the parabola be given as

$$y^2 = 4ax$$ ...(1)

and equation to any chord be, $x\cos\alpha + y\sin\alpha = p$. ...(2)

The equation to the lines joining the origin to the points of intersection of (1) and (2) is obtained by making (1) homogeneous with the help of (2).

So $$y^2 = \frac{4ax(x\cos\alpha + y\sin\alpha)}{p}$$

$$\Rightarrow \quad 4a\cos\alpha\, x^2 + 4a\sin\alpha.xy - py^2 = 0.$$

Now as the angle between the lines is 45°, so

$$\tan 45^\circ = \frac{\pm\sqrt{(4a^2\sin^2\alpha + 4ap\cos\alpha)}}{4a\cos\alpha - p}$$ ...(3)

as $$\tan\alpha = \frac{\pm 2\sqrt{(h^2 - ab)}}{a + b}.$$

Clearly the polar co-ordinates of the foot of the perpendicular from the vertex on chord (2) are (p, α).

Hence the required locus is obtained by generalising (3) for p and α.

So squaring, generalising and simplifying (3), the required locus is

$$16\,(a\sin^2\theta + r\cos\theta) = 16\,a^2\cos^2\theta + r^2 - 8\text{ are }\cos\theta$$

or $$r^2 - 24ar\cos\theta + 16a^2\cos 2\theta = 0.$$ **Proved.**

Example 40:

Find the locus of the middle points of chords of the parabola which are of given length l.

Solution:

Let B and C be any two points $\left(at_1^2, 2at_1\right)$ and $\left(at_2^2, 2at_2\right)$ respectively on the parabola $y^2 = 4ax$; then

$$\text{Length BC} = \sqrt{\{(at_2^2 - at_1^2)^2 + (2at_2 - 2at_1)^2\}}$$
$$= a(t_2 - t_1)\sqrt{\{(t_1 + t_2)^2 + 4\}}$$

$\Rightarrow$ $$BC = a\sqrt{\{(t_1 - t_2)^2 - 4t_1t_2\}}\sqrt{\{(t_1 - t_1)^2 + 4\}} = l \qquad ...(1)$$

[by hypothesis]

Again, if P (h, k) be the mid-point of BC, then

$$\frac{2h}{a} = t_1^2 + t_2^2,\text{ and } t_1 + t_2\ \frac{k}{a}\text{ and as in last question,}$$

$$2t_1t_2 = \left\{\left(t_1 + t_2\right)^2 - \left(t_1^2 + t_2^2\right)\right\} = \frac{y_1^2}{a^2} - \frac{2x_1}{a} = \frac{y_1^2 - 2ax_1}{a^2}$$

Now by (1), $l^2 = a^2\{(t_2 + t_1)^2 - 4t_1t_2\}\{(t_1 + t_2)^2 + 4]$

Putting the values of t_1 and t_2 etc, we get

$$l^2 = a^2\left\{\frac{k^2}{a^2} - 2\frac{\left(k^2 - 2ah\right)}{a^2}\right\}\left(\frac{k^2}{a^2} + 4\right)$$

$\Rightarrow$ $$l^2 = \frac{\left(4ah - k^2\right)}{a^2}\left(k^2 + 4a^2\right).$$

Simplifying and generalising, we get the required locus as

$$(y^2 - 4ax)(y^2 + 4a^2) + a^2 l^2 = 0$$ **Ans.**

Example 41:

Two parabolas have the same axis and tangents are drawn to the second from points on the first; prove that the locus of the middle points of the chords of contact with the second parabola all lie on a fixed parabola.

Solution:

$$y^2 = 4x \qquad ...(1)$$

and $$y = 4b\,(x + c) \qquad ...(2)$$

be the equations of the parabolas and let PQ be the polar, the co-ordinates of ends P and Q to be $(at_1^2, 2at_1)$ and $(at_2^2, 2at_2)$ respectively; then the equation of polar PQ is $(t_1 + t_2)\,y = 2x + 2at_1t_2$ ...(3)

and if (x_1, y_1) be the pole, then polar of (x_1, y_1) w.r.t. (1) is

$$yy_1 = 2a\,(x + x_1) \qquad ...(4)$$

Since (3) and (4) represent the same line, the coefficient will be proportional. So comparing the coefficients of x, y and constant term, we get

$$\frac{y_1}{t_1 + t_2} = \frac{2a}{2} = \frac{2ax_1}{2at_1t_2}$$

or $x_1 = at_1t_2$ and $y_1 = a(t_1 + t_2)$.

Hence the co-ordinates of the pole are $(at_1t_2,\ a\,(t_1 + t_2)\}$.

As this point lies on (2), we have

$$a^2(t_1 + t_2)^2 = 4b\,(at_1t_2 + c) = \frac{2b\left(2a^2t_1t_2 + 2ac\right)}{a}. \qquad ...(5)$$

If (h, k) be the co-ordinates of the middle point of the chord PQ, then

$$2h/a = t_1^2 + t_2^2 \text{ and } k/a = t_1 + t_2.$$

Hence $k^2 - 2ah = 2at_1t_2$

Substituting the values in (5)

$$a^2(k^2/a^2) = (2b/a)\,(k^2 - 2ah + 2ac).$$

Generalising, we have $ay^2 = 2b\,(y^2 - 2ax + 2ac)$ which is a fixed parabola.

Example 42(a):

Find the locus of the middle points of chords of the parabola which subtend a constant angle α at the vertex.

Section:

Let A(0, 0) be the vertex of the parabola $y^2 = 4ax$ and (h, k) be the co-ordinates of the middle point of the chord BC, where B and C are respectively $(at_1^2, 2at_1)$ and $(at_2^2, 2at_2)$; then

$$2h/a = t_1^2 + t_2^2 \qquad ...(1)$$

and $$k/a = t_1 + t_2 \qquad ...(2)$$

Now the slope of BA $= \dfrac{2at_1 = 0}{at_1^2 - 0} = \dfrac{2}{t_1} = m_1$ (say).

Similarly slope of CA $= \dfrac{2}{t_1} = m_1$ (say).

As the angle between AB and AC is given as a.

$$\pm \tan \alpha = \frac{m_2 - m_1}{1 + m_1 m_2} = \frac{\dfrac{2}{t_2} - \dfrac{2}{t_1}}{1 + \dfrac{4}{t_1 t_2}} = \frac{2(t_1 - t_2)}{t_1 t_2 + 4}$$

$$= \frac{2\sqrt{\left\{(t_1 + t_2)^2 - 4t_1 t_2\right\}}}{t_1 t_2 + 4}. \qquad \ldots(3)$$

Putting the values of $(t_1 + t_2)$ and $2t_1t_2 = \{(t_1 + t_2)^2 - (t_1^2 + t_2^2)\}$, from (1) and (2), we get

$$\pm \tan \alpha = \frac{\sqrt{\left\{\dfrac{k^2}{a^2} - 2\left(\dfrac{k^2 - 2ah}{a^2}\right)\right\}}}{\left(\dfrac{k^2 - 2ah}{2a^2} + 4\right)}.$$

$$= \frac{2\sqrt{\left(4ah - k^2\right)} \times 2a}{\left(k^2 - 2ah + 8a^2\right)} = \frac{4a\sqrt{\left(4ah - k^2\right)}}{k^2 - 2ah + 8a^2}$$

$\Rightarrow \qquad (k^2 - 2ah + 8a^2)^2 \tan^2 \alpha = 16a^2 (4ah - k^2)$ (on squaring).

Generalising, the required locus is

$$(8a^2 + y^2 - 2ax)^2 \tan^2 \alpha = 16a^2 (4ax - y^2).$$

Example 42(b):

Find the locus of a point O when the three normals drawn from it are such that two of them make equal angles with the given line $y = mx + c$.

Solution:

Let the line $y = mx + c$ make an angle θ with the axis; then $m = \tan \theta$. Let θ_1 and θ_2 be the angles of inclination of two normals having slopes m_1 and m_2 respectively; then

$\theta_1 - \theta = \theta - \theta_2 \qquad$ [by hypothesis]

$\therefore \qquad \theta_1 + \theta_2 = 2\theta$ or $\tan(\theta_1 + \theta_2) = \tan 2\theta$

$$\Rightarrow \quad \frac{\tan\theta_1 + \tan\theta_2}{1 - \tan\theta_1 \tan\theta_2} = \frac{2\tan\theta}{1-\tan^2\theta}$$

$$\frac{m_1 + m_2}{1 - m_1 m_2} = \frac{2m}{1-m^2} = \lambda \text{ (say)}.$$

Putting the values from equations (1) and (3)

$$m_1 + m_2 = -m_3 \text{ and } m_2 m_1 = \frac{-k}{am_3}.$$

$$\frac{-m_3}{1 + k/am_3} = \lambda \text{ or } \frac{-am_3^2}{k + am_3} = \lambda$$

$$\Rightarrow \quad am_3^2 + alm_3 + k\lambda = 0. \qquad \ldots(1)$$

Again, $(m_1 + m_2)m_3 + m_1m_2 = \frac{2a - h}{a}$.

Similarly, we get the other two terms. Substituting in (1) we get

$$\frac{1}{am_3}\left(am_3^3 + 4k\right).\frac{1}{am_1}\left(am_1^3 + 4k\right).\frac{1}{am_2}\left(am_2^3 + 4k\right) = \lambda$$

$$\Rightarrow \quad (4k + am_1^3)(4k + am_2^3)(4k + am_3^3) = \lambda a^3 m_1 m_2 m_3$$

$$\Rightarrow \quad 64k^3 + 16k^2 a\,\Sigma m_1^3 + 4ka^2\,\Sigma m_1^3 m_2^3 + a^2 m_1^3 m_2^3 m_3^3$$

$$= \lambda a^3 m_1 m_2 m_3. \qquad \ldots(2)$$

Now as $m_1 + m_2 + m_3 = 0$

$$\Rightarrow \quad m_1^3 + m_2^3 + m_3^3 = 3m_1m_2m_3 = -\frac{3k}{a}$$

$$\therefore \quad \Sigma m_1^3 = -\frac{3k}{a}.$$

Again $\quad m_1m_2 + m_1m_3 + m_2m_3 = \frac{2a - h}{a}$;

$$\therefore \quad \{\Sigma(m_1m_2)\}^3 = \left(\frac{2a-h}{a}\right)^3$$

$$\text{L.H.S.} = \Sigma m_1^3 m_2^3 + 3\,(m_1m_2 + (m_2m_3) \times (m_2m_3 + m_1m_2) \times (m_3m_1 + m_1m_2)$$

$$= \Sigma m_1^3 m_2^3 + 3m_1m_2m_3\,(m_1 + m_2)(m_2 + m_3)(m_3 + m_1)$$

$$= \Sigma m_1^3 m_2^3 + 3m_1m_2m_3\, -(m_3) -(m_1) -(m_2)$$

$$= \Sigma m_1^3 m_2^3 - 3m_1^2 m_2^2 m_3^2.$$

Hence $\Sigma m_1^2 m_2^2 = \frac{3k^2}{a^2} + \left(\frac{2a-h}{a}\right)^3$

Now putting different values in (2), we have

$$64k^3 + 16k^2a\left(-\frac{3k}{a}\right) + 4ka^2\left[\left(\frac{2a-h}{a}\right)^2 + \frac{3k^2}{a^2}\right] + a^3\left(-\frac{k^3}{a^3}\right) = \lambda a^3\left(-\frac{k}{a}\right)$$

or $$64k^2 - 48k^3 + \frac{4k}{a}\,[(2a-h)^3 + 3ak^2] - k^3 + \lambda a^2k = 0$$

$$\Rightarrow \quad 15k^3 + \frac{4k}{a}\,[(2a-h)^3 + 3ak^2] - k^3 + \lambda a^2k = 0$$

$$\Rightarrow \quad 15k^3 + \frac{4k}{a}\,(2a-h)^2 + 12k^3 + \lambda a^2k = 0$$

$$\Rightarrow \quad 27k^2a + 4(2a-h)^3 + \lambda a^3 = 0.$$

Generalising, the locus of (h, k) is

$$27ay^2 + 4(2a-x)^3 + \lambda a^3 = 0.$$ **Ans.**

$$\Rightarrow \quad -m_3^2 - \frac{k}{am_3} = \frac{2a-h}{a}$$

$$\therefore \quad am_3^2 + (2a-h)m_3 + k = 0. \quad \ldots(2)$$

On subtracting (2) from (1) after multiplying (4) by m_3, we have

$$a\lambda m_3^2 + (k\lambda - 2a - h)\,m_3 - k = 0 \quad \ldots(3)$$

Applying cross-multiplication to (2) and (3) to eliminate m_3, generalise the eliminant, which is the required to us and it is a parabola.

Example 43:

Find the locus of a point O when the three normals drawn from it are such that two of them make complementary angles with the axis.

Solution:

Let the equation to the parabola by $y^2 = 4ax$; then the equation to any normal to it will be $y = mx - 2am - am^3$.

If this normal passes through (h, k), the co-ordinates will satisfy, hence $k = mh - 2am - 2am - am^3$ or $am^3 + (2a-h)\,m + k = 0$.

Let m_1, m_2 and m_3 be the roots of this equation, then

$$m_1 + m_2 + m_3 = 0, \quad \ldots(1)$$

$$m_1 m_2 + m_2 m_3 + m_3 m_1 = \frac{2a-h}{a}$$

and $$m_1 m_2 m_3 = -\frac{k}{a}. \quad \ldots(3)$$

Let one of the normals make an angle θ with axis of x, then the other will make an angle (90° – θ) by hypothesis.

So $m_1 = \tan\theta$, and $m_2 \tan(90 - \theta) = \cot\theta$.

So $m_1 m_2 = 1$. ...(4)

Now we have to eliminate m_1, m_2 and m_3 from (1), (2), (3) and (4).

Now from (3) and (4), $m_3 = -\dfrac{k}{a}$. ...(5)

By (2), $m_1 m_2 + m_3(m_1 + m_2) = \dfrac{2a-h}{a}$

or $m_1 m_2 + m_3(-m_3) = \dfrac{2a-h}{a}$ [by (1), $m_1 + m_2 + m_3 = 0$] ...(6)

Putting the value of m_1m_2 from (4) and m_2 from (5) we have from (1),

$$1-\frac{k^2}{a^2}=\frac{2a-h}{a} \text{ or } k^2 = a(h-a).$$

Generalising, the locus of (h, k) is $y^2 = a(x - a)$. **Ans.**

Example 44(a):

If OP and OQ make complementary angles withe axis, then the tangent of R is parallel to SO.

Solution:

Let the equation of any normal OP be $y = m_1x - 2am_1 - am_1^3$, and the equation of the other normal OQ be $y = m_2x - 2am_2 - am_2^3$. Let OP be inclined at an angle of θ and OQ at an angle of ϕ. Then $\theta + \phi = 90^\circ$ or $\phi = (90^\circ - \theta)$ and if $m_1 = \tan\theta$ and $m_2 = \tan(90^\circ - \theta) = \cot\theta$.

$\therefore$ $m_1m_2 = 1$. ...(1)

The co-ordinates of the focus S of the parabola will be (a, 0) and those of O are (h, k) (given).

Hence the slope of SO $= \dfrac{k-0}{h-a} = \dfrac{k}{h-a}$. ...(2)

The normal at R is $y = m_3x - 2am_3 - am_3^3$. Hence the slope of the tangent at R is $= -1/m_3$.

Again, $m_1 + m_2 + m_3 = 0$, $m_1m_2 + m_2m_3 + m_1m_3 = \dfrac{2a-h}{a}$

and $m_1 m_2 m_3 = -\dfrac{k}{a}$.

Hence $m_3(m_1 + m_2) + m_1 m_2 = \dfrac{2a-h}{a}$

or $-m_3^2 + 1 = \dfrac{2a-h}{a} \quad \therefore \quad -m_3^2 = \dfrac{a-h}{a}$. ...(3)

Again $m_3 = \dfrac{-k}{am_1 \ m_2} = \dfrac{-k}{a}\left[\text{as } m_1 \ m_2 = 1 \text{ by } (1)\right]$. ...(4)

Dividing equation (4) by equation (3), we get $\dfrac{-1}{m_3} = \dfrac{k}{a-1}$.

Hence the slope of the tangent at R is same as the slope of SO.

$\therefore$ SO is parallel to the tangent at R.

Example 44(b):

The normals at three points P, Q and R of the parabola $y^2 = 4ax$ meet in a point O whose coordinates are h and k; prove that the point O and the orthocentre of the triangle formed by the tangents at P, Q and R are equi-distant from the axis.

Solution:

Let P, Q and R be respectively $\left(am_1^2, -2am_1\right)$, $\left(am_2^2, -2am_2\right)$ amd $\left(am_3^2, -2am_3\right)$. Then tangent at P is

$$-ym_1 = x + am_1^2 \qquad ...(1)$$

and tangent at Q is $-ym_2 = x + am_2^2$ ...(2)

Solving (1) and (2) simultaneously, the co-ordinates of the point of intersection are $x = am_1m_2$ and $y = -a(m_1 + m_2)$.

Tangent at R is $-ym_3 = x + am_3^2$. ...(3)

The equation to the line through $[amm_2 - a(m_1 + m_2)]$ and perpendicular to (3) will be $y + a(m_1 + m_2) = m_3(x - am_1m_2)$

or $\qquad y = m_3x - am_1m_2m_3 - a(m_1 + m_2)$

or $\qquad y = m_3x + k + am_3$ ...(4)

Similarly the equation of the other perpendicular will be

$$y = m_2x + k + am_2. \qquad ...(5)$$

Solving (4) and (5), we get the co-ordinates of point of intersection *i.e.*, ortho-centre as $(-a, k)$. As the ordinate of the ortho-centre and the point (h, k) are same, hence they are equi-distance from the x-axis.

Example 45:

Find the locus of a point O when the three normals drawn from it are such that two of them make angles with the axis the product of whose tangents is 2.

Solution:

$$m_1 m_2 = 2 \qquad \text{[By hypothesis]...(1)}$$

$$m_1\, m_2\, m_3 = \frac{-k}{a}. \qquad ...(2)$$

By (1) and (2), $m_3 = \dfrac{-k}{2a}$. ...(3)

Again, we have

$$m_1\, m_2 - m_3^2 = \frac{2a-h}{a}.$$

Putting the values from (1) and (3), we have $2-\dfrac{k^2}{4a^2}=\dfrac{2a-h}{a}$,

whence $k^2 = 4ah$.

Generalising, the locus of (h, k) is $y^2 = 4ax$.

Example 46:

Find the locus of a point O when the three normals drawn from it are such that the area of the triangle formed by their feet is constant.

Solution:

Suppose the co-ordinates of the feet of the normals A, B and C be $\left(am_1^2 + 2am_1\right)$, $\left(am_2^2, -2am_2\right)$ and $\left(am_3^2, +2am_3\right)$ respectively.

Hence the area of the triangle ABC is given as

$$= \frac{1}{2}\left[am_1^2\left(-2am_2 + 2am_3\right) + am_2^2\left(-2am_3 + 2am_1\right) + am_3^2\left(-2am_1 + 2am_2\right)\right]$$

$$-a^2\left[m_1^2\left(m_2 - m_3\right) + m_2^2\left(m_3 - m_1\right) + m_3^2\left(m_1 - m_2\right)\right]$$

$\Rightarrow \quad a^2 (m_1 - m_2)(m_2 - m_3)(m_3 - m_1) = \text{constant (by hypothesis)}$

$\therefore \quad (m_1 - m_2)^2 (m_2 - m_3)^2 (m_3 - m_1)^2 = \text{constant} = \lambda \text{ (say)}$ ...(1)

But $(m_1 - m_2)^2 = (m_1 + m_2)^2 - 4m_1 m_2 = (-m_3)^2 + \dfrac{4k}{am_3}$

$$= \frac{1}{am_3}\left(am_3^3 + 4k\right)$$

Example 47(a):

The circle circumscribing the triangle PQR goes through vertex and its equation is $2x^2 + 2y^2 - 2x(h + 2a) - ky = 0$.

Solution:

Let the co-ordinates of P, Q and R are $\left(am_1^2,-2a\right)$ $\left(am_2^2,-2am_2\right)$ and $\left(am_3^2,-2am_3\right)$ respectively and let the equation of circle passing through P, Q and R be $x^2 + y^2 + 2gx + 2fy + c = 0$.

Since the circle passes through P, Q and R, substituting the ordinates one by one, we have

$$a^2m_1^4 + 3a^2m_1^2 + 2agm_1^2 - 4afm_1 + c = 0$$

$$\Rightarrow \quad am_1\left(am_1^3 + 4am_1 + 2gm_1 - 4f\right) + c = 0$$

and as $\quad am_1^3 + (2a - h)\, m_1 + k = 0$

We know that

$am_1^3 = (h - 2a)\, m_1 - k$; putting in (1), we get

$am_1\ \{(h - 2a)\, m_1 - k + 4am_1 + 2gm_1 - 4f\} + c = 0$

$\Rightarrow a\ \{(h + 2a + 2g)\ m_2^2 - (k + 4f)\ m_1\} + c = 0.$

Similarly by the points Q and R, we get

$$a\ \{(h + 2a + 2g)\ m_2^2 - (k + 4f)\ m_2 + c = 0$$

and $\quad a\ \{(h + 2a + 2g)\ m_3^2 - (k + 4f)\ m_3\} + c = 0$

Subtracting (3) from (2), we get

$$a\ \{(h + 2a + 2g)\left(m_1^2 - m_2^2\right) - (k + 4f)\ (m_1 - m_2) = 0$$

$$\Rightarrow \quad (h + 2a + 2g)\ (m_1 + m_2) - (k + 4f) = 0 \ (\text{as } m_1 \neq m_2)$$

and similarly from (3) and (4), we get

$$(h + 2a + 2g)\ (m_3 + m_2) - (k + 4f) = 0$$

$$(h + 2a + 2g)\ (m_3 + m_1) - (k + 4f) = 0$$

Adding (4), (5) and (6),

$$2\ (h + 2a + 2g)\ (m_1 + m_2 + m_3) - 3\ (k + 4f) = 0$$

But as $(m_1 + m_2 + m_3) = 0$ (by Q. No. 1), so $k + 4f = 0$, $2f = \dfrac{-k}{2}$.

Subtracting (6) from (5), we get

$$h + 2a + 2g = 0 \text{ as } m_1\ '\ m_3 \text{ or } 2g = -(h + 2a).$$

Putting these values in (2), we get $c = 0$.

$$\Rightarrow \quad x^2 + y^2 - x\ (h + 2a) - \frac{k}{y} = 0$$

$$\Rightarrow \quad 2x^2 + 2y^2 - 2x\ (h + 2a) - ky = 0.$$

This passes through as the co-ordinates (0, 0) satisfy it.

Example 47(b):

Find the locus of a point O when the three normals drawn from it are such that the sum of the three angles made by them with the axis is constant.

Solution:

Let θ_1, θ_2 and θ_3 be the angles of inclination of normals with axis of x; then $\theta_1 + \theta_2 + \theta_3$ = constant (by hypothesis) = λ (say).

So $\tan(\theta_1 + \theta_2 + \theta_3) = \tan\lambda$

$$\Rightarrow \quad \frac{\tan\theta_1 + \tan\theta_2 + \tan\theta_3 - \tan\theta_1 \tan\theta_2 \tan\theta_3}{1-(\tan\theta_1 \tan\theta_2 + \tan\theta_2 \tan\theta_3 + \tan\theta_1 \tan\theta_3)} = \tan\lambda$$

$$\Rightarrow \quad \frac{m_1 + m_2 + m_3 - m_1 m_2 m_3}{1-(m_1 m_2 + m_2 m_3 + m_1 m_3)} = \tan\lambda .$$

Putting the values from equations No. (1), (2) and (3), we get

$$\frac{0-(-k/a)}{1-\dfrac{2a-h}{a}} = \tan\lambda$$

or $\dfrac{a}{h-a}$ = tan l or $k = (h - a)\tan\lambda$.

Generalising, the locus of (h, k) is $y = (x - a)\tan\alpha$.

Example 48:

Find the locus of a point O when the three normals drawn from it are such that one bisects the angle between the other two.

Solution:

Let θ_1, θ_2 and θ_3 be the angles which the three normals make with x-axis; we have $\tan\theta_1 = m_1$, $\tan\theta_2 = m_2$ and $\tan\theta_3 = m_3$. Now one normal bisects the angle between two. we must have $\theta_1 - \theta_2 = \theta_2 - \theta_3$;

$$\Rightarrow \quad 2\theta_2 = \theta_1 + \theta_3$$

$$\Rightarrow \quad \tan 2\theta_2 = \tan(\theta_1 + \theta_3)$$

$$\Rightarrow \quad \frac{2\tan\theta_2}{1-\tan^2\theta_2} = \frac{\tan\theta_1 + \tan\theta_3}{1-\tan\theta_1 \tan\theta_3}$$

$$\Rightarrow \quad \frac{2m_2}{1-m_2^2} = \frac{m_1+m_3}{1-m_1 m_3} = \frac{-m_3}{1-m_1 m_3}$$

$$\Rightarrow \quad m_2^2 + 2m_1m_3 = 3 . \qquad ...(1)$$

Again, we have

$$m_1m_2 + m_2m_3 + m_1m_3 = \frac{2a-h}{a}$$

$\Rightarrow$ $$(m_1 + m_3)\, m_2 + m_1\, m_3 = \frac{2a-h}{a},$$

and as $m_1 + m_2 + m_3 = 0$

we get $-m_2^2 + m_1\, m_3 = \dfrac{2a-h}{a}.$ ...(2)

From (1) and (2), by addition,

$$3m_1\, m_3 = \frac{2a-h}{a} + 3 = \frac{5a-h}{a} \text{ or } m_1\, m_3 = \frac{5a-h}{3a}$$

Putting this value in (4), we get $m_2^2 = \dfrac{2h-a}{3a}$

$$m_1\, m_2\, m_3 = \frac{-k}{a} \text{ or } m_1^2 m_3^2 = \frac{k^2}{a^2}.$$

Putting the values, we have $\dfrac{(5a-h)^2}{9a^2} \cdot \dfrac{(2h-a)}{3a} = \dfrac{k^2}{a^2}.$

Generalising and simplifying, the locus of (h, k) is

$$(x - 5a)^2\, (2x - a) = 27ay^2.$$

Example 49:

Through a point P are drawn tangents PQ and PR to a parabola and circles are drawn through the focus to touch the parabola in Q and R respectively; prove that the common chord of these circles passes through the centroid of the triangle PQR.

Solution:

If the co-ordinates of Q and R be $(at_1^2, 2t_1)$ and $(at_2^2, 2t_2)$ respectively, then

the equation of tangent at Q is $t_1y = x + at_1^2$,

the equation of the tangent at R is $t_2y = x + at_2^2$.

So the co-ordinates of the point of intersection P will be $\{at_1t_2 . a\,(t_1 + t_2)\}$. Let the co-ordinates of the centroid of the ΔPQR be (h, k).

Then $h = \dfrac{a}{3}\left(t_1^2 + t_2^2 + t_1 t_2\right)$ and $k = \dfrac{a}{3}\left(2t_1 + 2t_2 + t_1 + t_2\right)$

or $k = a\,(t_1 + t_2)$

the equation of the circle which passes through focus (a, 0) and touches the parabola at point $\left(at_1^2, 2at_1\right)$ is

$$x^2 + y^2 - ax\left(3t_1^2 + 1\right) - ay\left(3t_1 - t_1^3\right) + 3a^2t^2 = 0. \quad ...(1)$$

Similarly other circle touching the parabola at $\left(at_2^2, 2at_2\right)$ is

$$x^2 + y^2 - ax\left(3t_2^2 + 1\right) - ay\left(3t_2 - t_2^3\right) + 3a^2t_2^2 = 0 \quad ...(2)$$

Now the common chord of the circles is obtained by subtracting (2) from (1), so its equation is

$$-ax\left(3t_1^2 - 3t_2^2\right) - ay\left(3t_1 - 3t_2 - t_1^3 + t_2^3\right) + 3a^2\left(t_1^2 - t_2^2\right) = 0$$

$$\Rightarrow \quad 3ax\ 3ax\left(t_1 + t_2\right) - ay\left(t_1^2 + t_2^2 + t_1t_2 - 3\right) = 3a^2\left(t_1 + t_2\right)$$

$$\Rightarrow \quad kx - y(h - a) = ak.$$

This line is satisfied by (h, k). **Hence proved.**

Example 50(a):

The sum of the intercepts which the normals cut off from the axis is 2 (h + a).

Solution:

Let P, Q and R be the three points on the parabola having the co-ordinates as $\left(am_1^2, -2am_1\right)$, $\left(am_2^2, -2am_2\right)$ and $\left(am_3^2, -2am_3\right)$ respectively. Then equations of normal at these points are respectively

$$y = m_1x - 2am_1 - am_1^3, \quad ..(1)$$

$$y = m_2x - 2am_2 - am_2^3 \quad ...(2)$$

and

$$y = m_3x - 2am_3 - am_3^3. \quad ...(3)$$

Solving these with the axis of the parabola *i.e.*, y = 0, we get the intercepts on the x-axis respectively as $2a + am_1^2$, $2a + am_2^2$ and $2a + am_3^2$.

Sum of the intercepts made by the normals on the axis

$$= \left(2a + am_1^2\right) + \left(2a + m_2^2\right) + \left(2a + am_3^2\right)$$

$$= 6a + a\left(m_1^2 + m_2^2 + m_3^2\right)$$

$$= 6a + a\{(m_1 + m_2 + m_3)^2 - 2(m_1m_2 + m_2m_3 + m_3m_1)\}$$

$$= 6a + a\left\{0 - \frac{2(2a - h)}{a}\right\} = 2h + 2a = 2(h + a).$$

Example 50(b):

The normals at three points P, Q and R of the parabola $y^2 = 4ax$ meet in a point O whose coordinates are h and k; prove that the centroid of the triangle POR lies on the axis.

Solution:

Let the co-ordinates of P, Q and R be $\left(am_1^2,-2am_1\right)$, $\left(am_2^2,-2am_3\right)$ and $\left(am_3^2,-2am_3\right)$ respectively. Now, if O ≡ (x, y) be the co-ordinates of the centroid of ΔPQR, then we have

$$y=\frac{-2am_1-2am_2-2am_3}{3}=\frac{-2a\left(m_1+m_2+m_3\right)}{3}=0$$

Hence the locus of the centroid is y = 0 or x-axis which is axis of the parabola. **Proved.**

Example 50(c):

The sum of the squares of the sides of the triangle PQR is equal to 2 (h – 2a) (h + 10a).

Solution:

Let the co-ordinates of P, Q and R be $\left(am_1^2,-2am_1\right)$, $\left(am_2^2,-2am_2\right)$ and $\left(am_3^2,-2am_3\right)$ respectively.

Then $PQ^2 = a^2\left(m_1^2-m_2^2\right)^2 + (2am_1 - 2am_2)^2$

$$= a^2\left(m_1^2-m_2^2\right)^2 + 4a^2(m_1 - m_2)^2$$
$$= a^2(m_1 - m_2)^2\{(m_1 + m_2)^2 + 4\}$$
$$= a^2\left(m_1^2-m_2^2-2m_1m_2\right)\left(m_3^2+4\right)$$

(as $m_1 + m_2 + m_3 = 0$)

$$= a^2\left\{\left(m_1^2m_3^2+m_2^2m_3^2\right)+4\left(m_1^2+m_2^2\right)-2m_1m_2m_3^2-8m_1m_2\right\}$$

Similarly , we can write the value of PR^2 and QQ^2.

Hence $PQ^2 + QR^2 + PR^2$

$$= a^2\left(2\Sigma m_1^2m_2^2-8\Sigma m_1^2+2m_1m_2m_3\Sigma m_1-8\Sigma m_1m_2\right)$$

Now, $Sm_1 = 0$, we have

$$\Sigma m_1^2=\left(\Sigma m_1\right)^2-2\Sigma m_1m_2=-2\left(\frac{2a-h}{a}\right)$$

and $\Sigma\, m_1^2\, m_2^2 = (\Sigma\, m_1\, m_2)^2 - (2m_1\, m_2\, m_3\, \Sigma\, m_1) = \left(\frac{2a-h}{a}\right)^2$

$$a^2\left[2\left(\frac{2a-h}{a}\right) + 8\,(-2)\left(\frac{2a-h}{a}\right) - 8\left(\frac{2a-h}{a}\right)\right]$$

$$= 2a^2\left(\frac{2a-h}{a}\right)\left[\frac{2ah}{a} - 12\right] = 2\,(h-2a)\,(h+10a)\,.$$

Example 50(d):

Find the locus of a point O when the three normals drawn from it are such that the line joining the feet of two of them is always in a given direction.

Solution:

Let $A \equiv (am_1^2, -2am_1)$ and $B \equiv (am_2^2, -2am_2)$ be the co-ordinates of the feet of any two of the three normals. Now slope of the line joining AB is

$$= \frac{-2a\,(m_1 - m_2)}{a\,(m_1^2 - m_2^2)} = \frac{-2}{m_1 + m_2} = \text{constant.[by hypothesis]}$$

$\Rightarrow \quad m_1 + m_2 = \text{constant} = -\,l \text{ (say)}.$

As $m_1 + m_2 + m_3 = 0$ by Equation (1) of Q. No. 1.

so $\quad m_3 = -\,(m_1 + m_2) = \lambda.$

Again, we have

$$m_3\,(m_2 + m_1) + m_1\, m_2 = \frac{2a-h}{a} \text{ or } -m_3^2 - \frac{k}{a\lambda} = \frac{2a-h}{a}$$

$$\Rightarrow \quad -\lambda^2 - \frac{k}{a\lambda} = \frac{2a-h}{a}$$

$\Rightarrow \quad al^3 + (2a - h)\,\lambda + k = 0.$

Generalising the locus of (h, k) is $a\lambda^3 + (2a - x)\,\lambda + y = 0$ which is a normal at $(a\lambda^2, -a\lambda)$ to the parabola.

Example 51:

Prove that the normals at the points, where the straight line $lx + my = 1$ meets the parabola, meet on the normal at the point

$$\left(\frac{4am^2}{l^2}, \frac{4am}{l}\right)$$

of the parabola.

Solution:

The line is lx + my = 1. ...(1)

and the parabola is $y^2 = 4ax$.

Let (1) cut the parabola in P and Q such that

$$P \equiv \left(at_1^2, -2at_1\right) \text{ and } Q \circ \left(at_2^2, -2at_2\right).$$

Now equation of PQ is $(t_1 + t_2)\, y + 2x + 2at_1t_2 = 0$ ...(2)

Since the lines (1) and (2) represent the same line, so comparing the coefficients, we get

$$\frac{t_1 + t_2}{m} = \frac{2}{l} = \frac{2at_1\ t_2}{-1};\ \text{so } t_1 + t_2 = \frac{2m}{l} \qquad ...(3)$$

and $$t_1\ t_2 = \frac{1}{al}. \qquad ...(4)$$

Again, the normal at $(at^2, -2at)$ is $y = tx - 2at - at^3$, or $at^3 + t(2a - x) + y = 0$.

The roots of this equation will be t_1, t_2 and t_3, such that $t_1 + t_2 + t_3 = 0$.

So $$t_3 = -(t_1 + t_2) = \frac{-2m}{l} \quad \text{[by (3)]} \qquad ...(5)$$

As the third point on the parabola is $\left(at_3^2, -2at_3\right)$, so putting the value of t_3 from (5), we get the required point as $\left(\dfrac{4m^2}{l^2}, \dfrac{4am}{l}\right)$. **Proved.**

Example 52:

Show that the locus of the point of intersection of two tangents, which the tangent at the vertex form a triangle of constant area c^2, is the curve $x^2 (y^2 - 4ax) = 4c^4$.

Solution:

Let P ≡ (h, k) be the point of intersection of tangents of the parabola $y^2 = 4ax$, and if these cut an intercept AB on the y-axis, then $AB = \sqrt{(k^2 - 4ah)}$.

The length of perpendicular from P ≡ (h, k) on AB *i.e.*, y = 0 is clearly h. Hence the area of the triangle ABP

$$= \frac{1}{2}\sqrt{(k^2 - 4ah)}\,.\,h = c^2 \qquad \text{(by hypothesis).}$$

Squaring, simplifying and generalising, the locus of (h, k) is given as

$$x^2 (y^2 - 4ax) = 4c^4.$$ **Proved.**

Example 53(a):

Two tangents to a parabola meet at angle of 45°; prove that the locus of their point of intersection is the curve $y^2 - 4ax = (x + a)^2$.

If they meet at an angle of 60°, prove that the locus is $y^2 - 3x^2 - 10ax - 3a^2 = 0$.

Solution:

(a) Let m_1 and m_2 be the slopes of two tangents through P.

As the angle between them is 45°, we have

$$\tan 45^\circ = \pm\frac{m_1 - m_2}{1 + m_1 m_2} = \pm\frac{\sqrt{\left\{(m_1 + m_2)^2 - 4\, m_1\, m_2\right\}}}{1 + m_1\, m_2}.$$

Putting the values of m_1 and m_2, we get

$$1 = \pm\frac{\sqrt{\left(\frac{k^2}{h^2} - \frac{4a}{h}\right)}}{\left(1 + \frac{a}{h}\right)}$$

$$\Rightarrow \qquad 1 = \frac{\sqrt{(k^2 - 4ah)}}{h + a}.$$

Generalising after squaring and simplifying, we get the locus of (h, k) as $y^2 - 4ax = (x + a)^2$.

(b) Again, if the angle between the tangents if 60°, then

$$\tan 60^\circ = \pm\frac{m_1 - m_2}{1 + m_1\, m_2} = \pm\frac{\sqrt{\left\{(m_1 + m_2)^2 - 4\, m_1\, m_2\right\}}}{1 + m_1\, m_2}$$

$$\Rightarrow \qquad \sqrt{3} = \pm\frac{\sqrt{(k^2 - 4ah)}}{a + h}, \text{ [as in part (a)].}$$

Squaring, simplifying and generalising, the required locus is

$$y^2 - 3x^2 - 10ax - 3a^2 = 0.$$

Example 53(b):

From an external point P tangents are drawn to the parabola; find the equation to the locus of P when these tangents make angles q_1 and q_2 with the axis, such that $\tan q_1 + \tan q_2$ is constant (= b).

Solution:

Let the co-ordinates of P be (h, k) and the equation to the parabola be $y^2 = 4ax$.

Any tangent on the parabola is given by $y = mx + a/m$. If this passes through (h, k), the co-ordinates will satisfy. Hence $k = mh + a/m$

$\Rightarrow \qquad m^2h - mk + a = 0.$ ...(1)

[1] is a quadratic in m. Let its roots be m_1 and m_2, then $m_1 + m_2 = k/h$ and $m_1 m_2 = a/h$. Now, if the two tangents through P make angles θ_1 and θ_2 with axis of x_1 then $m_1 = \tan\theta_1$ and $m_2 = \tan\theta_2$.

$\therefore \qquad \tan\theta_1 + \tan\theta_2 = k/h.$ ...(2)

$m_1 m_2 = a/h$ ...(3)

$\Rightarrow \qquad \tan\theta_1 \tan\theta_2 = a/h$ ...(4)

By hypothesis, $\tan\theta_1 + \tan\theta_2 = b$

So from (2), $\quad k/h = b$ or $k = bh$

Generalising, the locus of (h, k) is $y = bx$. **Ans.**

Example 53(c):

Find the locus of the middle points of chords of the parabola which are such that the normals at their extremities meet on the parabola.

Solution:

Since the normals at $P \equiv \left(at_1^2, 2at_1\right)$ and $Q \equiv \left(at_2^2, 2at_2\right)$ meet on the parabola $y^2 = 4ax$, the necessary condition is $t_1t_2 = 2$ ...(1)

Now, if (h, k) be the mid-point of the chord PQ, then as in Q. No. 19, we have

$$\frac{2x_1}{a} = t_1^2 + t_2^2 \qquad \text{...(2)}$$

and $$\frac{y_1}{a} = t_1 + t_2. \qquad \text{...(3)}$$

Now, we have to eliminate 't_1' 't_2' from (1), (2) and (3).

Identically, we have $(t_1 + t_2)^2 - \left(t_1^2 + t_2^2\right) = 2t_1t_2$.

Putting the values from (1), (2) and (3),

$$\frac{k^2}{a^2} - \frac{2h}{a} = 2a \text{ or } k^2 - 2ah = 4a^2.$$

Generalising and rearranging the required locus is $y^2 = 2a(x + 2a)$.

Ans.

$$\Rightarrow \qquad t_3 = \frac{y - 2k}{4b - 2a} \ \{\text{by } (4)\}.$$

Multiplying (1) by t_1 and (2) by t_2 and subtracting we get

$$-a^2t_1t_2 + a^2t_3^2 + k^2 - 2akt_3 = -4bh.$$

$$\Rightarrow \qquad (y - 2k)^2 - 2ak\,(y - 2k)(4b - 2a) = (4b - 2a)^2(ax - 4bh - k)$$

Clearly this is a parabola.

Example 54:

Parabolas are drawn to touch two given straight lines which are inclined at an angle ω; if the chords of contact all pass through a fixed point, prove that (1) their directories all pass through another fixed point, and (2) their facial all lie on a circle which goes through the intersection of the two given straight lines.

Solution:

The equation of the directrix is written by

$$\frac{x + y\cos\omega}{a} + \frac{y + x\cos\omega}{b} = 1 \qquad ...(1)$$

and as the chord of contact passes through a fixed point whose co-ordinates are (h, k), we obtain

$$\frac{h}{a} + \frac{k}{b} = 1.$$

Therefore the directrix passes through the fixed point given by the following equation

$$x + y\cos\omega = h \text{ and } y + x\sec\omega = k.$$

And the focus is given by $ax = by = x^2 + y^2 + 2xy\cos\omega$, whence

$$a = \frac{x^2 + y^2 + 2\,xy\cos\omega}{x}$$

and

$$b = \frac{x^2 + y^2 + 2\,xy\cos\omega}{y}.$$

Substituting for 'a' and 'b' in (2) and simplifying, we have $x^2 + y^2 + 2xy\cos\omega - xh - yk = 0$, which is a circle which pass through the origin.

Example 55:

Prove that the sum of the angles which the three normals, drawn from any point O, make with the axis exceeds the angle which the focal distance of O makes with the axis by a multiple of π.

Solution:

Let y = mx – 2am – am^3 be the normal passing through O, whose co-ordinates are (h, k) (say); then k = mh – 2am – am^3 or am^3 + m (2a – h) + k = 0.

Then we have

$$m_1 + m_2 + m_3 = 0,\ \Sigma m_1\ m_2 = \frac{2a-h}{a} \text{ and } m_1\ m_2\ m_3 = \frac{-k}{a}.$$

If θ_1, θ_2 and θ_3 be the angles of inclination of these normals with axis, *i.e.*, x-axis then $m_1 = \tan\theta_1$, $m_2 = \tan\theta_2$ and $m_3 = \tan\theta_3$.

If S is the focus (a, 0), then the slope of $SO = \frac{k}{h-a} = \tan\alpha$ (say)

Again $\tan(\theta_1 + \theta_2 + \theta_3)$

$$= \frac{(m_1 + m_2 + m_3 - m_2\ m_2\ m_3)}{1-(m_1\ m_2 + m_2\ m_3 + m_3\ m_1)} = \frac{0-(-k/a)}{1-\left(\frac{2a-h}{a}\right)} = \frac{k}{h-a}$$

$\Rightarrow$ $\tan(\theta_1 + \theta_2 + \theta_3) = \tan\alpha.$

Hence $\theta_1 + \theta_2 + \theta_3 = n\pi + \alpha.$

Example 56:

If P be fixed, then QR is fixed in direction and the locus of the centre of the circle circumscribe PQR is a straight line.

Solution:

The slope of QR

$$= \frac{-2a\left[m_2 - m_3\right]}{a\left[m_2^2 - m_3^2\right]} = \frac{-2}{m_2 + m_3} = \frac{2}{m_1}$$

If m_1 is constant $2/m_1$ is also constant, hence the direction QR is fixed.

The centre of the circle given

$$(-g, -f) \text{ i.e.,} \left(\frac{h+2a}{a}, \frac{k}{4}\right).$$

So if (x, y) is the centre, $x = \frac{h+2a}{a}$ ...(1)

and $y = \frac{k}{4}$ ...(2)

Again the equation of normal at $P \equiv \left(am_1^2 - 2am_1\right)$ is

$$y = m_1x - 2am_1 - am_1^3. \quad ...(3)$$

As this passes through (h, k) and also m_1 = constant, then $k = m_1h - 2am_1 - am_1^3$. Eliminating h and k from (1), (2) and (3), we have $4y = m_1(2x - 2a) - 2am_1 - am_1^3$. This is the locus of centre (xm y) and it is clearly a straight line.

Example 57(a):

Prove that the equation to the circle, which power through the focus and touches the parabola $y^2 = 4ax$ at the point $(at^2, 2at)$, is $x^2 + y^2 - ax(3t^2 + 1) - ay(3t - t^3) + 3a^2t^2 = 0$.

Prove also that the locus of its centre is the curve

$$27ay^2 = (2x - a)(x - 5a)^2.$$

Solution:

If the circle touches the parabola $y^2 = 4ax$ at $(at^2, 2at)$, they must have a common tangent at that point, and hence a common normal. The centre of the circle must lie on that normal. Let (h, k) be the co-ordinates of the centre of the centre and r be the radius of the circle. Then its equation is

$$x^2 + y^2 - 2hx - 2ky + c = 0 \quad ...(1)$$

The equation to the normal at $(at^2, 2at)$ is

$$y = -tx + 2at + at^3 \quad ...(2)$$

As the centre (h, k) lies on the normal (2), hence

$$k = -th + 2at + at^3. \quad ...(3)$$

Focus of the parabola is (a, 0).

As the circle passes through (a, 0) and $(at^2, 2at)$, the distance of these points from the centre (h, k) must each be equal to the radius. So we have

$$r^2 = (h - a)^2 + k^2 = (h - at^2)^2 + (k - 2at)^2$$

$$\Rightarrow \quad -2ah + a^2 = -2aht^2 - 4akt + a^2t^4 + 4a^2t^2$$

$$\Rightarrow \quad 4kt = h(1 - t^2) + at^4 + 4at^2 - a \quad ...(4)$$

Solving (3) and (4), $2h = a(3t^2 + 1)$ ...(5)

and $\quad 2k = a(3t - t^3)$ ...(6)

Putting the values of h and k in (1), the equation to the circle is given by

$$x^2 + y^2 - a(3t^2 + 1)x - a(3t - t^3)y + c = 0 \quad ...(7)$$

As (7) passes through the focus (a, 0), the co-ordinates will satisfy it: Hence we have $a^2 - a^2(3t^2 + 1) + c = 0$;

$\therefore \qquad c = 3a^2t^2.$

Putting the values in (7) the circle given by

$$x^2 + y^2 - ax\,(3t^2 + 1) - ay\,(3t - t^3) + 3a^2t^2 = 0.$$

(ii) To get the locus of the centre, we have to eliminate t from (5) and (6).

Multiplying (5) by t and (6) by 3 and adding, we get $2th + 6k = 10at$; hence $t = 3k/(5a - h)$. Substituting the value of 't' in (5), we have $(2h - a)$

$$= 3a.\left(\frac{3k}{5a-h}\right)^2$$

Simplifying and generalising, the locus of centre (h, k) is

$$(2x - a)\,(x - 5a)^2 = 27ay^2.$$ **Hence proceed.**

Example 57(b):

Three normals are drawn to the parabola $y^2 = 4ax \cos \alpha$ from any point lying on the straight line $y = b \sin \alpha$. Prove that the locus of the orthocentre of the triangles formed by the corresponding tangents is the curve $\frac{x^2}{a^2} + \frac{y^2}{b^2} = 1$, the angle α being variable.

Solution:

The equation to the parabola is given as

$$y^2 = 4ax \cos \alpha \quad \text{(i)}$$

Equation to any normal to (i) is $y = mx - 2am \cos \alpha - am^3 \cos \alpha$.

If this normal passes through any point on $y = b\ \alpha$. Then we have

$$b \sin \alpha = mx - 2am \cos \alpha - am^3 \cos \alpha$$

$$\Rightarrow \qquad am^3 \cos \alpha + 2am \cos \alpha - mx + b \sin \alpha = 0. \quad \text{(ii)}$$

Let m_1, m_2 and m_3 be its roots; then we have

$$m_1 + m_2 + m_3 = 0,\ m_1 m_2 m_3 = \frac{b \sin \alpha}{a \cos \alpha} \text{ and } \Sigma m_1 m_2 = \frac{2a \cos \alpha - x}{a \cos \alpha}$$

The feet of the perpendicular of the normals are given as

$$\left(am_1^2 \cos \alpha, -2am_1 \cos \alpha\right);\ \left(am_2^2 \cos \alpha, -2am_2 \cos \alpha\right)$$

and $\left(am_3^2 \cos \alpha, -2am_3 \cos \alpha\right)$. Now the equation to the tangent at $\left(am_1^2 \cos \alpha, -2am_1 \cos \alpha\right)$ w.r.t. the parabola is

$$y\,(-2am_1 \cos \alpha) = 2a \cos \alpha \left(x + am_1^2 \cos \alpha\right)$$

$\Rightarrow \qquad -ym_1 = x + am_1^2 \cos\alpha.$ ...(3)

Similarly tangent at $\left(am_2^2 \cos\alpha, -2am_2 \cos\alpha\right)$ is given by

$$-ym^2 = x + am_2^2 \cos\alpha.$$

Solving the two, the co-ordinates of the point of intersection of the tangents are $\{am_1m_2 \cos\alpha, -a(m_1+m_2)\cos\alpha\}$

or $\left(\dfrac{b\sin\alpha}{m_3} . am_3 \cos\alpha\right)$ [on putting the values from (2)].

Similarly the other two vertices of the triangles formed by the tangents are given as

$$\left(\frac{b\sin\alpha}{m_2}, am_2\cos\alpha\right) \text{ and } \left(\frac{b\sin\alpha}{m_1}, am_1\cos\alpha\right)$$

The equation of the line perpendicular of line (3) through the vertex $\left(\dfrac{b\sin\alpha}{m_1}, am_1\cos\alpha\right)$ is given by

$y = m_1x + am_1 \cos\alpha - b\sin\alpha$ ...(4)

Similarly other perpendicular line is given as

$y = m_2x + am_2 \cos\alpha - b\sin\alpha.$ ...(5)

On solving (3) and (4), we get

$x = -a\cos\alpha,$ ...(6)

$y = -b\sin\alpha.$ ...(7)

The ortho-centre of the given circle is (x, y)

Hence eliminating α from (6) and (7), we get the required focus as

$$\cos^2\alpha + \sin^2\alpha = \frac{x^2}{a^2} + \frac{y^2}{b^2} = 1.$$

Example 57(c):

If the normals at the three points P, Q and R meet in a point and if PP', QQ' and RR' be chords parallel to QR, RP and PQ respectively, prove that the normals at P', Q', R' also meet in a point.

Solution:

Let P, Q and R be $\left(am_1^2, -2am_1\right)$, $\left(am_2^2, -2am_2\right)$ and $\left(am_3^2, -2am_3\right)$ respectively and P', Q' and R' be $\left(at_1^2, -2at_1\right)$, $\left(at_2^2, -2at_2\right)$ and $\left(at_3^2, 2at_3\right)$ respectively.

Then Slope of $PP' = \frac{-2am_1 + 2at_1}{am_1^2 - at_1^2} = \frac{-2}{m_1 + t_1}$

and Slope of $QR = \frac{-2am_2 + 2am_3}{am_1^2 - am_3^2} = \frac{-2}{m_2 + m_3}$

As PP' as parallel to QR, slopes must be equal.

Hence $m_1 + t_1 = m_2 + m_3 = -m_1$ $\{\because \Sigma m_1 = 0\}$

$\therefore$ $t_1 = -2m_1$.

Similarly $t_2 = -2m_2$ and $t_3 = -2m_1 - 2m_2 - 2m_3$.

$= -2(m_1 + m_2 + m_3) = 0$ (as $\Sigma m_1 = 0$) or $\Sigma t_1 = 0$.

Hence the normals at point P', Q' and R' meet in a point.

Example 58:

Show that three circles can be drawn to touch a parabola and also to touch at the focus a given straight line passing through the focus, and prove that the tangents at the point of contact with the parabola form an equilateral triangle.

Solution:

We know that the equation to the circle which passes through the focus (a, 0) and also touches the parabola $y^2 = 4ax$ at $(at^2, 2at)$ is $x^2 + y^2 - ax(3t^2 + 1) - ay(3t - t^3) + 3a^2t^2 = 0$ and equation of any line is $x^2 + y^2 - ax(3t^2 + 1) - ay(3t - t^3) + 3a^2t^2 = 0$ and equation of any line through (a, 0) is given by $y = \frac{\sin\alpha}{\cos\alpha}(x - a)$ $\left(\because \quad m = \tan\alpha = \frac{\sin\alpha}{\cos\alpha}\right)$

$x\sin\alpha - y\cos\alpha = a\sin\alpha$ and equation of perpendicular line will be

$$x\cos\alpha + y\sin\alpha = a\cos\alpha \quad ...(1)$$

Co-ordinates of the centre of the circle given are $\{a(3t^2 + 1)/2, a(3t - t^3)/2\}$.

Since this passes through the centre then we have

$$\frac{a(4t^2 + 1)}{2}\cos\alpha + \frac{a}{2}(3t - t^3)\sin\alpha = a\cos\alpha$$

$\Rightarrow$ $(3t^2 + 1)\cos\alpha + (3t - t^3)\sin\alpha = 2\cos\alpha$

$\Rightarrow$ $(3t - t^3)\sin\alpha = (1 - 3t^2)\cos\alpha$

$\therefore$ $\cot\alpha = \frac{3t - t^3}{1 - 3t^2}$.

Putting t = tan θ, we get $\cot\alpha = \dfrac{3\tan\theta - \tan^3\theta}{1-3\tan^2\theta} = \tan 3\theta$

$\Rightarrow \qquad \tan 3\theta = \tan\left(\dfrac{\pi}{2}-\alpha\right)$

then we have

$$3\theta = \left(\frac{\pi}{2}-\alpha\right), \left\{\pi+\left(\frac{\pi}{2}-\alpha\right)\right\} \text{ or } \left\{2\pi+\left(\frac{\pi}{2}-\alpha\right)\right\}$$

If θ_1, θ_2 and θ_3 be three values of θ, then

$$\theta_1 = \frac{1}{3}\left(\frac{\pi}{2}-\alpha\right), \theta_2 = \frac{\pi}{3}+\frac{1}{3}\left(\frac{\pi}{2}-\alpha\right) \text{ and } \theta_3 = \frac{2\pi}{3}+\frac{1}{3}\left(\frac{\pi}{2}-\alpha\right)$$

So $\qquad \theta_2 - \theta_1 = \dfrac{\pi}{3} = \theta_3 - \theta_2.$

Therefore the three normals are inclined at 60°. So the angles between the tangents are also 60°.

Therefore these form an equilateral triangle. **Hence proved.**

Example 59:

Two of the normals drawn from a point O to the curve know complementary angles with the axis; prove that the locus of O and the curve which is touched by its polar are parabolas such that their lateral recta and that of the original parabola form a geometrical progression. Sketch the three curves.

Solution:

Let the equation of the parabola be $y^2 = 4ax$ as the normal $y = mx - 2am - am^3$ passes through O, whose co-ordinates are

(h, k), we have $m_1 + m_2 + m_3 = 0$, ...(1)

$$\Sigma m_1 m_2 = \frac{ah-h}{a}$$

and $\qquad \Sigma m_1 m_2 m_3 = -\dfrac{k}{a}.$ ...(3)

Let the angles which the normals having slopes m_1 and m_2 make with x-axis be θ and ϕ; then θ + ϕ = 90° or ϕ = (90° – θ).

So $\qquad m_1 = \tan\theta_1$ and $m_2 = \tan(90° - \theta) = \cot\theta$

$\therefore \qquad m_1m_2 = 1.$ ...(4)

Again $m_3(m_2 + m_1) + m_1 m_2 = \dfrac{2a-h}{a}$ [by (2)]

But $\quad m_1 + m_2 = -m_3$ [by (1)]

So $\quad -m_3^2 + 1 = \dfrac{2a - h}{a}$ or $-\dfrac{k^2}{a^2} = \dfrac{2a - h}{a} - 1 = \dfrac{a - h}{a}$

$\Rightarrow \quad k^2 = a(h - a).$ ...(5)

Generalising, we obtain the locus of O as $y^2 = a(x - a)$ which as clearly a parabola having latus rectum equal to a.

Again the equation of the polar of (h, k) w.r.t., $y^2 = 4ax$ is

given by $yk = 2a(x + h) = 2a\ x + \left(\dfrac{k^2 + a^2}{a}\right)$ from (5)

$$\Rightarrow \quad y = \frac{2a}{k}(x + a) + 2k$$

$$\Rightarrow \quad y = \frac{2a}{k}(x + a) + \frac{4a}{2a/k}.$$

Changing the origin to (–a, 0), this becomes $y = \dfrac{2a}{k}x + \dfrac{4a}{2a/k}$ which is of the form $y = mx + 4a/m$, and hence is tangent to $y^2 = 4.4ax$.

Changing the origin back to the original one, parabola becomes

$$y^2 = 16a(x + a).$$

Hence the polar of (h, k) touches a parabola $y^2 = 16a(x + a)$.

Latus rectum of this parabola is 16a.

Clearly a, 4a, 16a are in G.P. **Hence proved.**

Example 60(a):

If the normals drawn from any power line x = 2a in points whose ordinates are in arithmetical progression, prove that the tangents of the angles which the normals make with the axis are in geometrical progression.

Solution:

Let the equation of the parabola be $y^2 = 4ax$. If the feet of the normals passing through a point be $(am_1^2, -2am_1)$, $(am_2^2, -2am_2)$ and $(am_3^2, -2am_3)$, equation of normal at $(am_1^2, -2am_1)$ is $y = m_1x - 2am_1 - am_1^3$.

Solving with the line x = 2a, we get the point of intersection as $(2a, -am^3)$. Similarly solving with the other normals the ordinates of the points of intersection with x = 2a are $-am_1^3, -am_2^3$ and $-am_3^3$.

These ordinates are in A.P.

So $\quad -am_1^3 - am_3^3 = -2am_2^3$

or $\quad m_1^3 + m_3^3 = 2m_2^3$. ...(1)

But $m_1 + m_3 = -m_2$, as $\Sigma m_1 = 0$ ...(2)

By (1) $2m_2^3 = (m_1 + m_3)\left(m_1^2 + m_3^2 - m_1 m_3\right)$

$$-2m_2^2 = \left(m_1^2 + m_3^2 - m_1 m_3\right) \text{ (as } m_1 + m_3 = -m_2) \quad ...(3)$$

and by (2) $m_2^2 = m_1^2 + m_3^2 + 2m_1 m_3$. ...(4)

Subtracting (3) from (4), $3m_2^2 = 3m_1 m_3$ or $m_2^2 = m_1 m_3$.

Example 60(b):

PG, the normal at P to a parabola, cuts the axis in G and is produced to Q so that GQ = 1/2PG; prove that the other normals which pass through Q intersect at right angles.

Solution:

Let the co-ordinates of P be $\left(am_1^2, -am_1\right)$.

Equation of normal at P will be given as

$$y = m_1 x - 2am_1 - am_1^3. \quad ...(1)$$

Solving (1) with x-axis *i.e.* y = 0, the abscissa of the point of intersection is $(2a + am_1^2)$.

The point Q is such that GQ = 1/2 PG. Hence Q divides PG in the ratio of 3: 1 externally. Let co-ordinates of Q be (h, k); then we have

$$k = \frac{3 \times 0 - 1\left(-2am_1\right)}{3 - 1} = am_1 \text{ or } m_1 = \frac{k}{a}.$$

But $\quad m_1 m_2 m_3 = \dfrac{-k}{a}$

Putting the value of m_1, we get $\dfrac{k}{a} m_2 m_3 = \dfrac{-k}{a}$ or $m_2 m_3 = -1$.

Hence the two normals through Q are perpendicular to each other.

Example 60(c):

A circle is described whose centre is the vertex and whose diameter is three-quarters of the latus rectum of a parabola; prove that the common chord of the circle and parabola bisects the distance between the vertex and the focus.

Solution:

Let the equation of the parabola be $y^2 = 4ax$

Its vertex is (0, 0) and latus rectum is 4a. Hence the centre of given circle will be (0, 0) and diameter will be 3/4.4a = 3a.
So its equation will

To get the point of intersection of (1) and (2), we solve them simultaneously. So putting the value of y^2 from (1) in (2), we get

$$x^2 + 4ax = \frac{9}{4}a^2$$

$$\Rightarrow \qquad 4x^2 + 16ax - 9a^2 = 0$$

$$\Rightarrow \qquad (2x - a)(2x + 9a) = 0, \text{ whence } x = a/2.$$

(neglecting –ve value as the parabola does not lie on that side).

As both the curves (1) and (2) are symmetrical about axis of x, the common chord will be parabola to axis of y.

Hence its equation will be x = a/2.

Clearly it cuts x-axis at (a/2, 0) which is mid-way between the vertex (0, 0) and the focus (a, 0).

Example 60(d):

PR and QR are chords of a parabola which are normals at P and Q. Prove that two of the common chords of the parabola and the circle circumscribing the triangle PRQ meet on the directrix.

Solution:

Suppose the co-ordinates of R on the parabola

$y^2 = 4ax$ be (h, k) and let $h = am_3^2$ and $k = -2am_3$.

Again, let the co-ordinates of P and Q be $(am_1^2, -2am_1)$ and $(am_2^2, -2am_2)$ respectively. Now, if any normal $y = mx - 2am - am^3$ passes through (h, k). Then we have

$$m_1 + m_2 + m_3 = 0, \; m_1 m_2 + m_2 m_3 + m_3 m_1 = \frac{2a - h}{h}$$

and $$m_1 m_2 m_3 = \frac{-k}{a}.$$

But as $m_3 = \dfrac{-k}{2a}$, so $m_1 m_2 = 2$ and $m_1 + m_2 = \dfrac{k}{2a}$. ...(1)

We have proved that the circle and parabola intersect in four points and that the line joining one pair of these four points and the line joining the other pair are equally inclined to the axis. Therefore, if S be the fourth point, then RQ and RS are equally inclined with the axis.

Again, the equation to PQ will be $-y\ (m_1 + m_2) = 2x + 2am_1m_2$

$\Rightarrow \quad -y\dfrac{k}{2a} = 2x + 4a$ [from (1)]

$\Rightarrow \quad -ky = 4ax + 8a^2.$ ...(2)

The equation to RS is $k\ (y - k) = 4a\ (x - h)$ or $yk = 4ax + k^2 - 4ah.$

But $k^2 = 4ah$. So $yk = 4ax.$ ...(3)

Adding (3) with (2), we have $8axk + 8a^2 = 0$ or $x + a = 0$. This is the equation of directrix. Hence the common chords PR and RS of parabola and circle meet in directrix.

Example 61:

Prove that the locus of the centre of the circle, which passes through the vertex of a parabola and through its intersections with a normal chord, is the parabola $2y^2 = ax - a^2$.

Solution:

Equation of normal at point $\left(at_1^2, 2at_1\right)$ of the parabola $y^2 = 4ax$ is $y = -t_1x + 2at_1 +$, this line intersects the parabola at point $\left(at_2^2, 2at_2\right)$, such that $t_2 + t_1 = -2/t_1$

The equation of any circle passing through (0, 0) is

$$x^2 + y^2 - 2gx - 2fy = 0.$$

If this passes through $\left(at_1^2, 2at_1\right)$, then

$$at_1^3 + 4at_1 - 2gt_1 - 4f = 0 \qquad ...(2)$$

Similarly, if the circle passes through $\left(at_2^2, 2at_2\right)$, then we have

$$at_2^3 + 4at_2 - 2gt_2 - 4f = 0. \qquad ...(3)$$

The locus of the centre (g, f) of the circle will be obtained by eliminating t_1, t_2 and t_3 from (1), (2) and (3).

Subtracting (3) from (2), we have

$$a\left(t_1^3 - t_2^3\right) + 4a\left(t_1 - t_2\right) - 2g\left(t_1 - t_2\right) = 0$$

$$\Rightarrow \quad a\left(t_1^2 + t_2^2 + t_1t_2\right) + 4a - 2g = 0$$

$$\Rightarrow \quad a\left(t_2^2 - 2\right) + 4a - 2g = 0 \qquad \text{[by (1)]}$$

$$\Rightarrow \quad at_2^2 + 2a - 2g = 0. \qquad ...(4)$$

Again multiplying (2) by t_2 and (3) by t_1 and then subtracting, we have

$$at_1t_2\left(t_1^2 - t_2^2\right) + 4f\left(t_1 - t_2\right) = 0$$

$\Rightarrow$ $at_1t_2(t_1 + t_2) + 4f = 0$ [by (1)]

$\Rightarrow$ $-2at_2 + 4f = 0$

Hence $t_2 = 2f/a$.

Substituting this value of t_2 in (4), we have

$$a(4f^2/a^2) + 2a - 2g = 0 \text{ or } 2f^2 + a^2 - ag = 0.$$

Gemeralising, the locus of the centre (g, f) is

$$2y^2 + a^2 - ax = 0 \text{ or } 2y^2 = ax - a^2.$$ **Hence Proceed.**

Example 62:

A parabola of latus rectum l, touches a fixed equal parabola, the axes of the two curves being parallel; prove that the locus of the vertex of the moving curve is a parabola of latus rectum 2l.

Solution:

Let (h, k) be the co-ordinates of the vertex of the moving parabola and its equation be $(y - k)^2 + 1(x - h) = 0$. The equation of the fixed curve be $y^2 - lx = 0$; then the tangents at the point (lt^3, lt) to the two curves are

$$-2yt + x + lt^2 = 0 \text{ and } (lt - k) + \frac{1}{2}x + \frac{l^2t^2}{2} + k^2 - lh - kl\,t = 0$$

As they are coincident $\frac{k - l\,t}{2t} = \frac{l/2}{1} = \frac{l^2t^2 + 2k - 2lh - 2klt}{2l\,t^2}$

from which $t = \frac{k}{2l}$ and $k^2 - l\,h = kl\,t$.

Eliminating t, we get $k^2 = 2l\,h$. Hence the locus of the vertex is $y^2 = 2l\,x$ which is a parabola having latus rectum as $2l$.

Example 63:

The two parabolas $y^2 = 4a(x - l)$ and $x^2 = 4a(y - l)$ always touch one another, the quantities l and l' being both variable; prove that the locus of their point of contact is the curve $xy = 4a^2$.

Solution:

Any point on parabola $y^2 = 4a(x - l)$ may be taken as $(l + at_1^2, 2at_1)$ and on $x^2\ 4a(y - l')$ as $(2at_2, l' + at_2^2)$. Let (x, y) be the common point, then

$$xy = 2at_2 \times 2at_1 \text{ or } xy = 4a^2t_1t_2. \quad ...(1)$$

Tangent at point 't_1' of the parabola $y^2 = 4a(x - l)$ will be

$$t_1 y = x + at_1^2 \quad ...(2)$$

Again the tangent at 't_2' of the parabola $x^2 = 4a(y - l')$ will be

$$xt_2 = y + 2at_2 \text{ or } y = xt_2 - 2at_2 \quad ...(3)$$

The tangents (2) and (3) will coincide, if

$$t_2 = 1/t_1 \text{ or } t_1 t_2 = 1. \quad ...(4)$$

By (1) and (4), we have $xy = 4a^2$, which is the locus of common point of the parabola given.

Example 64(a):

The sides AB and AC of a triangle ABC are given in position and the harmonic mean between the lengths AB and AC is also given; prove that the locus of the focus of the parabola touching the sides at B and C is a circle whose centre lies on the line bisecting the angle BAC.

Solution:

Let AB and AC be the axes and each equal to a and b respectively. Say k is the harmonic mean between a and b; then

$$\frac{1}{a} + \frac{1}{b} = \frac{2}{k} \quad ...(1)$$

Now the foci are given by $ax = by = x^2 + y^2 + 2ay \cos \omega$ whence

$$a = \frac{x^2 + y^2 + 2xy \cos \omega}{x} \quad ...(2)$$

and

$$b = \frac{x^2 + y^2 + 2xy \cos \omega}{y} \quad ...(3)$$

Putting the values of a and b from (2) and (3) in (1) and simplifying, we get

$$\frac{2}{k} = \frac{x + y}{x^2 + 2xy \cos\omega + y^2}$$

Hence the locus is $x^2 + 2xy \cos w + y^2 - \frac{k}{2}(x + y) = 0$, and it is clearly a circle.

Again as the abscissa and the ordinate of the centre are equal, so it lies on the bisector of the angle.

Example 64(b):

Prove that the sum of the angles which the four common tangents to a parabola and a circle make with the axis is equal to $n\pi + 2a$, where α is the angle which the radius from the focus to the centre of the circle makes with the axis and n is an integer.

Solution:

Let the equation of parabola be $y^2 = 4ax$. ...(1)

Equation to any tangent to (1) is $y = mx + a/m$. ...(2)

Solving (2) with any circle $x^2 + y^2 - 2fy + c = 0$, we get $x^2 (1 + m^2) + 2x (a - g - fm) + a^2/m^2 - 2f (a/m) + c = 0$.

If the line (1) touches the circle, then the roots of this equation must be equal and hence its discriminant will be zero.

So $\quad 4 (a - g - fm)^2 = 4 (1 + m^2) (a^2/m^2 - 2fm\ a/m + c)$

$\Rightarrow \quad m^4 (f^2 - c) + 2gfm^3 - (g^2 - 2ag - c) m^2 + 2fam - a^2 = 0.$

Let its roots be m_1, m_2, m_3 and m_4 and let $m_1 = \tan\theta_1$, $m_2 = \tan\theta_2$, $m_3 = \tan\theta_3$ and $m_4 = \tan\theta_4$; then $\Sigma m_1 = \dfrac{-2gf}{f^2 - c} = \Sigma \tan\theta_1 = s_1$ (say).

Similarly, $\Sigma \tan\theta_1 \tan\theta_2 = \dfrac{g^2 - 2ga - c}{f^2} = s_2$ (say).

$\Sigma \tan\theta_1 \tan\theta_2 \tan\theta_3 = \dfrac{-2fa}{f^2 - c} = s_4$ (say)

Now as $\tan(\theta_1 + \theta_2 + \theta_3 + \theta_4) = \dfrac{s_1 + s_3}{1 - s_2 + s_4}$

$$= \frac{\dfrac{-2gf}{f^2 - c} + \dfrac{2fa}{f^2 - c}}{1 - \dfrac{g^2 - 2ag - c}{f^2 - c} - \dfrac{a^2}{f^2 - c}}$$

$$\Rightarrow \qquad \tan(\Sigma\theta_1) = \frac{2f(g - a)}{(g - a)^2 - f^2}. \qquad ...(3)$$

Now as the co ordinates of the centre and focus are respectively (g, f) and (a, 0), then

$$\tan\alpha = \frac{f}{g - a}. \qquad ...(4)$$

$$\text{Again, } \tan 2\alpha = \frac{2\tan\alpha}{1 - \tan^2\alpha} = \frac{\dfrac{2f}{g - a}}{1 - \left(\dfrac{2f}{g - a}\right)^2} = \frac{2f(g - a)}{(g - a)^2 - f^2} \qquad \text{[by (4)]}$$

So $\quad \tan(\Sigma\theta_1) = \tan 2\alpha$.

Solving for most general value, we get

$\theta_1 + \theta_2 + \theta_3 + \theta_4 = n\pi + 2\alpha$, where n is an integer. **Proved.**

Example 65:

A parabola touches two given straight lines, which meet at O, in given points and a variable tangent meets the given lines in P and Q respectively; prove that the locus of the centre of the circumcircle of the triangle OPQ is a fixed straight line.

Solution:

Taking the given lines as axes, $(x/f) + (y/g) = 1$ as the equation of the variable tangent, we have $(f/a) + (g/b) = 1$.

Then the equation to OP will be $x + y \cos \omega = f/2$ and of the OQ, perpendicular to OP will be $y + x \cos \omega = g/2$.

The required locus will be the locus of the point of intersection of OP and OQ; so we have to eliminate g and f.

The required eliminant is $\dfrac{x + y \cos \omega}{a} + \dfrac{y + x \cos \omega}{b} = \dfrac{1}{2}$ which is a straight line. **Proved.**

Example 66:

Parabolas are drawn to touch two given rectangular axes and their foci are all at a constant distance c from the origin. Prove that the locus of the vertices of these parabolas is the curve $x^{2/3} + y^{2/3} = c^{2/3}$

Solution:

We have $b = \lambda x^{1/3}$, $a = \lambda y^{1/3}$

and
$$\lambda^2 = \frac{\left(x^{2/3} + y^{2/3}\right)^4}{x^{2/3}\, y^{2/3}} \qquad \text{...(1)}$$

As co-ordinates of focus are $\left\{\dfrac{ab^2}{a^2 + b^2}, \dfrac{a^2 b}{a^2 + b^2}\right\}$, square of the distance of focus from the origin (0, 0) is

$$\frac{a^2\, b^4}{\left(a^2 + b^2\right)^2} + \frac{a^2\, b^2}{\left(a^2 + b^2\right)^2} = \frac{a^2\, b^2}{\left(a^2 + b^2\right)} = c^2 \quad \text{(by hypothesis)}$$

$$= \frac{\lambda x^{2/3}\, y^{2/3}}{x^{2/3} + y^{2/3}} \quad \text{[by (1)]} \qquad \text{...(2)}$$

Eliminating λ from (1) and (2), we get $c^{2/3} = x^{2/3} + y^{2/3}$. **Proved.**

Example 67:

If a parabola, whose latus rectum is 4c, slide between two rectangular axes, prove that the locus of its focus is $x^2y^2 = c^2 (x^2 + y^2)$ *and that the curve traced out by its vertex is* $x^{2/3}\, y^{2/3} (x^{2/3} + y^{2/3}) = c^2$.

Solution:

(a) As $\omega = 90^\circ$, or $\cos\omega = 0$ so the focus is given by

$$x^2 + y^2 = ax = by$$

Also $\quad c^2 (a^2 + b^2)^3 = a^4b^4.$

Eliminating a and b from (1) and (2), we get

$$c^2(x^2 + y^2)^6 \left(\frac{1}{x^2} + \frac{1}{y^2}\right)^3 = \frac{(x^2 + y^2)^8}{x^2\, y^2}$$

whence, simplifying, $x^2y^2 = c^2 (x^2 + y^2)$,

(b) The co-ordinates of vertex are

$$x = \frac{ab^4}{(a^2 + b^2)^2},\ y = \frac{a^4b}{(a^2 + b^2)^2} \qquad \text{...(3)}$$

Hence $\quad \dfrac{x}{y} = \dfrac{b^3}{a^3}$ or $\dfrac{b}{x^{1/3}} = \dfrac{a}{y^{1/3}} = \lambda$ (suppose)

So $\quad a = \lambda y^{1/3}$ and $b = \lambda x^{1/3}$.

Putting the values of and b in (2),

$$c^2 = \frac{\lambda^2\, x^{4/3}\, y^{4/3}}{x^{2/3} + y^{2/3}} \text{ and from (3), } 1 = \frac{\lambda x^{1/3}\, y^{1/3}}{(x^{2/3} + y^{2/3})^2}$$

Eliminating λ, we get $c^2 = x^{2/3}\, y^{2/3}\, (x^{2/3} + y^{2/3})$.

Example 68:

The axes being rectangular, prove that the locus of the focus of the parabola $\left(\dfrac{x}{a} + \dfrac{y}{b} - 1\right)^2 = \dfrac{4xy}{ab}$, *a and b being variables such that a*

ab $= c^2$, *is the curve* $(x^2 + y^2)^2 = c^2xy$.

Solution:

The focus is given by $ax = by = x^2 + y^2$

and we have $\quad ab = c^2$ (by hypothesis).

To eliminate a and b from (1) and (2), we have

$$\frac{x^2 + y^2}{y} = b \text{ and } \frac{x^2 + y^2}{x} = a$$

So $\quad ab = \dfrac{(x^2 + y^2)^2}{xy} = c^2$ by (2)

$\Rightarrow \qquad (x^2 + y^2)^2 = c^2xy$ which is the required locus.

Example 69(a):

Parabolas are drawn to touch the axes, which are inclined at an angle ω, and their directories all pass through a fixed point (h, k). Prove that all the parabolas touch the straight line $\frac{x}{h+k\sec\omega}+\frac{y}{k+h\sec\omega}=1$.

Solution:

Let the equation of the parabola be

$$\sqrt{(x/a)}+\sqrt{(y/b)}=1 \qquad \text{...(1)}$$

The equation to the directories is

$$x(a + b\cos\omega) + y(b + a\cos\omega) = ab\cos\omega$$

Its (P's) distance from x-axis = y.

Its distance from origin *i.e.* (0, 0) = $\sqrt{[(x-0)^2+(y-0)^2}$

By hypothesis, $y=\frac{1}{2}\sqrt{(x^2+y^2)}$

$$\Rightarrow \qquad y=\frac{1}{4}(x^2+y^2) \text{ or } x^2 = 3y^2.$$

As this line passes through (h, k), we have

$$h(a + b\cos\omega) + k(b + a\cos\omega) = ab\cos\omega \qquad \text{...(2)}$$

The given line is $\frac{x}{h+k\cos\omega}+\frac{y}{k+h\cos\omega}=1$ ...(3)

If (1) touches (3), solving the two, we must get two coincident points.

By (1), $x = a/b\left(\sqrt{b}-\sqrt{y}\right)^2$

Putting in (3), we get $\frac{a\left(b+y-2\sqrt{by}\right)}{b\left(h+k\cos\omega\right)}+\frac{y}{k+h\cos\omega}=1$ ...(4)

or simplifying with the help of (2), we find (4) is a perfect square. Hence (3) touches (1).

Example 69(b):

Parabolas are drawn to touch two given straight lines which are inclined at an angle ω; if the chords of contact all pass through a fixed point, prove that (1) their directories all pass through another fixed point, and (2) their facial all lie on a circle which goes through the intersection of the two given straight lines.

Solution:

The equation of the directrix is written by

$$\frac{x+y\cos\omega}{a}+\frac{y+x\cos\omega}{b}=1 \qquad ...(1)$$

and as the chord of contact passes through a fixed point whose co-ordinates are (h, k), we obtain

$$\frac{h}{a}+\frac{k}{b}=1.$$

Therefore the directrix passes through the fixed point given by the following equation

$$x + y\cos\omega = h \text{ and } y + x\sec\omega = k.$$

And the focus is given by $ax = by = x^2 + y^2 + 2xy\cos\omega$, whence

$$a=\frac{x^2+y^2+2\,xy\cos\omega}{x}$$

and $$b=\frac{x^2+y^2+2\,xy\cos\omega}{y}.$$

Substituting for 'a' and 'b' in (2) and simplifying, we have $x^2 + y^2 + 2xy\cos\omega - xh - yk = 0$, which is a circle which pass through the origin.

Example 69(c):

If the normals at the three points P, Q and R meet in a point and if PP', QQ' and RR' be chords parallel to QR, RP and PQ respectively, prove that the normals at P', Q', R' also meet in a point.

Solution:

Let P, Q and R be $\left(am_1^2,-2am_1\right)$, $\left(am_2^2,-2am_2\right)$ and $\left(am_3^2,-2am_3\right)$ respectively and P', Q' and R' be $\left(at_1^2,-2at_1\right)$, $\left(at_2^2,-2at_2\right)$ and $\left(at_3^2, 2at_3\right)$ respectively.

Then Slope of $PP'=\dfrac{-2am_1+2at_1}{am_1^2-at_1^2}=\dfrac{-2}{m_1+t_1}$

and Slope of $QR=\dfrac{-2am_2+2am_3}{am_1^2-am_3^2}=\dfrac{-2}{m_2+m_3}$

As PP' as parallel to QR, slopes must be equal.

Hence $m_1 + t_1 = m_2 + m_3 = -m_1$ $\qquad \{\because \Sigma m_1 = 0\}$

$\therefore$ $t_1 = -2m_1$.

Similarly $t_2 = -2m_2$ and $t_3 = -2m_1 - 2m_2 - 2m_3$.

$= -2(m_1 + m_2 + m_3) = 0$ (as $\Sigma m_1 = 0$) or $\Sigma t_1 = 0$.

Hence the normals at point P', Q' and R' meet in a point.

Example 70(a):

The normal at the point $(at_1^2, 2at_1)$ meets the parabola again in the point $(at_2^2, 2at_2)$; prove that

$$t_2 = -t_1 - \frac{2}{t_1}.$$

Solution:

Let the equation to the parabola be $y^2 = 4ax$. ...(1)

The equation to the normal at any point $(at_1^2, 2at_1)$ to (1) will be

$$y - 2at_1 = \frac{-at_1}{2a}\left(x - at_1^2\right). \qquad ...(2)$$

If (2) passes through the point $(at_2^2, 2at_2)$, the co-ordinates will satisfy it, hence

$$2at_2 - 2at_1 = -t_1(at_2^2, 2at_1^2)$$

$$\Rightarrow \quad 2a(t_2 - t_1) = - . a.(t_2 - t_1)(t_2 + t_1)$$

$$\Rightarrow \quad 2 = -t_1 t_2 - t_1^2$$

$$\Rightarrow \quad t_1 t_2 = -2 - t_1^2$$

$$\Rightarrow \quad t_1 = -\frac{2}{t_1} - t_1.]$$

Proved.

Example 70(b):

If PQ be a normal chord to the parabola and it S be the focus, prove that the locus of the centroid of the triangle SPQ is the curve

$$36y^2(3x - 5a) - 18y^4 = 128a^4.$$

Solution:

Let the equation to the parabola by $y^2 = 4ax$ ...(1)

and say P and Q are any two points $(at_1^2, 2at_1)$ and $(at_2^2, 2at_2)$ such that PQ is normal at P. Then by of this example

$$t_2 = -\frac{2}{t_1} - t_1. \qquad ...(2)$$

Let the focus (a, 0) be S, as the centroid of any triangle having the vertices as (x_1, y_1), (x_2, y_2) and (x_3, y_3) is

$$\left(\frac{x_1 + x_2 + x_3}{3}, \frac{y_1 + y_2 + y_3}{3}\right)$$

so if the centroid of the triangle PQS will be (h, k) then

$$h = \frac{at_1^2 + at_2^2 + a}{3} \qquad ...(3)$$

or $$k = \frac{at_1 + 2at_2 + 0}{3} \quad ...(4)$$

The required locus will be obtained by eliminating t_1 & t_2 from (2), (3) and (4).

So putting the value of t_2 from (2) in (4); we get

$$3k = 2at_1 + 2a\left(-t_1 - \frac{2}{t_1}\right) = -\frac{4a}{t_1} \quad \text{or} \quad t_1 = -\frac{4a}{3k}.$$

Hence $$t_2 = +\frac{4a}{3k} + \frac{2.3k}{4a}$$

Putting the values of t_1 and t_2 in (2), we get

$$3h = a\left(\frac{4a}{3k}\right)^2 + a\left(\frac{4a}{3k} + \frac{2.3k}{4a}\right)^2 + a$$

$$\Rightarrow \quad 3h = \frac{16a^3}{9k^2} + \frac{16a^3}{9k^2} + \frac{9k^2}{4a} + 4a + a$$

$$\Rightarrow \quad 3h - 5a = \frac{32a^2}{9k^2} + \frac{9k^2}{4a}.$$

Simplifying further and generalising, we get

$$36ay^2 (3x - 5a) = 128a^4 + 81y^4$$

$$36ay^2 (3x - 5a) - 81y^4 = 128a^4$$

Example 71:

A point on a parabola, the foot of the perpendicular from it upon the directrix, and the focus are the vertices of an-equilateral triangle. Prove that the focal distance of the point is equal to the latus rectum.

Solution:

Let the equation to the parabola be $y^2 = 4ax$ ...(1)

and P be any point on it as $(at^2, 2at)$. The focus S of the parabola will be (a, 0). If PN is the perpendicular from P on the directrix, $x = -a$; clearly the co-ordinates of N are (– a, 2at0 and the distance

$$PN = (a + at^2) \quad ...(2)$$

Distance $$SN = \sqrt{[(-a - a)^2 + (2at - 0)^2]}$$
$$= \sqrt{(4a^2 + 2a^2t^2)} \quad ...(3)$$

If ΔPSN is an equilateral triangle, then

$$PN = NS \qquad [\because PN = PS \text{ for all po int s}]$$

Hence by (2) and (3), we get

$$(a + at^2) = \sqrt{(4a^2 + 2a^2t^2)}.$$

Squaring, we get $a^2 + a^2t^4 + 2a^2t^2 = 4a^2 + 4a^2t^2$

$$\Rightarrow \quad t^4 - 2t^2 - 3 = 0$$

$$\Rightarrow \quad (t^2 - 3)(t^2 + 1) = 0.$$

If $t^2 + 1 = 0$, t^2 is imaginary so, $t^2 - 3 = 0$ or $t = \pm\sqrt{3}$.

Hence the co-ordinates of P become $(3a, 2a\sqrt{3})$ or $(3a, -2a\sqrt{3})$.

Distance PS $= \sqrt{\{(3a - a)^2 + (2a\sqrt{3})^2\}} = \sqrt{(16a^2)} = 4a$

$= \text{latus rectum.}$ **Hence proved.**

Example 72:

Prove that on the axis of any parabola there is a certain point K which has the property that, if a chord PQ of the parabola be drawn through it, then

$$\frac{1}{PK^2} + \frac{1}{QK^2}$$

is the same for all positions of the chord.

Solution:

Let the equation to the parabola be $y^2 = 4ax$. ...(1)

It axis is $y = 0$. So let there be any point K on the axis as $(h, 0)$.

Equation of any line passing through $(h, 0)$ is

$$y = mx - mh. \quad \text{...(2)}$$

To get the point of intersection of (1) and (2) we solve them. So putting the value of y from (2) in (1), we get

$$(mx - mh)^2 = 4ax$$

$$\Rightarrow \quad m^2a^2 - 2x(m^2h + 2a) + m^2h^2 = 0$$

$$\Rightarrow \quad x = \frac{2(m^2h + 2a) \pm \sqrt{\left\{4(m^2h + 2a)^2 - 4.m^2.m^2h^2\right\}}}{2m^2}$$

$$= \frac{1}{m^2}\left\{(m^2h + 2a) \pm \sqrt{(4a^2 + 4ahm^2)}\right\}.$$

The corresponding values of y are

$$\frac{1}{m}\left[2a \pm \sqrt{(4a^2 + 4ahm^2)}\right]$$

Hence the co-ordinates of P and Q are respectively

$$\left[\frac{(m^2 h+2a)+\sqrt{(4a^2+4ahm^2)}}{m^2}, \frac{2a+\sqrt{(4a^2+4ahm^2)}}{m}\right]$$

and
$$\left[\frac{(m^2 h+2a)-\sqrt{(4a^2+4ahm^2)}}{m^2}, \frac{2a-\sqrt{(4a^2+4ahm^2)}}{m}\right]$$

So distance PK^2

$$=\left[\left\{\frac{(m^2 h+2a)+\sqrt{(4a^2+4ahm^2)}}{m^2}-h+\frac{2a+\sqrt{(4a^2+4ahm^2)}}{m}-0\right\}^2\right]$$

$$=\left[\left\{\frac{2a+(4a^2+4ahm^2)}{m^2}\right\}^2+\left\{\frac{2a+(4a^2+4ahm^2)}{m}\right\}^2\right]$$

$$=\frac{\left\{2a+\sqrt{(4a^2+4ahm^2)}\right\}^2\times(1+m^2)}{m^4}$$

Similarly, $OK^2 = \dfrac{\left\{2a-\sqrt{(4a^2+4ahm^2)}\right\}^2\times(1+m^2)}{m^4}$

So $\dfrac{1}{PK^2}+\dfrac{1}{Q^2}=\dfrac{m^2}{1+m^2}\left[\left\{\dfrac{1}{2a+\sqrt{(4a^2+4ahm^2)}}+\right\}^2\right.$

$$\left.+\left\{\frac{1}{2a-\sqrt{(4a^2+4ahm^2)}}\right\}^2\right]$$

$$=\frac{m^4}{1+m^2}\left[\frac{\left\{2a-\sqrt{(4a^2+4ahm^2)}\right\}^2+\left\{2a+\sqrt{(4a^2+4ahm^2)}\right\}^2}{\left\{4a^2-(4a^2+4ahm^2)\right\}^2}\right]$$

$$= \frac{m^4}{1+m^2}\left[\frac{4a^2 + 4a^2 + 4ahm^2}{16a^2 h^2 m^2}\right] = \frac{2a+hm^2}{4ah^2\left(1+m^2\right)}.$$

If we put h = 2a, we get

$$\frac{1}{PK^2}+\frac{1}{QK^2}=\frac{2a+2am^2}{4a(2a)^2\left(1+m^2\right)}=\frac{2a\left(1+m^2\right)}{16a^3\left(1+m^2\right)}=\frac{1}{8a^2}$$

which is constant for all values of m. Hence for the point (2a, 0) the given property holds good. **Hence proved.**

Example 73:

Prove that the semi-latus-rectum is a harmonic mean between the segments of any focal chord.

Solution:

Let the equation of the parabola be $y^2 = 4ax$. ...(1)

The co-ordinates of focus are (a, 0). Let P be any point $(at_1^2, 2at_1)$ and Q be another point $(at_2^2, 2at_2)$, such that the line PQ passes through the focus S, then the equation to PQ will be

$$y - 2at_1 = \frac{2at_2 - 2at_1}{at_2^2 - at_1^2}\left(a - at_1^2\right)$$

$$\Rightarrow \qquad y - 2at_1 = \frac{2}{t_2 + t_1}\left(a - at_1^2\right)$$

If this line passes through the focus (a, 0), we must have

$$0 - 2at_1 = \frac{2}{t_2 + t_1}\left(a - at_1^2\right)$$

$$\Rightarrow \qquad t_1 t_2 = -1 \qquad \text{...(2)}$$

Distance SP = $\sqrt{[(at_1^2 - a)^2 + (2at_1)^2]} = (at_1^2 + a)$

Distance SQ = $\sqrt{[(at_2^2 - a)^2 + (2at_2)^2]} = (at_2^2 + a)$

The harmonic mean of SP and SQ will be

$$= \frac{2SP.SQ}{SP+SQ} = \frac{2\left(at_1^2 + a\right)\left(at_2^2 + a\right)}{\left(at_1^2 + a\right) + \left(at_2^2 + a\right)}$$

$$= \frac{2a^2\left(t_1^2 + 1\right)\left(\frac{1}{t_1^2} + 1\right)}{a\left(2 + t_1^2 + \frac{1}{t_1^2}\right)} \qquad \left[\because = -\frac{1}{t_1} \text{ by } (2)\right]$$

$$= \frac{2a\frac{1}{t_1^2}\left(t_1^2+1\right)^2}{\frac{1}{t_1^2}\left(t_1^2+1\right)^2} = 2a = \text{semi-latus rectum.}$$ **Hence proved.**

Example 74(a):

Prove that the normal chord at the point whose ordinate is equal to its abscissa subtends a right angle at the focus.

Solution:

Let the equation of the parabola be $y^2 = 4ax$ ...(1)

and say P is any point on (1) for which the abscissa is equal to the ordinate is (h, h). Putting these co-ordinates in (1), we get $h^2 = 4ah$ hence either $h = 0$ or $h = 4a$.

Hence the point is either origin at which the normal is x-axis or (4a, 4a) at which the normal is

$$y - 4a = -\frac{4a}{2a}(x-4a)$$

or $$y = -2x + 12a. \quad ...(2)$$

Putting the value of y in (1), we get

$$(-2x + 12a)^2 = 4ax$$

$\Rightarrow$ $$x^2 - 13ax + 36a^2 = 0$$

or $$(x - 9a)(x - 4a) = 0$$

whence either $x = 9a$ or $4a$

The corresponding values of y are $-6a$ and $4a$.

Hence the other point of interaction of the normal at $P \equiv (4a, 4a)$ to the parabola (1) is (9a, 6a) say Q.

Slope of the line SP where S is the focus (a, 0) is

$$\frac{4a-0}{2a-a} = \frac{4}{3} = m_1 \text{ (say)} \quad ...(3)$$

$$\text{slope of SQ} = \frac{-6a-0}{+9a-a} = -\frac{3}{4} = m_2 \text{ (say)} \quad ...(4)$$

Clearly $m_1 \times m_2 = -1$, so the SP and SQ are at right angles.

Hence proved.

Example 74(b):

Prove that the area of the triangle formed by the normals to the parabola at the points $(at_1^2, 2at_1)$, $(at_2^2, 2at_2)$ *and* $(at_3^2, 2at_3)$ *is*

$$\frac{a^2}{2}(t_2-t_3)(t_3-t_1)(t_1-t_2)(t_1+t_2+t_3)^2$$

Solution:

Let the equation of the parabola be $y^2 = 4ax$ and let these be any three points $(at_1^2, 2at_1)$, $(at_2^2, 2at_2)$ and $(at_3^2, 2at_3)$.

The equation of the normals at these points to the parabola are respectively

$$y = -t_1x + 2at_1 + at_1^2, \qquad ...(1)$$

$$y = -t_2x + 2at_2 + at_2^2, \qquad ...(2)$$

and $$y = -t_3x + 2at_3 + at_3^2, \qquad ...(3)$$

If (1) and (2) meet in A whose co-ordinates are (x_1, y_1), (2) and (3) in B whose co-ordinates are (x_2, y_2) and (3) and (1) in C whose co-ordinates are (x_3, y_3), then solving in pairs , we get

$$x_1 = 2a + a(t_1^2 + t_1t_2 + t_2^2),\ y_1 = -at_1t_2(t_1 + t_2),$$

$$x_2 = 2a + a(t_2^2 + t_2t_3 + t_3^2),\ y_2 = -at_2t_3(t_2 + t_3),$$

and $$x_3 = 2a + a(t_3^2 + t_3t_2 + t_1^2),\ y_3 = -at_3t_1(t_3 + t_1).$$

Then area of the triangle ABC will be

$$\frac{1}{2}\begin{vmatrix} a(2+t_1^2+t_1t_2+t_2^2) & -at_1t_2(t_1+t_2) & 1 \\ a(2+t_2^2+t_2t_3+t_3^2) & -at_2t_3(t_2+t_3) & 1 \\ a(2+t_3^2+t_3t_1+t_1^2) & -at_3t_1(t_3+t_1) & 1 \end{vmatrix}$$

Subtracting lst row from (2) and (3) and factorizing, we get

$$=-\frac{1}{2}a^2\begin{vmatrix} 2+t_1^2+t_1t_2+t_2^2 & t_1t_2(t_1+t_2) & 1 \\ (t_3-t_1)(t_3+t_2+t_1) & t_2(t_3-t_1)(t_1+t_2+t_3) & 0 \\ (t_3-t_2)(t_1+t_2+t_3) & t_1(t_3-t_2)(t_1+t_2+t_3) & 0 \end{vmatrix}$$

$$=-\frac{1}{2}a^2\begin{vmatrix} (t_3-t_1)(t_1+t_2+t_3) & t_3(t_2-t_1)(t_1+t_2+t_3) \\ (t_3-t_2)(t_1+t_2+t_3) & t_1(t_3-t_2)(t_1+t_2+t_3) \end{vmatrix}$$

$$=-\frac{1}{2}a^2(t_1+t_2+t_3)^2(t_2-t_1)(t_3-t_2)\begin{vmatrix} 1 & t_2 \\ 1 & t_1 \end{vmatrix}$$

$$=\frac{1}{2}(t_1-t_2)(t_2-t_3)(t_3-t_1)(t_1+t_2+t_3)^2$$

Proved.

Example 74(c):

The normal at a point P of a parabola meets the curve again in Q and T is the pole of PQ; show that T lies on the diameter passing through the other end of the focal chord passing through P, and that PT is bisected by the directrix.

Solution:

Let the equation of the parabola be $y^2 = 4ax$ and the co-ordinates of the point P be $(at_1^2, 2at_1)$ and of Q be $(at_2^2, 2at_2)$.

The equation of the line PQ the normal at P, will be

$$y = -t_1x + 2at_1 + at_1^3 \qquad ...(1)$$

If $T \equiv (\alpha, \beta)$ be the pole of PQ, then PQ will be the polar of T. Hence the equation of PQ must be

$$y\beta = 2a(x + \alpha). \qquad ...(2)$$

As (1) and (2) represent the same line PQ, so the coeffs. must be proportional, therefore,

$$\frac{\beta}{1} = \frac{2a}{-t_1} = \frac{2a\alpha}{2at_1 + at_1^2}. \qquad ...(3)$$

Let PR be the focal chord through P, where the co-ordinates of R are $(at_3^2, 2at_3)$. Then, the equation of PR will be

$$y - 2at_3 = \frac{2at_1 - 2at_3}{at_1^2 - at_3^2}\left(x - at_3^2\right)$$

or $$y - 2at_3 = \frac{2}{t_1 + t_3}\left(x - at_3^2\right)$$

As PR passes through the focus (a, 0), the co-ordinates will satisfy this equation.

Hence $$0 - 2at_3 = \frac{2}{t_1 + t_3}\left(a - at_3^2\right)$$

or $$t_3 = -\frac{1}{t_1}$$

So the co-ordinates of R become

$$\left(\frac{a}{t_1^2}, -\frac{2a}{t_1}\right)$$

The equation of the diameter through R will be

$$y = -\frac{2a}{t_1}. \qquad ...(4)$$

By relation no (3),

$$\alpha = \left(-2a - at_1^2\right) \text{ and } \beta = -\frac{2a}{t_1}.$$

Hence the co-ordinates of T are $(-2a - at_1^2, -2a/t_1)$.

As these co-ordinates satisfy (4), T lies on (4) i.e. the diameter through R.

The equation of the directrix of the parabola

$$x + a = 0. \quad ...(5)$$

Abscissa of the mid-point of PT is

$$\frac{1}{2}\left(at_1^2 - 2a - at_1^2\right) = -a.$$

As the abscissa satisfies (5), the equation of the directrix, the line PT is bisected by the directrix. **Hence proved.**

Example 75:

If T be any point on the tangent at any point P of a parabola, and if TL be perpendicular to the focal radius SP and TN be perpendicular to the directrix, prove that SL = TN.

Hence obtain a geometrical construction for the pair of tangents drawn to the parabola from any point T.

Solution:

Let the equation to the parabola be $y^2 = 4ax$ ...(1)

and say $(at^2, 2at)$ is any point P on it. The equation to the tangent at P will be

$$y.2at = 2a(x + at^2)$$

or

$$ty = x + at^2$$

If T is any point (x_1, y_1) on this line, then

$$ty_1 = x_1 + at^2. \quad ...(2)$$

Slope of the focal chord SP $= \dfrac{2at - 0}{at^2 - a} = \dfrac{2t}{t^2 - 1}$.

If TL is perpendicular to SP, then the equation of TL is

$$y - y_1 = -\frac{t^2 - 1}{2t}(x - x_1)$$

$$\Rightarrow \quad 2ty - 2ty_1 + x(t^2 - 1) - x_1(t^2 - 1) = 0.$$

As SL is perp. to TL, the length of the perp.

$$SL = \frac{-2ty_1 + a(t^2 - 1) - x_1(t^2 - 1)}{\sqrt{\left[4t^2 + (t^2 - 1)^2\right]}} \quad [\text{as} \equiv (a, 0)]$$

$$= \frac{-x_1 - at^2 - a - x_1 t^2}{(t^2+1)} = \frac{(x_1+a)(1+t^2)}{(1+t^2)}$$

Omitting the sign, we get SL = $(x_1 + a)$. ...(3)

If TN is the length of the perpendicular from T $\equiv (x_1, y_1)$ on the directrix $x = -a$, then clearly TN = $(x_1 + a)$. ...(4)

by (3) and (4), we get SL = TN **Proved.**

Again

(i) If T is any point, from T draw perpendicular TN on the directrix AB.

(ii) Join TS where S is the focus.

(iii) Draw a circle on TS as diameter.

(iv) With S as centre, distance equal to TN, cut off two arcs from the circles say at L and M.

(v) Join SL and produce to meet the parabola at Q.

(vi) Join SM and produce to meet the parabola at Q.

(vii) Join TP and TQ.

(viii) TP and TQ are required tangents.

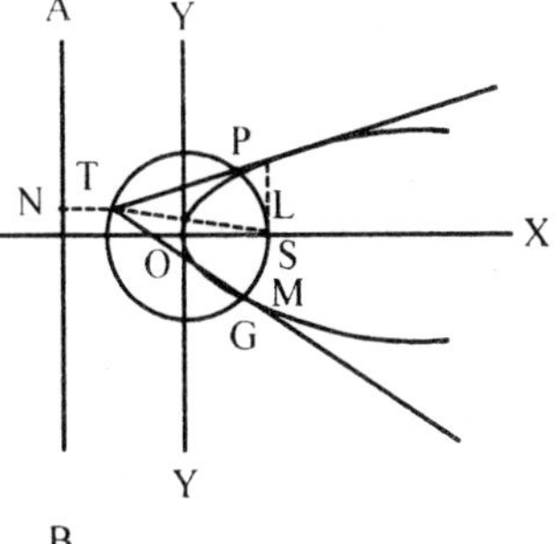

Example 76(a):

Parabola touches the sides of a triangle ABC in the points D. E and F respectively, prove that Bb and Cc are parallel.

Solution:

Let the parabola $y^2 = 4ax$ touch the sides BC, CA and AB of a triangle ABC at D, E and F respectively, and suppose the co-ordinates of the points D, E and F be $(at_1^2, 2at_1)$, $(at_2^2, 2at_2)$ and $(at_3^2, 2at_3)$ respectively.

So the tangents at D, E and F will represent the sides BC, CA and AB respectively.

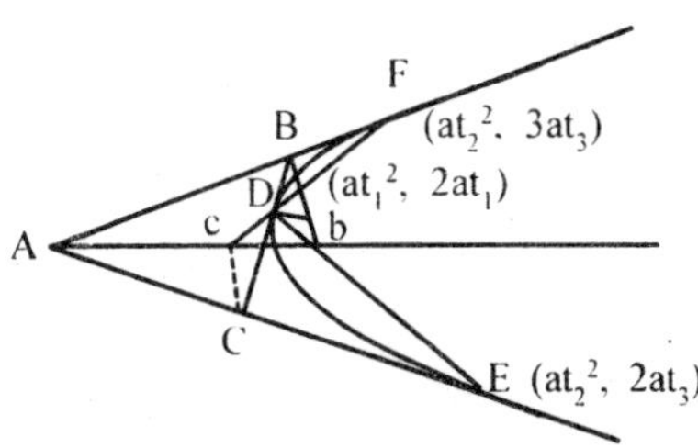

Tangents at the points D, E and F are respectively

$$t_1 y = x + at_1^2, \quad ...(1)$$

$$t_2 y = x + at_2^2, \quad ...(2)$$

and $$t_3 y = x + at_3^2, \quad ...(3)$$

On solving (2) and (3), (3) and (1) and (1) and (2) in pairs, we get the co-ordinates of A, B and C as $\{at_2t_3, a(t_2 + t_3)\}$;$\{at_1t_3, a(t_1 + t_3)\}$ and $\{at_1t_2, a(t_1 + t_2)\}$ respectively. Now the diameter through the point A will be $y = a(t_1 + t_2)$ and equation of DE is $y(t_1 + t_2) = 2x + 2at_1t_2$ solving these equation, we have the co-ordinates of point of intersection b as $\{(a/2)\{(t_1^2 + t_2t_3 + t_1t_3 - t_1t_2), a(t_1 + t_2)\}$.

Again the equation of DF is $y(t_1 + t_3) = 2x + at_1t_3$; this cuts the diameter through A in c

By solving the two equation the co-ordinates of c are

$$\{(a/2)\{(t_2^2 + t_2t_3 + t_1t_2 - t_1t_3), a(t_2 + t_3)\}.$$

If slope of Bb m_1 and that of Cc be m_2, then

$$m_1 = \frac{2a(t_2 - t_1)}{a(t_2^2 + t_2t_3 - t_1t_3 - t_1t_2)}$$

$$= \frac{2a(t_2 - t_1)}{a(t_1 + t_3)(t_2 - t_1)} = \frac{2}{(t_2 + t_3)}$$

Similarly slope of Cc is $m_2 = \dfrac{2}{(t_2 + t_3)}$

Since $m_1 = m_2$ hence the lines Cc and Bb are parallel. **Proved.**

Example 76(b):

T is the pole of the chord PQ; prove that the perpendicular from P, T and Q upon any tangent to the parabola are in geometrical progression.

Solution:

Let the equation of the parabola of $y^2 = 4ax$ and P and Q be any two points on it having the co-ordinates as $(at_1^2, 2at_1)$ and $(a_2^2, 2at_2)$ respectively; then the equation of the tangent at P will be and at Q will be

$$t_1 y = x + at_1^2 \quad ...(1)$$

$$t_2 y = x + at_2^2. \quad ...(2)$$

On solving (1) and (2); we get the co-ordinates of T as

$$(at_1t_2, a(t_1 + t_2)\}.$$

Any tangent of the parabola is y = mx + (a/m). ...(3)

Length of the perpendicular TM from P on (3) is given by

$$PL = \frac{mat_1^2 - 2at_1 + a/m}{\sqrt{(1+m^2)}} = \frac{a(mt_1 - 1)^2}{m\sqrt{(1+m^2)}}$$

Length of the perpendicular TM from T on (3) is given by

$$TM = \frac{mat_1t_2 - a(t_1 + t_2) + a/m}{\sqrt{(1+m^2)}} = \frac{a(mt_1 - 1)(mt_2 - 1)}{m\sqrt{(1+m^2)}}$$

and perpendicular QN from Q on (3) (as in the case of PL) is given by

$$QN = \frac{a(mt_1 - 1)^2}{\sqrt{m(1+m^2)}}$$

$$\text{Now } LP \times QN = \frac{a^2(mt_2 - 1)^2(mt_1 - 1)^2}{m^2(1+m^2)} = TM^2$$

Hence PL, TM, CN are in GP **Proved.**

Example 77:

If from the vertex of a parabola a pair of chords be drawn at right angles to one another and with these chords as adjacent sides a rectangle be made, prove that the locus of the further angle of the rectangle is the parabola $y^2 = 4a(x - 8a)$.

Solution:

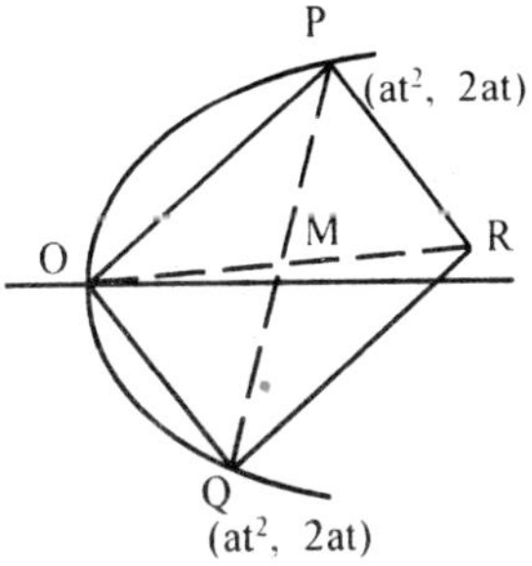

Let the equation of the parabola be $y^2 = ax$ and O be the vertex of it. If OP and OQ be the two adjacent sides of the rectangle OPRQ and the co-ordinates of P and Q be $(at_1^2, 2at_1)$ and $(at_2^2, 2at_2)$ resp; then the equation of the line PQ will be

$$y - 2at_1 = \frac{2at_2 - 2at_1}{at_2^2 - at_1^2}(x - at_1^2)$$

$$\Rightarrow \qquad y(t_1 + t_2) = 2x + 2at_1t_2$$

$$\Rightarrow \qquad \{y(t_1 + t_2) - 2x\}/2at_1t_2 = 1 \qquad ...(1)$$

The equation of the line joining the origin to the points of intersection of the parabola and (1) are obtained by making $y^2 = 4ax$ homogenous with the help of (1); so, we get

$$y^2 = 4a.x.\frac{y(t_1+t_2)-2x}{2at_1t_2}$$

$$2at_1t_2y^2 - 4a\,(t_1 + t_2)\,xy + 8ax^2 = 0.$$

If the lines are at right angles, then the sum of the coeff. of x^2 and y^2 should be zero, hence

$$2at_1t_2 + 8a = 0$$

or $$t_1t_2 + 4 = 0 \qquad ...(2)$$

If M is the mid-point of PR, then its co-ordinates will be

$$\left(\frac{at_1^2+at_3^2}{2}, \frac{2at_1+2at_3}{2}\right)$$

Let the co-ordinates of the reqd, point R be (h, k) then M will be the mid-point of OA; also as CPRQ is a rectangle, so, we get

$$\frac{h+0}{2} = \frac{at_1^2+at_2^2}{2} \quad \text{or} \quad h = a\left(t_1^2+t_2^2\right) \qquad ...(3)$$

and $$\frac{k+0}{2} = \frac{at_1+2at_2}{2} \quad \text{or} \quad k = 2a\left(t_1+t_2\right) \qquad ...(4)$$

To get the reqd. locus, we have to eliminate t_1 and t_2 from (2), (3) and (4), so squaring (4), we get

$$k^2 = 4a^2\,(t_1^2 + t_2^2 + 2t_1t_2)$$

$$= 4a^2\left\{\frac{h}{a}+2.(-4)\right\},$$

$$\{\text{as } t_1^2 + t_2^2 = h/a \text{ by (1)}$$

and $$2t_1t_2 = -4 \text{ by (2)}\}$$

$\Rightarrow$ $$k^2 = 4ah - 32a^2$$

Generalising, we get the reqd. locus as

$$y^2 = 4a\,(y - 8a).$$ **Proved.**

Example 78:

Prove that the locus of the poles of chords which subtend a right angle at a fixed point (h, k) is

$$ax^2 - hy^2 + (4a^2 + 2ah)\,x - 2aky + a\,(h^2 + k^2) = 0.$$

Solution:

Let $(at_1^2, 2at_1)$ and $(at_2^2, 2at_2)$ be the ends of the chord PQ, with respect to parabola $y^2 = 4ax$, and (x_1, y_1) be its, pole.

Equation of PQ will be $y - 2at_1 = \dfrac{2at_2 - 2at_1}{at_2^2 - at_1^2}\left(x - at_1^2\right)$

$\Rightarrow \quad y(t_1 + t_2) = 2ax + at_1t_2$ ...(1)

Again polar of (x_1, y_1) w.r.t. the parabola $y^2 = 4ax$ is

$yy_1 = 2a(x + x_1)$ ...(2)

As (1) and (2) represent the same straight line, their coeffs. will be proportional. Hence comparing the coeffs.

$$\frac{y_1}{t_1 + t_2} = \frac{2a}{2} = \frac{2ax_1}{2at_1t_2}$$

whence $x_1 = at_1t_2$ ...(3)

and $y_1 = a(t_1 + t_2)$. ...(4)

Let (h, k) be any point A. The slopes of PA and PB are respectively

$$\frac{k - 2at_1}{h - at_1^2} \text{ and } \frac{k - 2at_2}{h - at_2^2}$$

If PAQ = 90°, we have $\dfrac{k - 2at_1}{h - at_1^2} \times \dfrac{k - 2at_2}{h - at_2^2} = -1$

or $k^2 - 2ak(t_1 + t_2) + 4a^2t_1t_2 + h^2 - a(t_1^2 + t_2^2)h + a\, a^2t_1^2t_2^2 = 0$ ...(5)

From (3) and (4), $a^2(t_1^2 + t_2^2) = y_1^2 - 2ax_1$. ...(6)

Eliminating t_1 and t_2 from (3), (4), (5) and (6), we have

$$k^2 - 2ky_1 + 4ax_1 + h^2 - h\left(\frac{y_1^2 \quad 2ax_1}{a}\right) + x_1^2 = 0$$

Generalising and simplifying, the locus of the pole (x_1, y_1) is given by

$$ax^2 + hy^2 + (4a^2 + 2ah)x - 2aky + a(h^2 + k^2) = 0.$$

Proved.

Example 79:

If r_1 and r_2 be the lengths of radii vectors of the parabola which are drawn at right angles to one another from the vertex, prove that

$$r_1^{4/3} r_2^{4/3} = 16a^2 (r_1^{2/3} + r_2^{2/3}).$$

Solution:

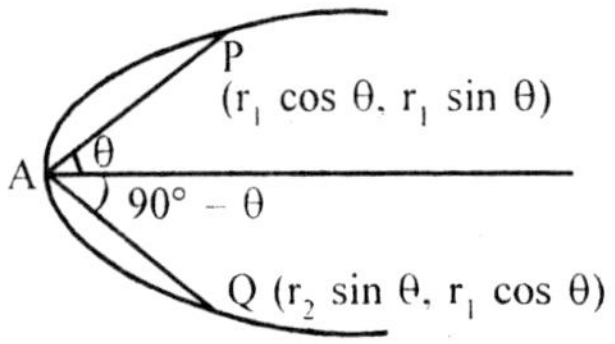

Let the parabola be $y^2 = 4ax$ ...(1)and let P and Q be any two points on it such that the co-ordinates of P and Q are $(r_1 \cos\theta, r_1 \sin\theta)$ and $(r_2 \sin\theta, r_2 \cos\theta)$ respectively.

$(\because \angle PAQ = 90^\circ)$.

Since the points lie on (1), the co-ordinates will satisfy it.

So $r_1^2 \sin^2\theta = 4 ar_1 \cos\theta$

$\Rightarrow \qquad r_1 = 4a \cos\theta/\sin^2\theta$

and $r_1^2 \cos^2\theta = 4ar_2 \sin\theta$

$\Rightarrow \qquad r_2 = 4a \sin\theta/\cos^2\theta.$

$$\text{Now} \qquad r_1 r_2 = \frac{16a^2 \cos\theta \sin\theta}{\sin^2\theta \cos^2\theta} = \frac{16a^2}{\sin\theta \cos\theta}$$

$$\Rightarrow \qquad r_1^{2/3} r_2^{4/3} = \frac{\left(16a^2\right)^{4/3}}{\sin^{4/3}\theta \cos^{4/3}\theta} \qquad ...(2)$$

$$\text{and } r_1^{2/3} r_2^{2/3} = (4a)^{2/3} \left\{ \frac{\cos^{2/3}\theta}{\sin^{4/3}\theta} + \frac{\sin^{2/3}\theta}{\cos^{4/3}\theta} \right\}$$

$$\text{or} \qquad r_1^{2/3} r_2^{2/3} = \frac{(4a)^{2/3}}{\sin^{4/3}\theta \cos^{4/3}\theta} \left(\sin^2\theta + \cos^2\theta\right)$$

$$= \frac{\left(16a^2\right)^{1/3}}{\sin^{4/3}\theta \cos^{4/3}\theta} \qquad ...(3)$$

Now dividing (2) by (3), we have

$$\frac{r_1^{4/3} r_2^{4/3}}{\left(r_1^{2/3} + r_2^{2/3}\right)} = 16a^2; \text{ or } r_1^{4/3} r_2^{4/3} = 16a^2 \left(r_1^{2/3} = r_2^{2/3}\right)$$

Proved.

Example 80(a):

A parabola is drawn such that each vertex of a given triangle is the pole of the opposite side; show that the focus of the parabola lies on the nine-point circle of the triangle and that the orthocentre of the triangle formed by joining the middle points of the sides lies on the directrix.

Solution:

The circle circumscribing the triangle formed by joining the middle points of the sides of the given triangle will be the nine point circle. As the circle circumscribing the triangle formed by the three tangents on the parabola always passes through the focus of the parabola (Ref. Art.179), hence, we have to prove that the lines joining the mid-points of the given triangle are the tangents on the parabola.

Let the equation of the parabola be $y^2 = 4ax$. ...(1)

Suppose the given triangle is ABC and let PQ be the chord of contact of A w.r.t. (1). Again, as BC is given to the polar of A w.r.t. (1), BC and PQ must lie on the same line. Hence the line joining the mid. points of AP and AQ also passes through the mid. points of AB & AC.

Suppose the co-ordinates of P and Q be $(at_1^2, 2at_1)$ and $(at_2^2, 2at_2)$ respectively so that the co-ordinates of the point of intersection of the tangents at P and Q i.e. A are $[at_1t_2, a(t_1 + t_2)]$

$\therefore$ Co-ordinates of mid-point of AP will be

$$\left\{\frac{at_1(t_1+t_2)}{2}, \frac{a(3t_1+t_2)}{2}\right\}$$

Similarly, the co-ordinates of mid, point of AQ will be

$$\left\{\frac{at_2(t_1+t_2)}{2}, \frac{a(t_1+3t_2)}{2}\right\}$$

The line joining the mid-points of AP and AQ is

$$y-\frac{a(3t_1+t_2)}{2}=\frac{\frac{a(3t_1+t_2)}{2}-a\frac{(t_1+3t_2)}{2}}{\frac{at_1(t_1+t_2)}{2}-\frac{at_2(t_1+t_2)}{2}}\left\{x-\frac{at_1(t_1+t_2)}{2}\right\}$$

$$\Rightarrow \qquad y-\frac{2}{t_1+t_2}x+\frac{a(t_1+t_2)}{2}$$

This is a tangent on the parabola as it is of the form $y = mx + a/m$

Hence proved.

Again the ortho-centre of the triangle formed by the tangents is on directrix (art. 181) and we have proved that the lines joining the mid. points of the sides of the triangle are tangents on the parabola.

Hence the ortho-centre of the triangle formed by joining the mid. Point of the sides of the triangle lies on directrix of the parabola. **Proved.**

Example 80(b):

Prove that the equation to the circle passing through the points $(at_1^2, 2at_1)$ and $(at_2^2, 2at_2)$ and the intersection of the tangents to the parabola at these points is

$$x^2 + y^2 - ax\,[(t_1 + t_2)^2 + 2] - ay\,(t_1 + t_2)\,(1 - t_1t_2) + a^2t_1t_2\,(2 - t_1t_2) = 0.$$

Solution:

Let $y^2 = 4ax$ be the parabola on which the points P $(at_1^2, 2at_1)$ and Q $(at_2^2, 2at_2)$ lie. The equation of the tangent at P will be

$$t_1y = x + at_1^2. \qquad ...(1)$$

and at Q will be $\quad t_2y = x + at_2^2. \qquad ...(2)$

Solving (1) and (2), we get the co-ordinates of T, the point of intersection of tangents as $[at_1t_2, a\,(t_1 + t_2)$. Now all these three points lie on the circle.

Let the equation of the circle be

$$x^2 + y^2 + 2gx + 2fy + c = 0. \qquad ...(3)$$

As T, P and Q lie on (3), the co-ordinates will satisfy it, hence taking one by one, we get

$$a^2t_1^2t_2^2 + a^2\,(t_1 + t_2)^2 + 2agt_1t_2 + 2af\,(t_1 + t_2) + c = 0, \qquad ...(4)$$

$$a^2t_1^4 + 4a^2t_1^2 + 2agt_1^2 + 4aft_1 + c = 0 \qquad ...(5)$$

$$a^2t_2^4 + 4a^2t_2^2 + 2agt_2^2 + 4aft_2 + c = 0 \qquad ...(6)$$

Now we have to solve these 3 equations.

From the sum of (5) and (6), subtracting twice equation (1), we get

$$a\,(t_1 + t_2)^2 + 2a + 2g = 0$$

or $\quad 2g = -\,[\,(t_1 + t_2)^2 + 2]$

Again subtracting (6) from (5); we get

$$a\,(t_1 + t_2)\,(t_1^2 + t_2^2) + 4a^2\,(t_1 + t_2) + 2g\,(t_1 + t_2)^2 = -\ 4f.$$

Substituting for g, we have $2f = -\,(t_1 + t_2)\,(1 - t_1t_2)$.

Substituting for 'g' and 'f' in (5), we have

$$a^2t_1^4 + 4a^2t_1^2 - a^2t_1t_2(t_1^2 + t_2^2 + 2t_1t_2 - 2a^2t_1^2 - 2a^2t_1^2 - (1 - t_1t_2) \times (t_1 + t_2) + c = 0$$

Whence $c = a^2t_1t_2\,(2 - t_1t_2)$

Putting these values in (3), we get the equation to the circle as

$$x^2 + y^2 - ax\,(t_1 + t_2)^2 + 2] - ay\,(t_1 + t_2)\,(1 - t_1t_2) + a^2t_1t_2\,(2 - t_1t_2) = 0.$$

Example 80(c):

Through the vertex A of the parabola $y^2 = 4ax$ two chords AP and AQ are drawn and the circles on AP and AQ as diameters intersect in R. Prove that, if θ_1, θ_2 and ϕ be the angles made with the axis by the tangents at P and Q and by AR, then

$$\cot \theta_1 + \cot \theta_2 + 2 \tan \phi = 0.$$

Solution:

Let the equation of the parabola by $y^2 = 4ax$. Its vertex (say A) will be (0, 0).

Let P and Q be any two points on the parabola as

$$(at_1^2, 2at_1) \text{ and } (at_2^2, 2at_2)$$

Then the equation of the circle drawn on AP as diameter will be

$$x(x - at_1^2) + y(y - 2at_1) = 0$$

$$\Rightarrow \quad x^2 + y^2 - axt_1^2 - 2ayt_1 = 0 \quad ...(1)$$

Similarly the equation of the circle having AQ as diameter is

$$x^2 + y^2 - axt_2^2 - 2ayt_2 = 0 \quad ...(2)$$

(1) and (2) intersect at R and pass through A : so AR will be the common chord.

Equation of common chord AR will be

$$(x^2 + y^2 - axt_1^2 - 2ayt_1) - (x^2 + y^2 - axt_2^2 - 2ayt_2) = 0$$

$$\Rightarrow \quad x(t_1 + t_2) + 2y = 0$$

Hence the slope of AR $= -\dfrac{t_1 + t_2}{2} = \tan\phi$ (by hypothesis). ...(3)

Again tangent at P is $t_1 y = x + at_1^2$.

$\therefore$ Its slope $= 1/t_1 = \tan\theta_1$ (by hypothesis).

So $t_1 = \cot\theta_1$ ...(4)

Tangent at Q is $t_2 y = x + at_2^2$.

$\therefore$ Its slope $= 1/t_2 = \tan\theta_2$ (by hypothesis).

So $t_2 = \cot\theta_2$ ...(5)

By (3), (4) and (5)

$$2\tan\phi = -(t_1 + t_2) = -\cot\theta_1 - \cot\theta_2$$

$$\Rightarrow \quad \cot\theta_1 + \cot\theta\theta_2 + 2\tan\phi = 0.$$ **Proved.**

Example 80(d):

Prove that the orthocentres of the triangles formed by three tangents and the corresponding three normals to a parabola are equidistant from the axis.

Solution:

Let the parabola by $y^2 = 4ax$ ...(1)

and $(am_1^2, -2am_1)$, $(am_2^2, -2am2)$ and $(am_3^2, -2am_2)$ be the co-ordinates of the points P, Q and R respectively. Then the equations of the tangents at P, Q and R w.r.t. the parabola (1) are

$$-m_1y = x + am_1^2, \quad ...(2)$$

$$-m_2y = x + am_2^2, \quad ...(3)$$

$$-m_3y = x + am_2^2, \quad ...(4)$$

Solving (2) and (3), we have the co-ordinates of the point of intersection say A as $\{am_1m_2, -a(m_1 + m_2)\}$.

Now equations to the line through A and perpendicular to (3) will be

$$y + a(m_1 + m_2) = m_3(am_1m_3) \quad ...(5)$$

Similarly the equation to the line through B the point of intersection (3) and (4) and perpendicular to (2) is

$$y + a(m_2 + m_3) = m_1(x - am_2m_3) \quad ...(6)$$

On solving (5) and (6), we get the ordinate of the orthocentre of the triangle formed by (2), (3) and (4) as

$$-a(m_2 + m_2 + m_3 + m_1m_2m_3)$$

Again equations to the normal at P, Q and R are respectively

$$y = m_1x - 2am_1 - am_1^3, \quad ...(7)$$

$$y = m_2x - 2am_2 - am_2^3, \quad ...(8)$$

and $$y = m_3x - 2am_3 - am_3^2, \quad ...(9)$$

Solving (8) and (7), we have the co-ordinates of one of the vertice of the triangle formed by these three normals as

$$\{2a + a(m_1^2 + m_2^2 + m_1m_2), am_1m_2(m_1 + m_2)\}$$

Equation to the line through this point and perpendicular to (9) is

$$y - am_1m_2(m_1 + m_2) = -\frac{1}{m_2}\left(x - 2a - am_1^2 - am_2^2 - am_1m_2\right) \quad ...(10)$$

Similarly the equation to the other perpendicular will be

$$y - am_2m_3(m_2 + m_3) = -\frac{1}{m_1}\left(x - 2a - am_2^2 - am_3^2 - am_2m_3\right) \quad ...(11)$$

On solving (10) and (11), we get the ordinate of the point of intersection as $-a(m_1 + m_2 + m_3 + m_1m_2m_3)$.

This is also the ordinate of the point of intersection or ortho-centre of the triangle formed by tangents.

Hence the ortho-centres are equidistant from the axis i.e. x-axis.

Example 81:

A series of chords is drawn so that their projections on a straight line which is inclined at an angle α to the axis are all of constant length c; prove that the locus of their middle point is the curve

$$(y^2 - 4ax)\ (y \cos \alpha + 2a \sin \alpha)^2 + a^2c^2 = 0.$$

Solution:

Let the equation of the parabola be $y^2 = 4ax$ ...(1)

and of the straight line having (x_1, y_1) as its mid-point be

$$\frac{x - x_1}{\cos\theta} = \frac{y - y_1}{\sin\theta} = r. \qquad ...(2)$$

By (2), we get $x = x_1 + r \cos \theta$ and $y = y_1 + r \sin \theta$.

Solving (1) and (2), we get

$$(y_1 + r \sin \theta)^2 = 4a\ (x_1 + r \cos \theta)$$

$$\Rightarrow \quad r^2 \sin^2 \theta + 2r\ (y_1 \sin \theta - 2a \cos \theta) + (y_1{}^2 - 4ax_1) = 0 \qquad ...(3)$$

(3) is quadratic in r.

The roots of this quadratic equation will be equal but of opposite sign ax (x_1, y_2) is the mid-point of the chord. So the coeff. of r must be zero.

Hence $\quad 2\ (y_1 \sin \theta - 2a \cos \theta) = 0$

$$\Rightarrow \quad \cot\theta = \frac{y_1}{2a} \qquad ...(4)$$

Again as the coeff. of r is zero, (iii) becomes

$$r^2 \sin^2 \theta = 4ax_1 - y_1^2 \quad \text{or} \quad r = \frac{\sqrt{\left(4ax_1 - y_1^2\right)}}{\sin\theta} \qquad ...(5)$$

Now length of the chord will be 2r. Let this chord and the given line make angle θ and α respectively with the axis.

Therefore, angle between the lines $= \theta - \alpha$, and the projection of the chord on the given line will be

$$2r \cos (\theta - \alpha) = c, \text{ (by hypothesis).}$$

So
$$2\sqrt{\left(\frac{4ax_1 - y_1^2}{\sin\theta}\right)} \cos(\theta - \alpha) = c, \qquad \text{[by (5)]}$$

$$\Rightarrow \quad 2\sqrt{\left(4ax_1 - y_1^2\right)}\ \frac{\cos\theta \cos\alpha + \sin\theta \sin a}{\sin\theta} = c$$

$$\Rightarrow \quad 2\sqrt{(4ax_1 - y_1^2)}\ (\cot \theta \cos \alpha + \sin \alpha) = c.$$

Putting the value of cot q from (4), we get

$$2\sqrt{\left(4ax_1 - y_1^2\right)}\left(\frac{y_1 \cos\alpha}{2a} + \sin\alpha\right) = c$$

$\Rightarrow$ $\sqrt{(4ax_1 - y_1^2)}\,(y.\cos\alpha + 2a\sin\alpha) = ac.$

Squaring, we have

$(4ax_1 - y_1^2)(y_1 \cos\alpha + 2a\sin\alpha)^2 = a^2c^2$

$\Rightarrow$ $(y_1^2 - 4ax_1)(y_1 \cos\alpha + 2a\sin\alpha)^2 = a^2c^2 = c.$

Generalising this for (x_1, y_1), we get

$(y^2 - 4ax)(y\cos\alpha + 2a\sin\alpha)^2 = a^2c^2 = c.$ **Proved.**

Example 82:

Prove that all circles described on focal chords as diameters touch the directrix of the curve and that all circles on focal radii as diameters touch the tangent at the vertex.

Solution:

(a) Let the equation of the parabola be $y^2 = 4ax$.

The co-ordinates of the focus S will be (a, 0). Let PSQ be any focal chord, where the co-ordinates of P and Q are $(at_1^2, -2at1)$ and $(at_2^2, -2at_2)$ respectively.

Equation of PQ is $y(t_1 + t_2) = 2x + 2at_1t_2$. As it passes through (a, 0) the co-ordinates will satisfy it, hence

$0 = 2a + 2at_1t_2$ or $t_1t_2 = -1$ $\therefore$ $t_2 = -1/t_1$.

So the co-ordinates of Q become $(a/t_1^2 - 2a/t_2)$.

Equation of circle drawn with PQ as diameter is

$(x - at_1^2)(x - a/t_1^2) + (y - 2at_1)(y + 2a/t_1) = 0$ ...(1)

The equation of directrix is $x = -a$. ...(2)

Solving (1) and (2) simultaneously, we have

$(-a - at_1^2)(-a - a/t_1^2) + (y - 2at_1)(y + 2a/t_1) = 0$

$\Rightarrow$ $y^2 - 2ay(t_1 - 1/t_1) + a^2(t_1 - 1/t_1)^2 = 0$

$$\Rightarrow \quad \left\{y - a\left(t_1 - \frac{1}{t_1}\right)\right\}^2 = 0$$

As this is a perfect square, it will give us only one value of y. So (2) cuts (1) at two coincident points. Hence (2) is tangent to (1). **Proved.**

(b) Equation of circle drawn with line PS as diameter is

$$(x - at_1^2)(x - a) + (y - 2at_1)(y - 0) = 0$$

{as P is $(at_1^2, 2at_1)$ and S is $(a, 0)$}

$$\Rightarrow \quad x^2 + y^2 - ax(1 + t_1^2) - 2at_1y + a^2t^2 = 0 \quad ...(3)$$

Equation to the tangent at vertex is $x = 0$.

Solving (3) and (4), we get

$$y^2 - 2ayt_1 + a^2t^2 = 0 \quad \text{or} \quad (y - at_1)^2 = 0$$

which is again a perfect square; hence the tow roots of y are coincident; therefore, the y-axis touches the circles, i.e. the tangent at the vertex touches (3). **Proved.**

Example 82(a):

A circle and a parabola intersect in four points; show that the algebraic sum of the ordinates of the four points is zero.

Show also that the line joining one pair of these pour points and the line joining the other pair are equally inclined to the axis.

Solution:

Let the equation to the circle be

$$x^2 + y^2 + 2gx + 2fy + c = 0 \quad ...(1)$$

and of the parabola be $y^2 = 4ax$. ...(2)

On solving (1) and (2), we have [by (2) $x = y^2/4a$]

$$\frac{y^4}{16a^2} + y^2 + \frac{gy^2}{2a} + 2fy + c = 0 \quad ...(3)$$

This is a fourth-degree equation in y. Let its roots be y_1, y_2, y_3 and y_4. Then sum of the roots

$$= -\frac{\text{coeff. of } y^3}{\text{coeff. of } y^4}.$$

Hence $\quad y_1 + y_2 + y_3 + y_4 = \dfrac{0}{1/16a^2} = 0$

Hence the sum of the ordinates of four points in which the two curves intersect is zero. Proved.

Again, let the points of intersection by P, Q, R and S having the co-ordinates as (x_1, y_1), (x_2, y_2), (x_3, y_3) and (x_4, y_4) resp. As all these points lie on the curve $y^2 = 4ax$, we have

Subtracting, we get $y_1^2 - y_2^2 = 4a(x_1 - x_2)$

$$\Rightarrow \quad \frac{y_1 - y_2}{x_1 - x_2} = \frac{4}{y_1 + y_2}. \quad ...(4)$$

Now slope of the line PQ is say

$$m_1 = \frac{y_1 - y_2}{x_1 - x_2} = \frac{4a}{y_1 + y_2}. \qquad \text{[by (4)] ...(5)}$$

Similarly the slope of RS is say

$$m_2 = \frac{y_3 - y_4}{x_3 - x_4} = \frac{4a}{y_3 + y_4}. \qquad \text{...(6)}$$

As $y_1 + y_2 + y_3 + y_4 = 0$, so $y_3 + y_4 = -(y_1 + y_2)$

Putting in (6), $m_2 = \dfrac{4a}{-(y_1 + y_2)} = -m_1$

So m_1 and m_2 are equal in magnitude; hence pQ and RS are equally inclined to axis.

Example 82(b):

TP and TQ are tangents to the parabola and the normals are P and Q meet at a point R on the curve; prove that the centre of the circle circumscribing the triangle TPQ lies on the parabola

$$2y^2 = a(x - a).$$

Solution:

Let the parabola be $y^2 = 4ax$ and co-ordinates of P and Q be $(at_1^2, -2at_1)$ and $(at_2^2, -2at_2)$ respectively. At Q will be $t_2y = x + at_2^2$. Solving these two co-ordinates of the point of intersection T are $\{at_1t_2, a(t_1 + t_2)\}$. Now normal at P is $y = -t_1x + 2at_1 + at_1^3$.

Let this normal pass through R $(at_3^2, -2at_3)$ on the parabola. Then

$$t_1 + \frac{2}{t_1} = -t_3 \qquad \text{...(1)}$$

Similarly if normal at Q passes through R, $t_2 + \dfrac{2}{t_2} = -t_3$ ...(2)

So $t_1 + \dfrac{2}{t_2} = t_2 + \dfrac{2}{t_2} = -t_3$

Subtracting, we get $(t_1 - t_2) + \dfrac{2}{t_1t_2}(t_2 - t_1) = 0$

$$\Rightarrow \quad (t_1 - t_2)\left(1 - \frac{2}{t_1t_2}\right) = 0$$

So either $t_1 = t_2$ which means P and Q coincide which is impossible

$$\Rightarrow \quad 1 - \frac{2}{t_1t_2} = 0$$

$$\Rightarrow \quad t_1t_2 = 2. \qquad \text{...(3)}$$

Again let the centre of the circle passing through P, Q and T be (x, y): we have

$$2x = a(t_1 + t_2)^2 + 2a \qquad ...(4)$$

and $$2y = a(t_1 + t_2)(1 - t_1t_2) \qquad ...(5)$$

Now, we have to eliminate $'t'_1$ and $'t'_2$ from these relations to find the locus of the centre of the circle.

By (5) and (3), we get

$$2y = a(t_1 + t_2)(1 - 2) = -a(t_1 + t_2); \quad \text{or} \quad t_1 + t_2 = \frac{-2y}{a}$$

Putting in (4), we get

$$2x = a\left(\frac{-2y}{a}\right)^2 + 2a \quad \text{or} \quad 2x = \frac{4y^2}{a} + 2a.$$

$$2y^2 = a(x - a).$$ **Proved.**

Example 83:

Prove that the locus of the middle points of all tangents drawn from points on the directrix to the parabola is

$$y^2(2x + a) = a(3x + a)^2.$$

Solution:

Let P $(at^2, 2at)$ be any point on the parabola $y^2 = 4ax$, whose directrix is the line

$$x + a = 0. \qquad ...(1)$$

Now tangent at P is $ty = x + at^2$. ...(2)

Solving (1) and (2), the co-ordinates of point of intersection are

$$\left(-a, \frac{at^2 - a}{t}\right)$$

Now, if (x, y) be the co-ordinates of the middle point of the tangent, then

$$2x = at^2 - a \qquad ..(3)$$

and $$2y = \frac{at^2 - a}{t} + 2at$$

or $$2ty = 3at^2 - a.$$

Also $$6x = 3at^2 - 3a$$

Hence $$2ty = 6x + 2a$$

or $$ty = 3x + a \qquad ...(4)$$

Eliminating t between (3) and (4), we have

$$(2x + a)\, y^2 = a\, (3x + a)^2$$

Example 84:

A circle is described on a focal chord as diameter; if m be the tangent of the inclination of the chord to the axis, prove that the equation to the circle is

$$x^2 + y^2 - 2ax\left(1 + \frac{1}{m^2}\right) - \frac{4ay}{m} - 3a^2 = 0.$$

Solution:

Let the equation to the parabola be $y^2 = 4ax$.

If PSQ is any focal chord where S is the focus (a, 0) and P is any point $(at^2, 2at)$, the co-ordinates of Q will be $(a/t^2, -2a/t)$.

Now equation of the circle drawn with PQ as diameter is

$$(x - at^2)\,(x - a/t^2) + (y - 2at)\,(y + 2a/t) = 0$$

$\Rightarrow$ $$x^2 + y^2 - ax\,(t^2 + 1/t^2) - 2ay\,(t - 1/t) - 3a^2 = 0 \qquad ...(1)$$

The slope of PQ = m (by hypothesis).

Hence $$m = \frac{2at + 2a/t}{at^2 - a/t^2} = \frac{2t}{t^2 - 1}$$

$\Rightarrow$ $$2/m = (t - 1/t). \qquad ...(2)$$

Squaring, we get

$$\frac{4}{m^2} = t^2 + \frac{1}{t^2} - \quad \text{or} \quad t^2 + \frac{1}{t^2} = \left(\frac{4}{m^2} + 2\right).$$

Now eliminating t from equations (1) and (2), we have

$$x^2 + y^2 - ax\,(4/m^2 + 2) - 2ay\,(2/m) - 3a^2 = 0$$

$\Rightarrow$ $$x^2 + y^2 - 2ax\,(1 - 2/m^2) - (4ay/m) - 3a^2 = 0$$

Proved.

Example 85(a):

Circles are drawn through the vertex of the parabola to cut the parabola orthogonally at the either point of intersection. Prove that the locus of the centres of the circles is the curve

$$2y^2\,(2y^2 + x^2 - 12ax) = ax\,(3x - 4a)^2$$

Solution:

Let P $(at^2, 2at)$ be any point on the parabola $y^2 = 4ax$. The vertex A will be (0, 0). The equation of the tangent at p is

$$ty = x + at^2 \quad ...(1)$$

and the co-ordinates of the middle point of AP are $(at^2/2, at)$.

The slope of AP will be $\dfrac{2at-0}{at^2-0} = \dfrac{2}{t}$

The equation to the line through the point $\left(\dfrac{at^2}{2}, at\right)$

and perpendicular to AP will be $(y - at) = \dfrac{-t}{2}\left(x - \dfrac{at^2}{2}\right)$

$$\Rightarrow \quad t\left(x - \frac{at^2}{2}\right) + 2(y - at) = 0 \quad \text{or} \quad tx + 2y = \frac{at^3}{2} + 2at. \quad ...(2)$$

Now the locus of the point of intersection of lines (1) and (2) will give us the locus of the centre of the which passes through the vertex A of the parabola.

from (1), $\dfrac{t^2y}{2} - \dfrac{tx}{2} = \dfrac{at^3}{2}$ and multiplying by $\dfrac{t}{2}$. ...(3)

Hence from (2) and (3), we have

$$t^2y + t(4a - 3x) - 4y = 0 \quad ...(4)$$

Also (1) can be written as $t^2a - ty + x = 0$. ...(5)

Now we have to eliminate 'a' from (4) and (5) to get focus of the centre of the circle.

On cross-multiplication,

$$\frac{t^2}{x(4a-3y)-4y^2} = \frac{t}{-4ya-xy} = \frac{1}{-y^2-a(4a-3x)}$$

$$t = \frac{y(x+4a)}{y^2-a(3x-4a)} \text{ and } t^2 = \frac{4y^2+x(3x-4a)}{y^2-a(3x-4a)}$$

hence $$\frac{y^2(x+4a)^2}{\left[y^2-a(3x-4a)\right]^2} = \frac{4y^2+x(3x-4a)}{y^2-a(3x-4a)}.$$

$$\Rightarrow \quad y^2(x + 4y)^2 = [\{4y^2 + x(3x - 4a)\}\{y^2 - 2(3x - 4a)\}].$$

On simplifying, we get the locus of centre as

$$2y^2(2y^2 + x^2 - 12ax) = ax(3x - 4a)^2$$ **Proved.**

Example 85(b):

LOL' and MOM' are two chords of a parabola passing through a point O on its axis. Prove that the radical axis of the circles described on LL' and MM' as diameters passes through the vertex of the parabola.

Solution:

Let the equation of the parabola be $y^2 = 4ax$.

If O is any point on its axis, its co-ordinates may be taken as (b, 0). As LOL' and MCM' are two chords, suppose the co-ordinates or L be $(at_1^2, 2at_1)$, of L' be $(at_2^2, 2at_2)$, of M be $(at_3^2, 2at_3)$ and of M' be $(at_4^2, 2at_4)$.

The equation of LL' is $y(t_1 + t_2) = 2x + 2at_1t_2$; this passes through (b, 0); the co-ordinates will satisfy t; hence

$$0 = 2b + 2at_1t_2$$

or $\quad t_1t_2 = -b/a$

Similarly, $\quad t_3t_4 = -b/a.$

So $\quad t_1t_2 = t_3t_4$

or $\quad t_1t_2 - t_3t_4 = 0.$ ...(1)

Now equation to the circles drawn with LL' as diameter is

$$(x - t_1^2)(x - at_2^2) + (y - 2at_1)(y - 2at_2) = 0 \quad ...(2)$$

and the equation of the circle on MM' as diameter is

$$(x - t_3^2)(x - at_4^2) + (y - 2at_3)(y - 2at_4) = 0 \quad ...(3)$$

Equation of radical axis of (3) and (2) is obtained by subtracting (3) from (2).

Subtracting (3) from (2), we get the constant term as

$$a^2(t_1^2t_2^2 + t_3^2t_4^2) + 4a^2(t_1t_2 + t_3t_4) = \lambda \text{ (say)}. \quad ...(4)$$

By (1) and (4), we get $\lambda = 0$; hence the equation of the radical axis has no constant term; therefore this radical axis passes through the origin, the vertex of the parabolas. **Proved.**

Example 85(c):

The sides of a triangle touch a parabola and two of its angular points lie on another parabola with its axis in the same prove that the locus of the third angular point is another parabola.

Solution:

Suppose the points of contact of the side triangle on the parabola be $(at_1^2, 2at_1)$, $(at_2^2, 2at_2)$, and $(at_3^2, 2at_3)$.

Solving the tangents at these points, the co-ordinates of two are $\{at_2t_3,$, $a(t_2 + t_3)\}$ and $\{at_1t_3, a(t_1 + t_3)$.

Let these vertices lie on the other parabola $(y - h)^2 = 4b(x - 1)$.

Hence $\qquad \{a(t_2 + t_3) - k\}^2 = 4b(at_2t_3 - h)$

and $\qquad \{a(t_3 + t_1) - k\}^2 = 4b(a_1t_3 - h)$.

Let the co-ordinates of third vertex be (x, y); we have

$$x = at_1t_2$$

and $\qquad y = a(t_1 + t_2)s$

To eliminate t_1, t_2 and t_3 from (1), (2), (3) and (4), we subtract from (1), we have $a(t_1 + t_2) + 2at_3 - 2k = 4bt_3$

Example 86:

Prove that the normal chord at the point whose ordinate is equal to its abscissa subtends a right angle at the focus.

Solution:

Let the equation of the parabola be $y^2 = 4ax$ ...(1)

and say P is any point on (1) for which the abscissa is equal to the ordinate is (h, h). Putting these co-ordinates in (1), we get $h^2 = 4ah$ hence either $h = 0$ or $h = 4a$.

Hence the point is either origin at which the normal is x-axis or (4a, 4a) at which the normal is

$$y - 4a = -\frac{4a}{2a}(x-4a) \text{ or } y = -2x + 12a. \qquad ...(2)$$

Putting the value of y in (1), we get

$$(-2x + 12a)^2 = 4ax$$

$\Rightarrow \qquad x^2 - 13ax + 36a^2 = 0$ or $(x - 9a)(x - 4a) = 0$

whence either $\qquad x = 9a$ or $4a$

The corresponding values of y are $-6a$ and $4a$.

Hence the other point of interaction of the normal at $P \equiv (4a, 4a)$ to the parabola (1) is (9a, 6a) say Q.

Slope of the line SP where S is the focus (a, 0) is

$$\frac{4a-0}{2a-a} = \frac{4}{3} = m_1 \text{ (say)} \qquad ...(3)$$

$$\text{slope of SQ} = \frac{-6a-0}{+9a-a} = -\frac{3}{4} = m_2 \text{ (say)} \qquad ...(4)$$

Clearly $m_1 \times m_2 = -1$, so the SP and SQ are at right angles.

Hence proved.

Similarly, the equation to any tangent to (2) is given by

$$y = m_1 (x + a') + \frac{a'}{m_1}. \qquad ...(5)$$

If (4) and (5) are perpendicular to each other, $m_1 = -1/m$ and if the point of intersection of (4) and (5) be (h, k), these co-ordinates will satisfy (4) and (5). So putting $x = h$, $y = k$ and $m_1 = -1/m$, we get respectively

$$k = m(h + a) + \frac{a}{m}$$

or $$km = m^2 h + m^2 a + a. \qquad ...(6)$$

and k = – $$k = -\frac{1}{m}(h + a) + \frac{a'}{-1/m}$$

$$\Rightarrow \quad km = -h - a' - a'm^2. \qquad ...(7)$$

The reqd locus of (h, k) is obtained by eliminating m from (6) and (7); so subtracting (7) from (6), we get

$$0 = h(1 + m^2) + a(1 + m^2) + a'(1 + m^2)$$

$$\Rightarrow \quad h + a + a' = 0. \qquad ...(8)$$

Generalising, we get $x + a + a' = 0$. **Proved.**

Again to get the points of intersection of (1) and (2), we solve them simultaneously. So subtracting (2) from (1), we get

$$0 = 4a(x + a) - 4a'(x + a')$$

$$\Rightarrow \quad (a - a') + (a^2 - a'^2) = 0$$

or $$(a - a')(x + a + a') = 0$$

As $a - a'$ ¹ 0, $x + a + a' = 0$ which is same as (8). **Hence proved.**

Example 87(a):

If perpendiculars be drawn on any tangent to a parabola from two fixed points on the axis, which are equidistant from the focus, prove that the difference of their squares is constant.

Solution:

Let the equation to the parabola be $y^2 = 4ax$. ...(1)

Its axis is x-axis and focus is (a, 0). So any point on the axis equidistant from the focus may be taken as (a + k, 0) and (a – k, 0) say P and Q respectively.

Equation to any tangent to (1) is

$$y = mx + \frac{a}{m}. \qquad ...(2)$$

Distance of P from (2), say PN = $\dfrac{m(a+k)+a/m}{\sqrt{(1+m^2)}}$

and the distance of Q from (2) say QM = $\dfrac{m(a-k)+a/m}{\sqrt{(1+m^2)}}$

Hence $PN^2 - QM^2 = \dfrac{[m(a+k)+a/m]^2}{1+m^2} - \dfrac{[m(a-k)+a/m]^2}{1+m^2}$

$$= \frac{1}{(1+m^2)}\left[m^2(a+k)^2 + \frac{a^2}{m^2} + 2m.\frac{a}{m}(a+k) - m(a-k)^2 - \frac{a^2}{m^2} - 2\frac{a}{m}.m(a-k)\right]$$

$$\frac{1}{(1+m^2)}[4m^2 ak + 4ak)$$

$= 4ak$ which is constant as k and a are constant. **Hence proved.**

Example 87(b):

Find the equation to that tangent to the parabola $y^2 = 7x$ which is parallel to the straight line $4y - x + 3 = 0$. Find also its point of contact.

Solution:

The parabola is given as $y^2 = 7x$. ...(1)

here $4a = 7$ or $a = 7/4$.

Equation to any tangent to (1) is

$$y = mx + \frac{a}{m} \quad ...(2)$$

The given line is $4y - x + 3 = 0$. ...(3)

Its slope = 1/4

If (2) is parallel to (3), m = 1/4. Substituting in (2), the equation of tangent is

$$y = \frac{1}{2}x + \frac{7}{4} \div \frac{1}{4}$$

$\Rightarrow$ $4y = x + 28$. **Ans.**

The point of contact is given by $\left(\dfrac{a}{m^2}, \dfrac{2a}{m}\right)$. So putting values of a and m, we get the point of contact as

$$\left\{\frac{7}{4}\div\left(\frac{1}{4}\right)^2,\ 2\times\frac{7}{4}\div\frac{1}{4}\right\}\quad \text{or}\quad (28,\ 14)$$ **Ans.**

Example 87(c):

A tangent to the parabola $y^2 = 4ax$ makes an angle of 60° with the axis; find its point of contact.

Solution:

The curve is $y^2 = 4ax$. ...(1)

the axis of (1) is x-axis. If a line makes an angle of 60° to the axis, its slope is tan 60o = $\sqrt{3}$.

Any tangent to (1) is $y = mx + \dfrac{a}{m}$.

Putting the values of m as $\sqrt{3}$, we get the equation to the tangent as

$$y=\sqrt{3x}+\frac{a}{\sqrt{3}}$$

and the point of contact will be

$$\left(\frac{a}{m^2},\frac{2a}{m}\right),\text{i.e.}\left(\frac{a}{3},\frac{a}{\sqrt{3}}\right)\text{as}\, m=\sqrt{3}.$$ **Ans.**

Example 88:

Find the equation of tangent and normal. At the ends of the latus rectum of the parabola $y^2 = 4a\ (a - a)$.

Solution:

The equation to the parabola is given as

$$y^2 = 4a\ (x - a) \qquad ...(1)$$

Changing the origin to (a, 0), the equation of the curve becomes

$$y^2 = 4ax. \qquad ...(2)$$

Co-ordinates of the ends of latus rectum are (a, 2a) and (a, – 2a).

The equation to the tangent at (a, 2a) is

$$y.2a = 2a\ (x + a) \quad \text{or} \quad y = x + a$$

Changing the axes to the original origin, the equation will become

$$y = (x - a) + a \quad \text{or} \quad y = x.$$ **Ans.**

Equation of the normal according the new origin is

$$y - 2a = -\frac{2a}{2a}(x-a) \quad \text{or} \quad y + x = 3a.$$

Equation to the normal according to the old origin is

$$y + (x - a) = 3a \qquad \text{or} \qquad y + x = 4a \qquad \textbf{Ans.}$$

Again, equation to the tangent at (a, – 2a) according to the new origin is y (– 2a) = 2a (x + a) or x + y + a = 0.

Equation to the normal at (a, – 2a) according to the new origin is

$$y - (-2a) = -\frac{-2a}{-2a}(x-a)$$

or $$y = x - 3a.$$

So equation to the normal at (a, – 2a) according to the old origin is

$$y = (x - a) - 3a \qquad \text{or} \qquad x - y = 4a \qquad \textbf{Ans.}$$

Example 89:

Find the equation to the tangents to the parabola $y^2 = 9x$ which goes through the point (4, 10).

Solution:

The parabola is given as $y^2 = 9x$. ...(1)

Equation to any tangent to (1) is

$$y = mx + \frac{9}{4m}\left(\text{as } a = \frac{9}{4}\right). \qquad ...(2)$$

If the tangent (2) passes through (4, 10), it will satisfy (2); hence

$$10 = m \,.\, 4 + \frac{9}{4m}$$

$$\Rightarrow \qquad 16m^2 - 40m + 9 = 0$$

$$\Rightarrow \qquad (4m = 9)(4m - 1) = 0, \text{ whence } m = \frac{9}{4} \text{ or } m = \frac{1}{4}.$$

If m = 9/4 substituting ın (2), the equation to the tangent is

$$y = \frac{1}{4}x + \frac{9}{4}.\frac{4}{9} \qquad \text{or} \qquad 4y = 9x + 4. \qquad \textbf{Ans.}$$

If m = 1/4, substituting in (2), we get

$$y = \frac{1}{4}x + \frac{9}{4}.\frac{4}{1} \qquad \text{or} \qquad 4y = x + 36. \qquad \textbf{Ans.}$$

Example 90:

Find the points of the parabola $y^2 = 4ax$ at which (i) the tangent and (ii) the normal is inclined at 30° to the axis.

Solution:

The parabola is given as $y^2 = 4ax$. ...(1)

Any tangent to (1) is $y = mx + (a/m)$ and the corresponding pcint of contact is $\left(\frac{a}{m^2}, \frac{2a}{m}\right)$; as the axis of the parabola is x-axis, if the tangent is inclined to its axis at 30°, then m=tan 30° $= \frac{1}{\sqrt{3}}$. So the point of contact is

$$\left\{a/\left(\frac{1}{\sqrt{3}}\right)^2, 2a/\frac{1}{\sqrt{3}}\right\} \text{ or } \left(3a, 2\sqrt{3}\,a\right)$$ **Ans.**

Equation to any normal to (1) is $y = mx - 2am - am^3$ and the corresponding point of intersection is $(am^2, -2am)$.

It the normal is inclined at an angle of 30° = $1/\sqrt{3}$.

So the corresponding point on the parabola is

$$\left\{a\left(\frac{1}{\sqrt{3}}\right)^2, -2a\frac{1}{\sqrt{3}}\right\}$$

$$\left(\frac{a}{3}, -\frac{2\sqrt{3}}{3}a\right)$$ **Ans.**

Example 90(a):

For what point of the parabola $y^2 = 4ax$ (1) the normal equal to twice the subtangent, (2) the normal equal to the difference between the subtangent and the subnormal

Solution:

The parabola is given as

$$y^2 = 4ax \quad ...(1)$$

If P be any point $(at^2, 2at)$ on the parabola, PT be the tangent at P meeting axis of x at T, PG be the normal meeting axis of x at G and PN be the perpendicular from P on X-axis, then TN is the subtangent, NG is the sub-normal and PG is the normal.

Equation to PT is $y.2at = 2a(x + at^2)$

$$\Rightarrow \quad ty = x + at^2. \quad ...(2)$$

Solving with X axis, i.e., putting

$y = 0$, we get $x = -at^2$.

So co-ordinates of T are $(-at^2, 0)$.

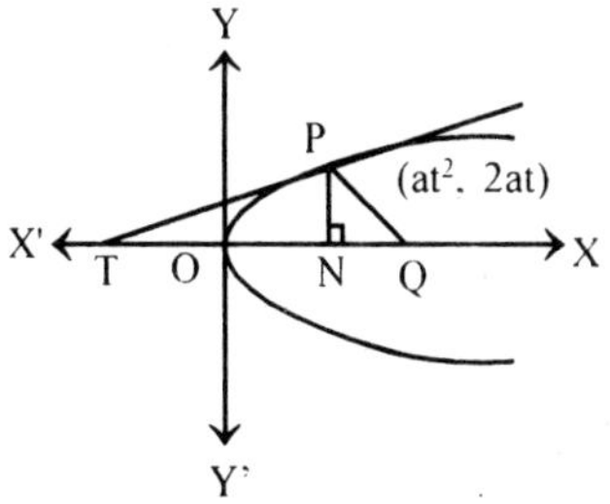

Co-ordinates of N are clearly $(at^2, 0)$

Equation to the normal PG is

$$y - 2at = -\frac{2at}{2a}\left(x - at^2\right)$$

$\Rightarrow$ $y + tx = 2at + at^3$

Again solving with X-axis i.e.

$y = 0$, we get

$x = 2a + at^2$.

So the co-ordinates of G are $(2a + at^2, 0)$.

Normal $PG = [(2a + at^2 - at^2)^2 + (0 - 2at)^2]^{1/2}$

$= 2a\sqrt{(1 + t^2)}$.

(i) The condition is given as

normal = 2 × sub-tangent

$\Rightarrow$ $PG = 2TN$ or $PG^2 = 4TN^2$

$\Rightarrow$ $4a^2(1 + t^2) = 4(2at^2)^2$. $[\because TN = at^2 + at^2 = 2at^2]$

$\Rightarrow$ $4t^4 - t^2 - 1 = 0$,

whence $t^2 = \dfrac{1 \pm \sqrt{(16+1)}}{2.4} = \dfrac{1 \pm 17}{8}$

$\Rightarrow$ $t = \dfrac{1}{4}\sqrt{\left\{2\sqrt{17} + 2\right\}}$.

(omitting -ve sign, as if t^2 is-ve, t will be imaginary),

So the co-ordinates of P are

$$\left[\frac{\sqrt{17}+1}{8}a, \frac{a}{2}\sqrt{\left\{2\sqrt{17}+2\right\}}\right]$$ **Ans.**

(ii) The condition is PG = TN – GN

$\Rightarrow$ $2a\sqrt{(1 + t^2)} = 2at^2 - 2a$

$[\because GN = OG - ON = 2a + at^2 - at^2 = 2a]$

$\Rightarrow$ $\sqrt{(1 + t^2)} = (t^2 - 1)$.

Squaring, we get $1 + t^2 = t^4 - 2t^2 + 1$,

whence $t^2 = 0$ or $t^2 = 3$.

If $t^2 = 0$, the co-ordinates of P become (0, 0) which is the vertex.

If $t^2 = 3$, the co-ordinates of P become $(3a, \sqrt{3}a)$. **Ans.**

Example 90(b):

Two equal parabolas have the same vertex and their axes are at right angles; prove that the common tangent touches each at the end at the and of a latus rectum.

Solution:

Taking the axis of the first parabola as x-axis and the axis of the other parabola as y-axis, their equation can be written as

$$y^2 = 4ax \qquad ...(1)$$

and $$x^2 = 4ay. \qquad ...(2)$$

(As the parabolas are equal, their latera recta are equal, say 4s).

Equation to any tangent to (1) is $y = mx + \dfrac{a}{m}$. ...(3)

To solve (2) and (3), we put the value of y from (3) in (2); we get

$$x^2 = 4a\left(mx + \frac{a}{m}\right)$$

$$\Rightarrow \qquad mx^2 - 4am^2x - 4a^2 = 0. \qquad ...(4)$$

If (3) touches (2), then (4) must be a perfect square, i.e. the discriminant of (4) must be zero. Hence

$$(-4am^2)^2 - 4.m(-4a^2) = 0$$

whence either $m = 0$

or $m^3 = -1$

or $m = -1$.

If $m = 0$, line is x-axis, which is not a tangent to (1). So putting $m = -1$ in (3), the equation to the common tangent is

$$y = -x - a$$

or $$y + x + a = 0 \qquad ...(5)$$

To get the point of contact, putting the value of m in (4),

$$-x^2 - 4ax - 4a^2 = 0$$

or $(x + 2a)^2 = 0$

whence $x = -2a$.

Putting this value in (2), $y = a$. So (5) touches (2) at $(-2a, a)$ which is one end of the latus rectum.

Similarly solving (1) and (5), we can show that the point of contact is $(a, -2a)$ which is the end of the latus rectum. **Proved.**

Example 90(c):

PN is an ordinate of the parabola; a straight line is drawn parallel to the axis to bisect NP and meets the curve in Q; prove that NQ meets the tangent at the vertex in a point T such that AT = (2/3) NP.

Solution:

Let the equation of the parabola be

$$y^2 = 4ax. \qquad ...(1)$$

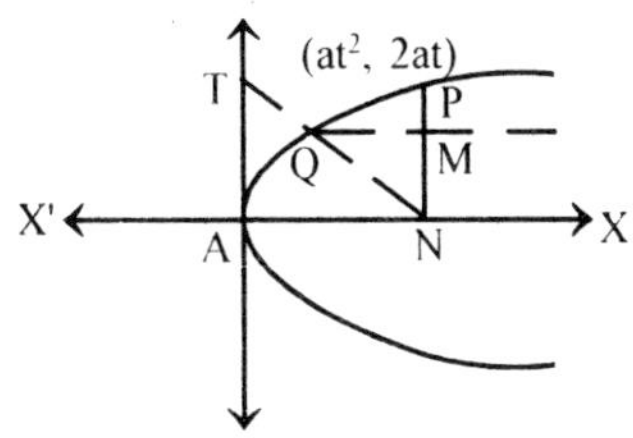

Tangent at the vertex is y-axis

or $\qquad x = 0 \qquad ...(2)$

Let P be any point $(at^2, 2at)$ on the parabola. So PN = 2at and co-ordinates of N are $(at^2, 0)$.

If M is the mid-point of PN, then NM = (1/2) PN = at.

As QM is parallel to x-axis, its equation is y = at. ...(3)

To solve (2) with (1), putting the value of y from (2) in (1), we get

$$a^2t^2 = 4ax \qquad \text{or} \qquad x = \frac{1}{4}at^2$$

So the co-ordinates of Q are $\left(\frac{1}{4}at^2, at\right)$

Equation of NQ will be

$$y - 0 = \frac{at - 0}{\frac{1}{4}at^2 - at^2}\left(a - at^2\right) \quad \text{or} \quad y = -\frac{4}{3t}\left(x - at^2\right) \qquad ...(4)$$

If NQ meets the tangent at the vertex i.e. x = 0 at T, then putting x = 0 in (3), we get $y = \frac{4at}{3}$.

So the co-ordinates of T are $= 0, \frac{4at}{3}t \quad \text{or} \quad AT = \frac{4a}{3}t.$

Clearly AT $= \frac{2}{3}.2at = \frac{2}{3}.PN.$ **Hence proved.**

Example 90(d):

A parabola is drawn touching the axis of x at the origin and having its vertex at a given distance k from this axis. Prove that the axis of the parabola is a tangent to the parabola $x^2 = -8k(y - 2k)$.

Solution:

Let the equation to any parabola in its most general form be

$$(ax + by)^2 + 2gx + 2fy + c = 0. \quad ...(1)$$

As the parabola passes through the origin, hence c = 0, x-axis i.e. y = 0 is a tangent. So solving with (1), we get

$$a^2a^2 + 2gx = 0.$$

If y = 0 is a tangent, this equation must be a perfect square. So g = 0.

Putting the values of c and g, (1) becomes

$$(ax + by)^2 + 2fy = 0. \quad ...(2)$$

The axis of the parabola (2) will be

$$ax + by = -\frac{bf}{\left(a^2+b^2\right)} \quad ...(3)$$

To find out vertex, we solve (2) and (3) simultaneously.

Putting the value of ax + by from (3) in (2), we get

$$\frac{b^2 f^2}{\left(a^2+b^2\right)} + 2fy = 0 \quad \text{or} \quad y = -\frac{b^2 f}{2\left(a^2+b^2\right)^2}.$$

So the ordinate of the vertex $-\dfrac{b^2 f}{2\left(a^2+b^2\right)^2} = k$ (by hypothesis).

Hence $\quad f = -\dfrac{2k\left(a^2+b^2\right)^2}{b^2}$

So (3) becomes

$$ax + by = -\, b \times \frac{-2k\left(a^2+b^2\right)^2}{b^2\left(a^2+b^2\right)} = \frac{2k\left(a^2+b^2\right)}{b}$$

$$\Rightarrow \quad x = -\frac{b}{a}y + \frac{2k}{ab}\left(a^2+b^2\right)$$

$$\Rightarrow \quad x = -\frac{b}{a} + 2k\left(\frac{a}{b} + \frac{b}{a}\right) \quad ...(4)$$

Equation to any tangent to the given parabola

$$x^2 = -\,8k\,(y - 2k) \quad \text{or} \quad x^2 = 4.\,(-\,2k)\,(y - 2k)$$

will be $\quad x = m\,(y - 2k) - \dfrac{2k}{m} \qquad (\because a = -2k)$

$$\Rightarrow \quad x = my - 2km - \frac{2k}{m}$$

$$\Rightarrow \quad x = my - 2k\left(m + \frac{1}{m}\right) \quad ...(5)$$

As it is true to any value of m, putting m = – b/a, (5) becomes

$$x = - \quad x=-\frac{b}{a}y-2k\left(-\frac{b}{a}-\frac{a}{b}\right)$$

$$\Rightarrow \qquad x=-\frac{b}{a}y+2k\left(\frac{b}{a}+\frac{a}{b}\right)$$

which is same as (4), the axis of the parabola. **Hence proved.**

Example 91:

PNP' is double ordinate of the parabola; prove that the locus of the point of intersection of the normal at P and the straight line through P' parallel to the axis is the equal parabola $y^2 = 4a\ (a - 4a)$.

Solution:

Let the equation to the parabola be $y^2 = 4ax$. ...(1)

If PQ be any double ordinate, let the co-ordinates of P be $(am^2, -2am)$; then the co-ordinates of Q will be $(am^2, 2am)$.

Equation to the normal at P is

$$y = mx - 2am - am^3. \qquad ...(2)$$

A line parallel to axis of parabola, i.e. x-axis through $Q \equiv (am^2, 2am)$ is

$$y = 2am \qquad ...(3)$$

The required locus will be obtained by eliminating m from (2) and (3). So putting the value of m from (3) in (2), we get

$$y=\frac{y}{2a}-2a\frac{y}{a}-a\left(\frac{y}{2a}\right)^3$$

Multiplying throughout by $8a^2$ and dividing by y, we get

$$8a^2 = 4ax - 8a^2 - y^2$$

$$\Rightarrow \qquad y^2 = 4ax - 16a^2$$

$$\Rightarrow \qquad y^2 = 4a\ (x - 4a) \text{ which is a parabola.}$$

Example 92:

Tangents are drawn to a parabola at points whose abscissa are in the ratio μ ; 1; prove that they intersect on the curve

$$y^2 = (\mu^{1/4} + \mu^{1/4})^2\ ax.$$

Solution:

Let the equation to the parabola be $y^2 = 4ax$. ...(1)

Let P and Q be two points on the parabola (1), such that abscissa of P is μat^2 and of Q is at^2; then the corresponding ordinates are 2at $\sqrt{\mu}$) and

2at. So the co-ordinats of P and Q are respectively (μat^2, 2at $\sqrt{\mu}$) and (at^2, 2at).

Equation of tangent to (1) at (μat^2, 2at $\sqrt{\mu}$)

$$y.2at\sqrt{\mu} = 2a\,(x + \mu at^2)$$

or $$ty\sqrt{\mu} = (x + \mu at^2). \qquad ...(2)$$

Similarly the equation to the tangent at Q is

$$ty = x + at^2. \qquad ...(3)$$

The required locus of the point of intersection will be obtained by eliminating t from (2) and (3).

Multiplying (3) by μ and subtracting (2), we get

$$ty\sqrt{\mu}\,(1 - \sqrt{\mu}) = x\,(1 - \mu)$$

or $$t = \frac{x}{y}.\frac{(1+\sqrt{\mu})}{\sqrt{\mu}} \qquad \left[\because 1-\mu = 1(1-\sqrt{\mu})(1+\sqrt{\mu}\right].$$

Substituting the value of t in (3) we get

$$y.\frac{x}{y}.\frac{\left(1+\sqrt{\mu}\right)}{\sqrt{\mu}} = x + a.\frac{x^2}{y^2}.\frac{\left(1+\sqrt{\mu}\right)^2}{\sqrt{\mu}}$$

Multiplying by μy^2, we get

$$xy^2\sqrt{\mu} + xy^2\mu = xy^2\mu + ax^2\,(1 + \sqrt{\mu})^2$$

or $$y^2\sqrt{\mu} = ax.\,(1 + \mu + 2\sqrt{\mu})$$

or $$y^2 = ax\left[\frac{1}{\sqrt{\mu}} + \sqrt{\mu} + 2\right]$$

or $$y^2 = [\mu^{1/4} + \mu^{-1/4}]^2\,ax$$

which is required locus. **Proved.**

Example 92(a):

If the tangents at the points (x', y") meet at the point (x_1, y_1) and the normals at the same points in (x_2, y_2) prove that

(1) $x_1 = \dfrac{y'y''}{4a}$ and $y_1 = \dfrac{y'+y''}{2}$.

(2) $x_2 = 2a\,\dfrac{y'^2 + y'y'' + y''^2}{4a}$ and $y_2 = -y'y''\dfrac{y'+y''}{8a^2}$,

and hence that

(3) $x_2 = 2a\cdot\dfrac{y_1'^2}{4a} - x_2$ and $y_2 = \dfrac{y_1 + y_1}{a}$.

Solution:

Let the equation to the parabola be $y^2 = 4ax$.

Equation to the tangent at (x', y') is $yy' = 2a(x + x')$. ...(1)

As the point (x', y') lies on the parabola, so $y'^2 = 4ax'$

$\Rightarrow \quad x' = y'^2/4a$.

So the equation to the tangent becomes

$$yy' = 2a\left(x + \frac{y'^2}{4a}\right)$$

$$\Rightarrow \quad yy'' = 2ax\ \frac{y'^2}{2} \qquad ...(2)$$

Similarly the equation to the tangent at $(x'', y''$ will be

$$yy'' = 2ax\ \frac{y''^2}{2} \qquad ...(3)$$

To solve (2) and (3), subtracting (3) from (2), we get

$$y(y' - y'') = \frac{1}{2}\left(y'^2 - y''^2\right)$$

$$\Rightarrow \quad y\ \frac{1}{2}(y' + y'').$$

Putting this value in (1), we have

$$\frac{1}{2}(y' + y'')\,y' = 2ax\ \frac{y'^2}{2}$$

$$\Rightarrow \quad x = \frac{y'y''}{4a}$$

As the point of intersection is given as (x_1, y_1), so

$$x_1 = \frac{y'y''}{4a}, y_1\ \frac{1}{2}(y' + y'')$$ **Proved.**

(2) Equation to the normal at (x', y') is given as

$$y - y' = -\frac{y'}{2a}(x - x').$$

Again substituting the value for x', i.e. $x' = y'2/4a$,

the equation becomes $y - y' = -\dfrac{y'}{2a}\left(x - \dfrac{y'^2}{4a}\right)$ ...(4)

Similarly the equation of the normal at (x'', y'') will be

$$y - y'' = -\frac{y''}{2a}\left(x - \frac{y''^2}{4a}\right). \qquad ...(5)$$

To solve (4) and (5) subtract (5) from (4), we get

$$y'' - y' = \frac{1}{2a}\left[(y''-y')x - \frac{1}{4a}\left(y''^2 - y'^3\right)\right]$$

$$\Rightarrow \qquad 1 = \frac{1}{2a}.x - \frac{1}{8a^2}\left(y''^2 + y''y' + y'^2\right) \text{ [dividing by (y'' – y')]}$$

$$\Rightarrow \qquad 1 = \frac{1}{8a^2}\left(y''^2 + y''y' + y'^2\right) = \frac{1}{2a}.x$$

$$\Rightarrow \qquad x = 2a + \frac{1}{4a}\left(y''^2 + y''y' + y'^2\right)$$

Putting the value in (4), we get

$$y - y' = -\frac{y'}{2a}\left[2a + \frac{1}{4a}\left(y''^2 + y''y' + y'^2\right) - \frac{y''^2}{4a}\right]$$

$$\Rightarrow \qquad y - y' = y' - \frac{y'}{8a^2}\left(y''^2 + y''y' + y'^2 - y'^2\right)$$

$$\Rightarrow \qquad \left[\frac{y''+y'}{8a^2}\right]$$

(3) Adding the values of x_1 and x_2 which we get in lest and 2nd part of this question, we get

$$x_2 = x_1 = 2a + \frac{1}{4a}\left(y''^2 + y''y' + y'^2\right) + \frac{y'y''}{4a}$$

$$= 2a + \frac{1}{4a}\left(y''^2 + 2y''y' + y'^2\right) = 2a + \frac{1}{4a}(y''+y')^2$$

$$= 2a + \frac{y_1^2}{a} \quad \left\{\because y_1 = \frac{1}{2}(y'+y'') \text{ by Q. No. 23(1)}\right\}$$

$$x_2 = 2a + \frac{y_1^2}{a} - x_1.$$

Again, $$y_2 = -\frac{y'y''(y'+y)}{8a^2}$$

$$= \frac{y'y''}{4a}\left(\frac{y'+y''}{2}\right).\frac{1}{a}$$

$$= \frac{x_1 y_1}{a} \quad \left\{\because x_1 = \frac{y'y''}{4a} \text{ and } y_1 = \frac{1}{2}(y'+y'')\right\}$$

Example 92(b):

Prove that the straight line $lx + my + n = 0$ touches the parabola $y^2 = 4ax$ if $ln = am^2$

Solution:

The straight line is given as

$$lx + my + n = 0 \qquad ...(1)$$

and the parabola is given as $y^2 = 4ax$. ...(2)

The solve (1) and (2), we put the value of x from (1) in (2). So we get

$$y^2 = 4a\left[-\frac{my+n}{l}\right]$$

$$\Rightarrow \qquad ly^2 + 4amy + 4an = 0. \qquad ...(3)$$

If (1) touches (2), the roots of (3) must coincide, i.e. (3) must be a perfect square; so its discriminant will be zero.

Hence $(4am)^2 - 4l.4an = 0$

$\Rightarrow \quad a^2 m^2 = a/n \quad$ or $\quad ln = am^2$ **Proved.**

Example 92(c):

The circle $x^2 + y^2 = 4ax$ and the parabola $y^2 = 4ax$.

Solution:

The equation to the circle is $x^2 + y^2 = 4ax$, ...(1)

and to the parabola is $y^2 = 4ax$. ...(2)

Equation to any tangent to (2) is $y = mx + a/m$ ...(3)

If (3) is a tangent to (1) also, the length of perpendicular from the centre of (1) on (3) must be equal to the radius of (1). Centre of (1) is (2a, 0) and radius is 2a,

Hence, we must have

$$\frac{m.2a - 0 + a/m}{\sqrt{(1+m^2)}} = 2a$$

$$\Rightarrow \qquad 2m^2 + 1 = 2m\sqrt{(1+m^2)}$$

Dividing by m^2, we get $2 + \frac{1}{m^2} = \frac{2}{m}\sqrt{(1+m^2)}$

Squaring both sides, $4 = \frac{1}{m^4} = \frac{4}{m^2} = \frac{4}{m^2} + 4$

$$\Rightarrow \qquad \frac{1}{m^4} = 0 \quad \text{or} \quad \frac{1}{m} = 0.$$

Dividing (3) m, we get $\frac{y}{m} = x + \frac{a}{m^2}$.

Putting the value of $1/m = 0$, we get the required equation as $x = 0$

Example 92(d):

From the proceeding question prove that, if tangents be drawn to the parabola $y^2 = 4ax$ from any point on the parabola $y^2 = a(x + b)$, then the normals at the points of contact meet on a fixed straight line.

Solution:

Let (x', y') and (x", y") be any two points on the parabola $y^2 = 4ax$ and let the tangents on (x', y') & (x", y") meet in (x_1, y_1), then by hypotesis, (x_1, y_1) lies on the parabola $y^2 = a(x + b)$. So the co-ordinates will satisfy it. Hence we get

$$y_1^2 = a(x_1 + b). \qquad ...(1)$$

Let the normals at (x', y') and (x", y") meet in (h, k); then by

$$h = 2a + \frac{y_1^2}{a} + x_1 \qquad ...(2)$$

and $$k = -\frac{x_1 y_1}{a} \qquad ...(3)$$

(putting h and k for x_2 and y_2 respectively).

The required locus will be obtained by eliminating (x_1, y_1) from (1), (2) and (3).

Putting the values of y_1^2 from (1) in (2), we have

$$h = 2a + \ h = 2a + \frac{1}{a}a(x_1 + b) - x_1$$

$\Rightarrow$ $$h = 2a + x_1 + b - x_1 = 2a + b.$$

Generalising, we get $x = 2a + b$ which is a fixed line. So the tangents meet on a fixed line.

Example 93:

Prove that the chord of the parabola $y^2 = 4ax$, whose equation is $y - x\sqrt{2} + 4a\sqrt{2} = 0$, is a normal to the curve and that its length is $6\sqrt{3}a$.

Solution:

The equation to the parabola is given as

$$y^2 = 4ax, \qquad ...(1)$$

and the equation to the chord is

$$y - x\sqrt{2} + 4a\sqrt{2} = 0. \qquad ...(2)$$

To solve (1) and (2) simultaneously, we put the value of y from (2) in (1); we get $(x\sqrt{2} - 4a\sqrt{2})^2 = 4ax$.

$\Rightarrow$ $$2x^2 + 32a^2 - 16ax = 4ax$$

$\Rightarrow \qquad x^2 - 10ax + 16a^2 = 0.$

whence either $x = 8a$ or $x = 2a$.

Putting in (2), the corresponding values of y are respectively $4a\sqrt{2}$ and $-2a\sqrt{2}$. Hence the points of intersection of (1) and 92) are $(8a, 4a\sqrt{2})$ and $(2a, -2a\sqrt{2})$

The distance between the points $(8a - 4a\sqrt{2}$ and $(2a - 2a\sqrt{2})$

$$\sqrt{\{(8a - 2a)^2 + (4a\sqrt{2} + 2a\sqrt{2})^2\}}$$

$$\sqrt{\{(36a^2 + 72a^2} = 6a\sqrt{3}.$$ **Ans.**

Hence the length of the chord $= 6a\sqrt{3}$.

Again, equation of the normal at $(8a, 4a\sqrt{2})$ is

$$y - 4a\sqrt{2} = \frac{4a\sqrt{2}}{2a}(x - 8a)$$

$$\Rightarrow \qquad 2x\sqrt{2} - y = 12a\sqrt{2}$$

And the normal at $(2a, -2a\sqrt{2})$is

$$y + 2a\sqrt{2} = \frac{-2a\sqrt{2}}{2a}(x - 2a)$$

$$\Rightarrow \qquad y - x\sqrt{2} + 4a\sqrt{2} = 0$$

Which is same as (2), hence the chord given by (2) is a normal at $(2a, -2a\sqrt{2})$ **Proved.**

Example 93(a):

It P, Q and R be three points on a parabola whose ordinates are in geometrical progression, prove that the tangents at P and R meet on the ordinate of Q.

Solution:

Let the equation of the parabolas $y^2 = 4ax$. ...(1)

and the co-ordinates of three points P, Q and R on it be $(at_1^2, 2at_1)$, $(at_2^2, 2at_2)$ and $(at_3^2, 2at_3)$ respectively. If the ordinates are in GP, then

$$2at_1^2 \times 2at_3 = (2at_2)^2 \quad \text{or} \quad t_1 t_3 = t_2^2$$

Equation to the tangent on (1) at $P \equiv (2at_1^2, 2at_1)$ is

$$y.2at_1 = 2a(x + at_1^2) \quad \text{or} \quad yt_1 = x + at_1^2. \qquad ...(3)$$

Similarly the equation to the tangent at $R \equiv (2at_3^2, 2at_3)$ to (1) is

$$yt_3 = = x + at_3^2. \qquad ...(4)$$

To get the value of x from (3) and (4), multiplying (3) by t_3, (4) by t_1 and subtracting, we get

$$0 = x(t_3 - t_1) - at_1t_3(t_3 - t_1)$$

$\Rightarrow$ $x = at_3t_1 = at_2^2$ [by (2), $t_1t_3 = t_2^2$]

The abscissa of Q is also given by $x = at_2^2$. **Hence proved.**

Example 93(b):

Prove that the distance between a tangent to the parabola and the parallel normal is a cosec θ sec^2 θ, where θ is the angle the either makes with the axis.

Solution:

Let any parabola be $y^2 = 4ax$. ...(1)

And tangent to (1) will be $y = mx + \dfrac{a}{m}$. ...(2)

If there be any normal which is parallel to (2), the slopes will be the same; so the equation of the normal parallel to (2) will be

$$y = mx - 2am - \theta m^3. \quad ...(3)$$

If ON' be the lengths of the perpendiculars from the origin O upto (2) and (3) respectively, clearly, the distance between (2) and (3) is ON – ON'. So the distance

$$ON - ON' = \frac{a/m}{\sqrt{(1+m^2)}} - \frac{-2am - am^2}{\sqrt{(1+m^2)}}$$

If θ is the angle which (1) or (2) make with x-axis, then

Putting the values, $m = \tan\theta$

$$ON - ON' = \frac{a/\tan\theta}{\tan\theta\sqrt{(1+\tan^2\theta)}} - \frac{-2\tan\theta - a\tan^2\theta}{\sqrt{(1+\tan^2\theta)}}$$

$$= \frac{a}{\tan\theta\sec\theta}\left[1 + 2\tan^2\theta + \tan^4\theta\right]$$

$$= \frac{a\cos^2\theta}{\sin\theta}.\left[1+\tan^2\theta\right]^2$$

$= a \operatorname{cosec}\theta. \cos^2\theta \sec^2\theta = a \operatorname{cosec}\theta \sec^2\theta$. **Proved.**

Example 94(a):

Prove that two parabolas, having the same focus and their axes in opposite directions, cut at right angles.

Solution:

Let us take the common focus as the origin, x-axis as the common axis (in opposite direction), the line perpendicular to the axis at 0 as y-axis, and the latus rectum of the parabolas as 4a and 4b, then their equations may be written as

$$y^2 = 4a(x + a)$$

or $$y^2 = 4ax + 4a^2 \quad ...(1)$$

and $$y^2 = -4b(x - b)$$

or $$y^2 = -4bx + 4b^2 \quad ...(2)$$

To get the points of intersection, we solve (1) and (2). So subtracting (2) from (1),

$$0 = 4(a + b)x + 4(a^2 - b^2)$$

or $$x = (b - a).$$

Putting this value in (1), $y^2 = 4b(b - a) + 4a^2 = 4ab$

$\Rightarrow$ $y = \pm 2\sqrt{(ab)}$.

Hence the points of intersection (say P and Q) are $\{b - a, 2\sqrt{(ab)}\}$ and $\{b - a, 2\sqrt{(ab)}\}$.

Tangent at P to (1) will be

$$y.2\sqrt{(ab)} = 2a(x + b - a) + 4a^2. \quad ...(3)$$

Tangent at P to (2) will be

$$y.2\sqrt{(ab)} = -2b(x + b - a) + 4b^2. \quad ...(4)$$

Slope of (3) is $\dfrac{2a}{2\sqrt{(ab)}} = \sqrt{\dfrac{a}{b}} = m_1$ (say)

and slope of (4) is $-\dfrac{2b}{2\sqrt{(ab)}} = \sqrt{\dfrac{b}{a}} = m_2$ (say)

Then $m_1 \times m_2 = \sqrt{\dfrac{a}{b}} \times -\sqrt{\dfrac{b}{a}} = -1$. So the two tangents are perpendicular to each other.

By symmetry, the tangents at Q are also perpendicular to each other. Hence the two curves (1) and (2) intersect at right angles.

Example 94(b):

Find the lengths of the normals drawn from the point on the axis of the parabola $y^2 = 8ax$ whose distance from the focus is 8a.

Solution:

The parabola is $y^2 = 8ax = 4.\ 2ax$. ...(1)

Hence the focus is (2a, 0). Clearly the co-ordinates of the point which is at a distance of 8a from (2a, 0) and is on the axis of the parabola i.e. x-axis will be (10a, 0).

Equation to any normal to the given parabola is

$$y = mx - 2 \,.\, 2a.m - 2am^3. \qquad ...(3)$$

If the normal passes through (10a, 0), the co-ordinates will satisfy. So $0 = 10am - 4am - 2am^3$.

or $\quad 2am\,(3 - m^3) = 0$

whence $\quad m = 0$

or $\quad m = \pm\sqrt{3}$.

If m = 0, it is x-axis; so length of the normal is 10a clearly.

The point of intersection of (2) and (1) is $(2am^2, -4am)$.

Putting the value of $m = \sqrt{3}$, the point is $(6a, -4a\sqrt{3})$.

So the length of the normal

$$= \sqrt{\{(10a - 6a)^2 + (0 + 4a\sqrt{3})^2} = \sqrt{(16a^2 + 48a^2)} = 8a.$$

Similarly, putting $m = -\sqrt{3}$ length will be 8a.

Example 94(c):

Prove that the locus of the middle point of the portion of a normal intersected between the curve and the axis is a parabola whose vertex is the focus and whose latus rectum is one quarter of that of the original parabola.

Solution:

Let $y^2 = 4ax$ be any parabola and point P on it be $(am^2, -2am)$.

Then the equation to the normal at $(am^2, -2am)$ for the given parabola will be $y = mx - 2am - am^3$. ...(1)

If (1) meets the axis of parabola i.e x-axis at $(x_1, 0)$, then these co-ordinates will satisfy (1). So $0 = mx_1 - 2am - am^2$

or $\quad x_1 = (2a = am^2)$.

Hence the point of intersection say G becomes

$$(2a + am^2, 0).$$

Let (h, k) be mid-point of PG; then

$$h = \frac{am^2 + \left(2a + am^2\right)}{2} = a + am^2 \qquad ...(2)$$

and $$k = \frac{0 - 2am}{2} - am \qquad ...(3)$$

The required locus will be obtained by eliminating m from (2) and (3).

So putting the values of m from (3) in (2), we get

$$h = a + a\left(-\frac{k}{a}\right)^2$$

Generalising and simplifying, we get the required locus as

$$y^2 = a (x - a)$$

This is a parabola whose vertex is (a, 0), i.e. focus of (1) and latus rectum is a, i.e. 1/4 of that of (1). **Hence proved.**

Example 95(a):

Show that the two parabolas

$$x^2 + 4a (y - 2b - a) = 0 \quad \text{and} \quad y^2 = 4b (x - 2a + b)$$

intersect at right angles at the common end of the latus rectum of each.

Solution:

The two parabolas are given as

$$x^2 + 4a (y - 2b - a) = 0 \quad \text{or} \quad x^2 + - 4a [y - (a + 2b)] \qquad ...(1)$$

$$\text{and } y^2 = 4b (x - 2a + b) \quad \text{or} \quad y^2 = 4b [x - (2a - b)]. \qquad ...(2)$$

For (1) changing the origin to {0, (a + 2b)}, we find the ends of the latus rectum, say P and Q as (2a, a) and (– 2a, a). Hence with respect to the original origin, the co-ordinates of P are [2a, a + 2b – a] or (2a, 2b).

Similarly the co-ordinates of one of the end points of the latus rectum in (2) is (2a, 2b).

Hence (2a, 2b) is the common point of (1) and (2) which is one of the end points of the latus rectum.

Now the equations to the tangent at (2a, 2b) to (1) and (2) are respectively

$$x.2a = - 2a [y - 2b - (a + 2b)] \qquad ...(3)$$

$$x.2b = 2b [x - 2a - (2a + b)]. \qquad ...(4)$$

The slope of (3) and (4) are respectively – 1 and 1.

Hence their product is – 1, or the tangents are perpendicular to each other.

So the two curves intersect at right angles

Example 95(b):

The normal at any point P meets the axis in G and the tangent at the vertex in G'; if A be the vertex and the rectangle AGQG' be completed, prove that the equation to the locus of Q is

Solution:

Let P be any point (at^2, 2at) on the parabola $y^2 = 4ax$. ...(1)

Equation of the normal PG at (at^2, 2at) tp (1) is

$$y = -tx + at + at^3 \quad ...(2)$$

[as the normal at $(am^2, -2am)$ is

$y = [mx - 2am - am^3]$.

If G is the point on x-axis, ordinate of G will be 0; so its abscissa will be $(2a + at^2)$.

Here $\qquad AG = 2a + at^2. \quad ...(3)$

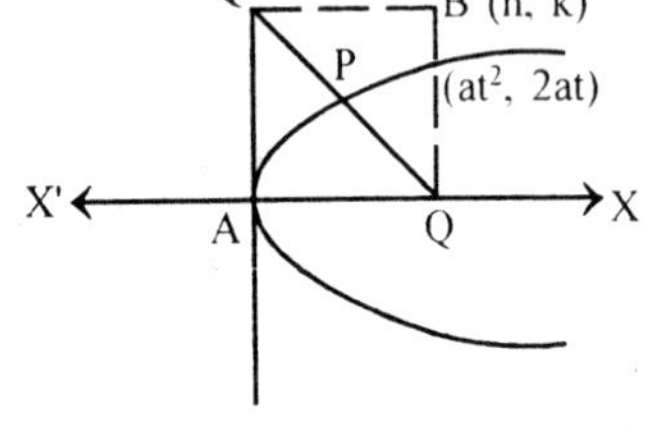

Similarly if PG meets tangent at the vertex i.e. y-axis at G'; abscissa of G' is zero, so ordinate will be [by putting $x = 0$ in (2)].

$$y = 2at + at^3$$

So $\qquad AG = 2at + at^3. \quad ...(4)$

If the required point Q be (h, k), clearly by the fig.,

$$h = AG = 2a + at^2 \text{ [by (3)]} \quad ...(5)$$

$$k = AG' = 2a + at^2 \text{ [by (3)]} \quad ...(6)$$

The required locus will be obtained by eliminating t from (5) and (6). So multiplying (5) by t and subtracting from (6); we get

$$k - ht = 0, \text{ so } t = \frac{k}{h}.$$

Putting in (5), we get $h = 2a + a\left(\frac{k}{h}\right)^2$

$\Rightarrow \qquad h^3 = 2ah^2 + ak^2$

Generalising we get $x^3 = 2ax^2 + ay^2$. **Proved.**

Example 96:

Prove that the parabolas $y^2 = ax$ and $x^2 = by$ cut one another at an angle $\tan^{-1} \dfrac{3a^{1/3}b^{1/3}}{2\left(a^{2/3}+b^{2/3}\right)}$.

Solution:

The parabolas are given as

$$y^2 = ax \quad ...(1)$$

and $\qquad x^2 = by \quad ...(2)$

To solve (1) and (2), putting the value of y from (2) in (1), we get

$$\frac{x^4}{b^2} = ax.$$

whence either $x = 0$ or $x = a^{1/2}b^{2/3}$. The corresponding values of y are 0 and $a^{1/3}b^{1/3}$. Hence the points of intersection of (1) and (2) are (0, 0) and $(a^{1/3}b^{2/3},\ a^{2/3}b^{1/3})$.

The tangents at (0, 0) to (2) and (1) are respectively the x and y axis, so the angle between them is 90°.

Now tangents at $(a^{1/3}b^{2/3},\ a^{2/3}b^{1/3})$ to (1) and (2) are respectively

$$y.\ a^{2/3}\ b^{1/3} = \frac{a}{2}.\left(x + a^{1/3}.b^{2/3}\right) \qquad ...(3)$$

and $$y.\ a^{1/3}\ b^{2/3} = \frac{b}{2}.\left(y + a^{2/3}.b^{1/3}\right) \qquad ...(4)$$

Slopes of (3) and (4) are respectively

$$\tan\theta = \frac{2.\dfrac{a^{1/3}}{b^{1/3}} - \dfrac{1}{2}.\dfrac{a^{1/3}}{b^{1/3}}}{1 + 2.\dfrac{a^{1/3}}{b^{1/3}}.\dfrac{a^{1/3}}{2b^{1/3}}} = \frac{\dfrac{3}{2}\dfrac{a^{1/3}}{b^{1/3}}}{\dfrac{\left(b^{2/3} + a^{2/3}\right)}{b^{2/3}}}$$

$$= \frac{3a^{1/3}\,b^{1/3}}{2\left(b^{2/3} + a^{2/3}\right)}$$

or $$\theta = \tan^{-1}\left\{\frac{3a^{1/3}\,b^{1/3}}{2\left(a^{2/3} + b^{2/3}\right)}\right\}$$ **Proved.**

Example 97:

Prove that the two parabolas $y^2 = 4ax$ and $y^2 = 4ax\ (x - b)$ cannot have a common normal, other than the axis, unless $b/(a - c) > 2$.

Solution:

The two parabolas are given as

$$y^2 = 4ax \qquad ...(1)$$

and $$y^2 = 4c\ (a - b). \qquad ...(2)$$

Equation to any normal to (1) is

$$y = mx - 2am - am^3. \qquad ...(3)$$

Equation to any normal to (2) is

$$y = m\ (x - a) - 2cm - am^3. \qquad ...(4)$$

If there is any common normal, then (3) and (4) must be identical. As the coefficients of x and y are equal, so the constant terms will also be equal; hence

$$-2am - am^3 = -bm - 2mc - cm^3$$

$\Rightarrow$ $$m[m^2(c-a) - 2a + b + 2c)] = 0.$$

So either $m = 0$ or $m^2 = \frac{2a-b-2c}{c-a}$.

If $m = 0$, the common normal is the x-axis.

If $m^2 = \frac{2a-b-2c}{c-a}$, then

$$m = \sqrt{\left(\frac{2(a-c)-b}{c-a}\right)} = \sqrt{\left(-2-\frac{b}{c-a}\right)}$$

If the value of m is real, then $-2-\frac{b}{c-a} > 0$

$$-\frac{b}{c-a} \geq 2 \quad \text{or} \quad \frac{b}{c-a} \geq 2.$$

Hence proved.

Example 98(a):

Two equal parabolas have the same focus and their axes are at right angles; a normal to one is perpendicular to a normal to the other; prove that the locus of the point of intersection of these normals is another parabola.

Solution:

Taking the common focus as origin and axis of the parabolas as axis of the co-ordinates, the equations of the parabolas may be written as

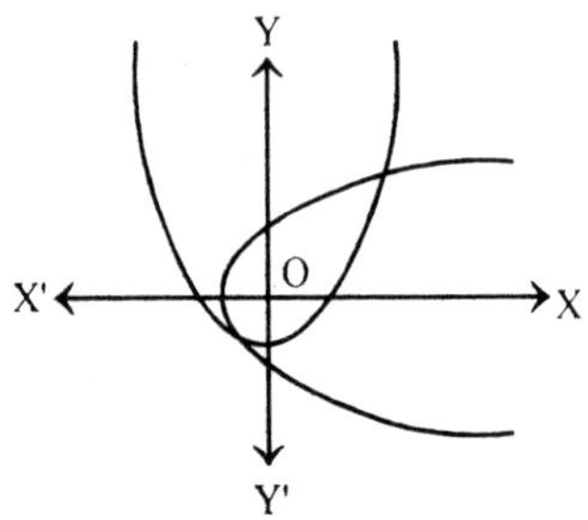

$$y^2 = 4a(x + a) \quad ...(1)$$

and $$x^2 = 4a(y + a) \quad ...(2)$$

where 4a is the latus rectum of each.

Equation to any normal to (1) is

$$y = m(x + a) - 2am - am^3. \quad ...(3)$$

Equation to any normal to (2) is

$$y = m'(y + a) - 2am' - am'^3. \quad ...(4)$$

But the two normal are at right angles, then

$$m.\frac{1}{m} = -1 \quad \text{or} \quad m' = -m \left[\text{as the slope of (4) is} \frac{1}{m'}\right]$$

Substituting $m' = -m$ in (4), we have

$$x = -m(y + a) + 2\ 2am + am^3 \qquad ...(5)$$

The locus of the point of intersection of the normals will be the eliminant of m from (3) and (5). So adding (3) and (5), after taking the terms to R.H.S. in both the cases, we get

$$0 = mx - x - y - my$$

$\Rightarrow$ $$x + y = m(x - y)$$

or $$m = \frac{x+y}{x-y}.$$

Putting the value of m in (3), we get

$$y = \frac{x+y}{x-y}.(x+a) - 2a.\frac{x+y}{x-y} - a\left(\frac{x+y}{x-y}\right)^3$$

$\Rightarrow$ $y(x-y)^3 = (x+y)[(x+a)(x-y)^2 - 2a(x-y)^2 - a(x+y)^2]$

$= (x+y)[(x-y)^2(x+a-2a) - a(x+y)^2]$

$= (x+y)[(x(x-y)^2 - 2a(x^2+y^2)]$

$\Rightarrow$ $2a(x^2+y^2)(x+y) = (x+y).\ x(x-y)^2 - y(x-y)^3$

$\Rightarrow$ $2a(x^2+y^2)(x+y) = (x-y)^2[x^2 + xy - xy + y^2]$

$\Rightarrow$ $2a(x+y) = (x-y)^2.$

This is the reqd. locus which is a parabola. **Proved.**

Example 98(b):

If $a^2 > 8b^2$, prove that a point can be found such that the two tangents from it to the parabola $y^2 = 4ax$ are normals to the parabola $x^2 = 4by$.

Solution:

Two parabolas are given as

$$y^2 = 4ax \qquad ...(1)$$

and $$x^2 = 4by \qquad ...(2)$$

Tangent to (1) $y = mx + \dfrac{a}{m}$. ...(3)

Equation to any normal sto (2) is

$$x = m'y - 2bm' - bm'^3$$

$\Rightarrow$ $$y = \frac{1}{m'}x + 2b + bm'^2 \qquad ...(4)$$

If the equations (3) and (4) are the same, then the coefficients must be identical. As the coefficients of y are equal in both, so coefficients of x and constant. terms will also be equal. Hence,

$$m = \frac{1}{m'} \quad \text{or} \quad m' = \frac{1}{m} \qquad ...(5)$$

and $$\frac{a}{m} = 2b + bm'^2 \qquad ...(6)$$

Putting the value of m' from (5) in (6), we get

$$\frac{a}{m} = 2b + \frac{b}{m^2} \quad \text{or} \quad 2bm^2 - am + b = 0. \qquad ...(7)$$

If the roots of (7) are real, then its discriminate must be greater then zero.

Therefore $a^2 - 4.\ 2b.b \geq 0$ or $a^2 \geq 8b^2$,

which is the given condition. **Hence proved.**

Example 99:

Prove that three tangents to a parabola, which are such that the tangents of their inclinations to the axis are in a given harmonical progression, form a triangle whose area is constant.

Solution:

Let the parabola be $y^2 = 4ax$ and say any three tangents on it are

$$y = m_1 x + \frac{a}{m_1} \qquad ...(1)$$

$$y = m_2 x + \frac{a}{m_2} \qquad ...(2)$$

and $$y = m_3 x + \frac{a}{m_3} \qquad ...(3)$$

Solving (1) and (2), we get

$$x = \frac{a}{m_1 m_2} \quad \text{and} \quad y = \frac{a}{m_1} + \frac{a}{m_2}$$

Hence the points of intersection of (1) and (2), (2) and (3) and (1) and (3) respectively, i.e., the co-ordinates of the angular points of the triangle are

$$\left(\frac{a}{m_1 m_2}, \frac{a}{m_1} + \frac{a}{m_2}\right), \left(\frac{a}{m_2 m_3}, \frac{a}{m_2} + \frac{a}{m_3}\right) \text{and} \left(\frac{a}{m_1 m_3}, \frac{a}{m_3} + \frac{a}{m_1}\right)$$

Hence the area of the triangle so formed is given by

$$\Delta = \frac{1}{2}\left[\frac{a}{m_1 m_2}\left(\frac{a}{m_2} + \frac{a}{m_3} - \frac{a}{m_3} - \frac{a}{m_1}\right) + \frac{a}{m_2 m_3}\left(\frac{a}{m_3} + \frac{a}{m_1} - \frac{a}{m_1} - \frac{a}{m_2}\right)\right.$$

$$\left. + \frac{a}{m_3 m_1}\left(\frac{a}{m_1} + \frac{a}{m_2} - \frac{a}{m_2} - \frac{a}{m_3}\right)\right]$$

$$=\frac{1}{2}a^2\left[\frac{a}{m_1 m_2}\left(\frac{a}{m_2}-\frac{a}{m_1}\right)+\frac{a}{m_2 m_3}\left(\frac{a}{m_3}-\frac{a}{m_2}\right)\frac{a}{m_3 m_1}\left(\frac{a}{m_1}-\frac{a}{m_3}\right)\right]$$

Factorizing by cyclic-order method,

$$\Delta=\frac{1}{2}a^2\left(\frac{1}{m_2}-\frac{1}{m_1}\right)\left(\frac{1}{m_3}-\frac{1}{m_2}\right)\left(\frac{1}{m_1}-\frac{1}{m_3}\right) \quad ...(4)$$

As by hypothesis, m_1, m_2 and m_3 are in H.P.

$\Rightarrow$ $\frac{1}{m_1},\frac{1}{m_2},\frac{1}{m_2}$ are in A.P.

$\Rightarrow$ $\frac{1}{m_2}-\frac{1}{m_1}=\frac{1}{m_3}-\frac{1}{m_2}=d$, (say) hence $\frac{1}{m_3}-\frac{1}{m_1}=2d$

Putting the values in (4), we get

$\Delta = -1/2.\ a^2\ d.\ d\ (-2d) = a^2\ d^3$ which is constant quantity as far as a and d are constant. **Proved.**

Example 100(a):

If a normal to a parabola make an angle ϕ with the axis, show that it will cut the curve again at an angle $\tan^{-1}$ (1/2) $\tan\phi$.

Solution:

Let the equation of the normal to any parabola

$$y^2 = 4ax \quad ...(1)$$

be $$y = mx - 2am - am^3 \quad ...(2)$$

So get the other point of contact, we solve (1) and (2) simultaneously.

Putting the value of x from (1) in (2), we get

$$y=\frac{y^2}{4a}-2am-am^3$$

$$\Rightarrow \quad my^2 - 4ay - 2am - am^3 = 0. \quad ...(3)$$

As (2) is the normal to (1) at $(am^2, -2am)$, so one point of intersection is $(am^2, -2am)$; hence one factor to (3) is $(y + 2am)$ and therefore the other factor is

$$(my - 4a - 2am^2).$$

So $$y=\frac{1}{m}\left(4a+2am^2\right).$$

Putting this value in (2),

$$\frac{4a}{m} + 2am = mx - 2am - am^3$$

$$\Rightarrow \qquad x = \frac{a}{m^2}\left(2+m^2\right)^2.$$

So the second point of intersection, say Q, of (1) and (2) is

$$\left[\frac{a}{m^2}\left(2+m^2\right)^2, \frac{2a}{m}\left(2+m^2\right)^2\right]$$

As the equation to a tangent at the point (x_1, y_1) to parabola (1) is $yy_1 = 2a(x + x_1)$, its slope is $2a/y_1$. Hence slope of the tangent at the second point of intersection is

$$(2a)/\frac{2a}{m}\left(2+m^2\right) = \frac{m}{2+m^2}$$

If the reqd. angle be q, then $\tan\theta = \dfrac{m - \dfrac{m}{2+m^2}}{1 + m.\dfrac{m}{2+m^2}}$

$$= \frac{2m + m^3 - m}{2 + m^2 + m^2} = \frac{m\left(1+m^2\right)}{2\left(1+m^2\right)} = \frac{m}{2}$$

So $\tan\theta = \dfrac{m}{2}$.

Given that the angle that the line (2) make with x-axis is ϕ, then

$$m = \tan\phi.$$

Hence $\qquad \tan\theta = \dfrac{1}{2}.\tan\theta$

or $\qquad \theta = \tan^{-1}\left[\dfrac{1}{2}.\tan\phi\right]$ **Proved.**

Example 100(b):

What is the equation to the chord of the parabola $y^2 = 8x$ which is bisected at the point (2, – 3)

Solution:

Equation to the parabola is given as $y^2 = 8x$. ...(1)

Any line passing through (2, – 3) may be given by

$$y + 3 = m(x - 2). \qquad ...(2)$$

Let (2) cut the parabola (1) at A and B whose co-ordinates are (x_1, y_1) and (x_2, y_2) respectively.

Putting the values of x from (2) in (1), we get

$$y^2 = 8\left[\frac{1}{m}(y + 3 + 2m)\right]$$

or $\qquad my^2 - 8y - 8(2m + 3) = 0.$ ...(3)

The ordinates of the points A and B i.e. y_1 and y_2 will be given by (3); so $y_1 + y_2 = 8/m$.

The ordinate of the middle point of AB will be $\frac{y_1+y_2}{2}=\frac{4}{m}$ and if (2, – 3) be the mid-point of the chord, we have

$$\frac{4}{m}=-3 \qquad \text{or} \qquad m=-\frac{4}{3}.$$

Putting the value in (2), we get the required equation as

$$y + 3 = -\frac{4}{3}(x-2)$$

or $$4x + 3y + 1 = 0.$$ **Ans.**

Example 100(c):

Prove that the area of the triangle formed by the tangents from the point (x_1, y_1) and the chord of contact is $y_1^2 - 4ax_1)^{3\ 2}$ / 2a.

Solution:

Let P be the point (x_1, y_1) outside the parabola

$$y^2 = 4ax, \qquad ...(1)$$

A and B the points of contact of the tangents from P on (1), as (α, β) and (α', β') respectively, then the equation of the chord of contact AB of the point P will be

$$yy_1 = 2a(x + x_1). \qquad ...(2)$$

Length of PN, the perpendicular from P on AB will be

$$\frac{2a(x_1+x_2)-y_1y_1}{\sqrt{(y_1^2+4a^2)}}=\frac{y_1^2-4ax_1}{\sqrt{(y_1^2+4a^2)}} \text{ (omitting-ve sign.)}$$

Area of the triangle PAB $= \frac{1}{2}AB\times PN$

$$=\frac{1}{2}\frac{\sqrt{(y_1^2+4a^2)}\sqrt{(y_1^2+4ax_1)}}{a}\times\frac{\sqrt{(y_1^2+4ax_1)}}{\sqrt{(y_1^2+4a^2)}}$$

[as AB = $\sqrt{(y_1^2 + 4a^2)}\ \sqrt{(y_1^2 + 4ax_1)}/a$, by the previous question]

$$=\frac{(y_1^2-4ax)^{3/2}}{2a}$$

Example 100(d):

A chord of a parabola passes through a point on the axis (outside the parabola) whose distance from the vertex is half the latus rectum; prove that the normals at its extremities meet on the curve.

Solution:

Let the equation of the parabola be $y^2 = 4ax$.

The point on the axis outside the parabola at a distance of half the latus rectum from the vertex will be (– 2a, 0) and let T_1 and T_2 be two points on the parabola as $(at_1^2, 2at_1)$ and $(at_2^2, 2at_2)$ respectively.

The line passing through T_1 and T_2 will be

$$y - 2at_1 = \frac{2at_1 - 2at_2}{at_1^2 - at_2^2}\left(a - at_1^2\right) \text{ or } y - 2at_1 = \frac{2}{t_1 + t_2}\left(x - at_1^2\right)$$

If this line passes through (– 2a, 0), we have

$$0 - 2at_1 = \frac{2}{t_1 + t_2}\left(-2a - at_1^2\right) \quad \text{or} \quad t_1 t_2 = 2 \qquad ...(1)$$

Equation of the normal at T_1 is

$$y = - t_1x + 2at_1 + at_1^3 \qquad ...(2)$$

and equation of the normal at T_2 is

$$y = - t_2x + 2at_2 + at_2^3 \qquad ...(3)$$

To solve (2) and (3), subtracting (2) from (3), we get

$$0 = (t_1 - t_2)\ x - 2a\ (t_1 - t_2) - a\ (t_1^3 - t_2^3)$$

$$\Rightarrow \qquad x = 2a + a\ (t_1^2 + t_1t_2^2 + t_1^2 \text{ as } t_1 - t_1 \neq 0.$$

Substituting this value of x in (2), we get

$$y = - [2a + a\ (t_1^2 + t_1t_2 + t_2^2) + 2at_1 + at_1^3$$

$$= - at_1t_2\ (t_1 + t_2).$$

Hence the point of intersection of the normals given by (2) and (3) is

$$[2a + a\ (t_1^2 + t_1t_2 + t_2^2),\ - at_1t_1\ (t_1 + t_2)].$$

The equation of the parabola is $y^2 - 4ax = 0$.

Substituting the co-ordinates of the point of intersection in the equation of the parabola, the L. H. S. becomes

$$[- at_1t_2\ (t_1 + t_2)]^2 - 4\ [2a + a\ (t_1^2 + t_1t_2 + t_2^2)]$$

$$4a^2\ \{t_1^2 + t_2^2\ 2t_1t_1\} - 4a^3\ (2 + t_1^2 + t_1t_2 + t_2^2\}$$

$$[\because t_1t_2 = 2 \text{by} (1)]$$

$$4a^2\ \{t_1^2 + t_2^2 + 4 - 2 - t_1^2 - 2 - t_2^2] = 0 \text{ R. H. S.}$$

As the point of intersection satisfies the equation of the parabola, it lies on the curve. **Hence proved.**

Example 101:

Prove that the length of the intercept on the normal at the point $(at^2, 2at)$ made by the circle which is described on the focal distance of the given point as diameter is a $\sqrt{(1 + t2)}$.

Solution:

The equation of the parabola is $y^2 = 4ax$ ...(1)

If the point $(at^2, 2at)$ be P, then the normal at P is,

$$y = -tx + 2at + at^2 \quad ...(2)$$

If PS is the diameter of ay circle,

□ ∠ P S

= 90° (being in a semi-circle).

The length of the intercept of the normal by the circle

$$= PN = \sqrt{(PS^2 - SN^2)} \quad ...(3)$$

Now $PS = \sqrt{(a - at^2)^2 + (0 - 2at)^2}$

$$= a(1 + t^2) \quad ...(4)$$

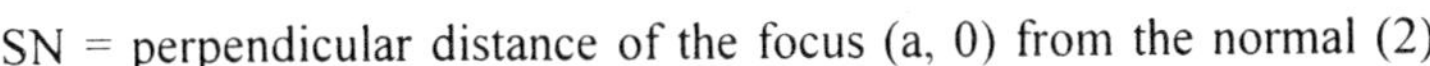

SN = perpendicular distance of the focus (a, 0) from the normal (2)

$$= \frac{-at + 2at + at^3}{\sqrt{(1+t^2)}} = \frac{at\,1+t^2}{\sqrt{(1+t^2)}} = at\sqrt{(1+t^2)} \quad ...(5)$$

Putting the values in (3) from (4) and (5), we get

$$PN = \sqrt{[\{a^2(1 + t^2)^2 - \{a^2t^2(1 + t^2)\}]}$$

$$= \sqrt{a^2(1 + t^2)(1 + ts^2 - t^2)} = a\sqrt{(1 + t^2)}.$$ **Proved.**

Example 102:

If a perpendicular be let fall from any point P upon its polar prove that the distance of the foot of this perpendicular from the focus is equal to the distance of the point P form the directrix.

Solution:

Let the equation of the parabola be $y^2 = 4ax$. ...(1)

Its focus will be (a, 0) and directrix will be

$$x + a = 0 \quad ...(2)$$

If P is any point (x_1, y_1), the equation of its polar to (1), will be

$$yy_1 = 2a(x + x_1) \quad ...(3)$$

Equation to the line passing through P and perpendicular to (3), will be

$$y - y_1 = -\frac{y_1}{2a}(x - x_1)$$

or $$2ay = -y_1x + x_1y_1 + 2ay_1. \quad ...(4)$$

Solving these equations, we get

$$x = -\frac{x_1 y_2^2 + 2ay_1^2 - 4a^2 x_1}{y_1^2 + 4a^2} \text{ and } y = \frac{4ay_1(x_2 + a)}{y_1^2 + 4a^2}$$

So the co-ordinates of the point of intersection of (3) and (4) i.e. the foot of the perpendicular from P on the polar, say N, are

$$\left[\frac{x_1 y_2^2 + 2ay_1^2 - 4a^2 x_1}{y_1^2 + 4a^2}, \frac{4ay_1(x_1 + a)}{y_1^2 + 4a^2}\right]$$

Distance between the focus (a, 0) and N is given by

$$\sqrt{\left[\left\{\frac{x_1 y_2^2 + 2ay_1^2 - 4a^2 x_1}{y_1^2 + 4a^2} - a\right\}^2 + \left(\frac{4ay_1(x_1 + a)}{y_1^2 + 4a^2} - 0\right)^2\right]}$$

$$= \frac{1}{y_1^2 + 4a^2}\sqrt{\left[\left(x_1 y_1^2 + ay_1^2 - 4a^2 x_1 - 4a^3\right)^2 + 16a^2 y_1^2 (x_1 + a)^2\right]}$$

$$= \frac{1}{y_1^2 + 4a^2}\sqrt{\left[\left(y_1^2 - 4a^2\right)^2 (x_1 + a)^2 + 16a^2 y_1^2 (x_1 + a)^2\right]}$$

$$= \frac{(x_1 + a)}{y_1^2 + 4a^2}\sqrt{\left(y_1^2 - 4a^2\right)^2 + 16a^2 y_1^2} = \frac{(x_1 + a)}{\left(y_1^2 + 4a^2\right)} \times \left(y_1^2 + 4a^2\right)$$

$$= x_1 + a \quad ...(5)$$

Distance of the point P (x_1, y_1) from the directrix $x + a = 0$ is clearly $(x_1 + a)$ which is equal to the distance given by (5). **Hence proved.**

Example 102(a):

Prove that the length of the chord joining the points of contact of tangents drawn from the point (x_1, y_1) is

$$\frac{\sqrt{y_1^2 + 4a^2}\sqrt{y_1^2 - 4a_1}}{a}.$$

Solution:

Let the equation to the parabola by $y^2 = 4ax$. ...(1)

and (x_1, y_1) be any point outside it.

The equation to the chord or contact from (x_1, y_1) to (1) is

$$yy_1 = 2a(x+x_1) \quad \text{or} \quad x = \frac{1}{2a}(yy_1 - 2ax_1) \qquad ...(2)$$

To find out the point of intersection, we solve (1) and (2)

$$y^2 = 4a\frac{1}{2a}(yy_1 - 2ax_1)$$

$$\Rightarrow \qquad y^2 - 2yy_1 + 4ax_1 = 0. \qquad ...(3)$$

Let the co-ordinates of the points of intersection of (1) and (2) be (α, β) and (α', β'). So b and b' will be given by (3). So $\beta + \beta' = 2y_1$ ...(4)

and $\beta\beta' = 4ax_1$. ...(5)

Again as (α, β) and (α', β') be on (2), the points will satisfy it. So

$$\alpha = \frac{1}{2a}(y_1\beta - 2ax_1)$$

$$\Rightarrow \qquad \alpha' = \frac{1}{2a}(y_1\beta' - 2ax_1)$$

Subtracting one from the other, we get

$$\alpha - \alpha' = \frac{y_1}{2a}(\beta - \beta'). \qquad ...(6)$$

The distance between the points (α, β) and (α', β') is

$$\sqrt{\{(\alpha-\alpha')^2 + (\beta-\beta')^2\}} = \sqrt{\frac{y_1^2}{4a^2}(\beta-\beta')^2 + (\beta-\beta')^2}$$

Putting the value of $(\alpha, - \alpha')$ from (6),

$$= \sqrt{\left[(\beta-\beta')^2\left(\frac{y_1^2}{4a^2}+1\right)\right]}$$

$$= \frac{1}{2a}\sqrt{(y_1^2+4a^2)}\sqrt{\{(\beta+\beta')^2 - 4\beta\beta'\}}$$

$$= \frac{1}{2a}\sqrt{(y_1^2+4a^2)}\sqrt{\{(2y_1)^2 + 4.4ax\}}$$

Putting the values from (5) and (4) we have

$$= \frac{1}{2a}\sqrt{(y_1^2+4a^2)\times 4(y_1^2 - 4ax_1)}$$

$$= \frac{\sqrt{(y_1^2+4a^2)}\cdot\sqrt{(y_1^2-4ax\ )}}{a} \qquad \textbf{Proved.}$$

Example 102(b):

P, Q and R three points on a parabola and the chord PQ cuts the diameter through R in V. Ordinates Pm and QN are drawn to this diameter. Prove that RM.RN = RV².

Solution:

Let the equation of the parabola referred to the diameter through R and the tangent at R as axis be $y^2 = 4px$ and let the points P and Q be $(pt_1^2, 2pt_1)$ and $(pt_2^2, 2pt_2)$ respectively.

Equation of the diameter RN is y = 0 and the ordinates through P and Q are $x = pt_1^2$ and $x = p_2t^2$. ...(1)

Hence RM.RN = $p_2t^2 . pt_2^2 = (pt_1t_2)^2$.

Also the equation of PQ is

$$y - 2pt_1 = \frac{2pt_2 - 2pt_1}{pt_2^2 - pt_1^2}\left(x - pt_1^2\right)$$

or $y(t_1 + t_2) = 2x + 2pt_1t_2$ ...(2)

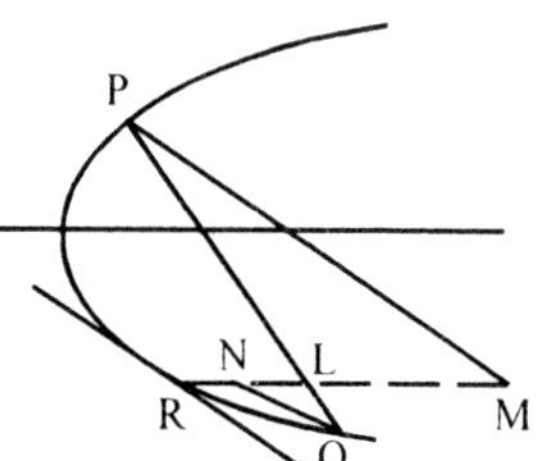

It meets x-axis i.e. y = 0 at L. So putting y = 0 in (2), $x = -pt_1t_2$; hence the co-ordinates of L are $(-pt_1t_2, 0)$.

So $RL = -pt_1t_2$

or $RL^2 = (pt_1t_2)^2$. ...(3)

By (1) and (3) we get RN.RM = RL^2. **Proved.**

Example 102(c):

A chord is a normal to a parabola and is inclined at an angle θ to the axis, prove that the area of the triangle formed by it and the tangents at its extremities is $4a^2 \sec^3 \theta \operatorname{cosec}^3 \theta$.

Solution:

Let the extremities of the normal chord be P and Q and the tangents at P and Q to the parabola say $y^2 = 4ax$ meet in T. Let the co-ordinates of T be (x_1, y_2).

PQ will be the chord of contact for T with respect to the parabola so area of triangle TPQ will be

$$[y_1^2 - 4ax_1]^{3/2}/2a \qquad ...(1)$$

The equation to the chord of contact of T will be

$$yy_1 = 2a(x + x_1)$$

$$\Rightarrow \qquad y = \frac{2ax}{y_1} + \frac{2ax_1}{y_1}. \qquad ...(2)$$

Equation to any normal to $y^2 = 4ax$ is

$$y = mx - 2am - am^2. \qquad ...(3)$$

So (2) and (3) must be identical. As the coefficients of y are equal, others must also equal, so

$$m = \frac{2a}{y_1} \quad \text{and} \quad -2am - am^3 = \frac{2ax_1}{y_1}.$$

whence $\quad y_1 = \frac{2a}{m} \quad$ and $\quad x_1 = \left(-2a - am^2\right).$

If the inclination of the chord of contact, i.e., the normal is θ; then

$$m = \tan\theta.$$

So $\; y_1 = \frac{2a}{m} = 2a\cot\theta \quad$ and $\quad x_1 = \left(-2a - am^2\theta\right).$

Substituting in (1), we get

Area of the triangle $= [4a^2 \cot^2\theta - 4a(-2a - a\tan^2\theta)]^{3/2}/2a$

$$= \frac{1}{2a}\left[4a^2\cot^2\theta + 8a^2\, 4a^2\tan^2\theta\right]^{3/2} = \frac{1}{2a}\left[4a^2\left(\cot^2\theta + 2 + \tan^2\theta\right)\right]^{3/2}$$

$$= \frac{1}{2a}\left(4a^2\right)^{3/2}\left\{(\cot\theta + \tan\theta)^2\right\}^{3/2}$$

$$= \frac{1}{2a}8a^3\left\{(\sec\theta + \operatorname{cosec}\theta)^2\right\}^{3/2}$$

$$\left[\because \cot\theta + \tan\theta = \frac{\cos^2\theta + \sin^2\theta}{\sin\theta.\cos\theta} = \sec\theta.\operatorname{cosec}\theta\right]$$

$= 4a^2 \sec^3\theta \,.\, \operatorname{cosec}^3\theta.$ **Proved.**

Example 102(d):

Two equal parabolas with axes in opposite directions touch at a point O. From a point P on one of them are drawn tangents PQ and PQ' to the other. Prove that QQ' will touch the first parabola in P' where PP' is parallel to the common tangent at O.

Solution:

Taking the common vertex O as origin, common axis as x-axis and the line perpendicular to x-axis through O as y-axis, the equations of the parabolas may be written as

$$y^2 = 4ax. \qquad ...(1)$$

and $y^2 = 4ax$, ...(2)

as the parabolas are in opposite direction.

Any point P is $(at^2, 2at)$. If PQ and PQ' are the tangents to (2), QQ' will be the chord of contact for $P \equiv (at^2, 2at)$ with respect to (2). So its equation will be

$$y.2at = -2a(x + at^2)$$

$$\Rightarrow \quad yt = -x - at^2. \qquad ...(3)$$

To get the points of intersection of (1) and (3), we put the value of x from (3) in (1); we get

$$y^2 = 4a(-yt - at^2)$$

$$\Rightarrow \quad y^2 = 4ayt + 4a^2t^2 = 0$$

$$\Rightarrow \quad (y + 2at)^2 = 0$$

This being a perfect square, the points of intersection of (1) and (3) coincide; so (3) is a tangent (1). **Hence proved.**

Again by (4) $y = -2at$. Putting this value in (3), we get $x = at^2$.

Hence the co-ordinates of the point of contact of (1) and (3), P', are $(at^2, -2at$. So the equation of PP' will be

$$y + 2at = \frac{-2at - 2at}{at^2 - at^2}\left(x - at^2\right)$$

or $\quad x - at^2 = 0 \quad$ and $\quad x = at^2$

This line is clearly parallel to y-axis which is the common tangent at the common vertex of (1) and (2). **Hence proved.**

Example 103:

The general equation to a system of parallel chords in the parabola $y^2 = (25/7)x$ is $4x - y + k = 0$. what is the equation to the corresponding diameter

Solution:

The equation of the parabola is given as

$$y^2 = \frac{25}{7}x \qquad ...(1)$$

and a line is given as $\quad 4x - y + k = 0$. ...(2)

To solve (1) and (2), we put the value of x from (2) in (1); we get

$$y^2 = \frac{25}{7}\,\frac{y - k}{4}$$

$$\Rightarrow \quad 28y^2 - 25y + 25k = 0.$$

If the ordinates of the point of intersection of (1) and (2) be y_1 and y_2 respectively; then y_1 and y_2 are the roots of (3).

So $$y_1 + y_2 = \frac{25}{28}$$

If the ordinate of the mid-point of the line joining the points whose ordinates are y_1 and y_2 is

$$\frac{y_1 + y_2}{2} = \frac{25}{28 \times 2} = \frac{25}{56}.$$

As the diameter of (1) corresponding to (2) is a line parallel to the axis of (1) i.e. y = C (x-axis) through any chord parallel to (2) will be passing through the point whose ordinate is 25/56 so its equation is y = 25/56 or 56y = 25.

Ans.

Example 104:

If ω be the angle which a focal of a parabola makes with the axis, prove that the length of the chord is 4a cosec² ω and that the perpendicular on it from the vertex is a sin ω.

Solution:

Let the equation of the parabola by $y^2 = 4ax$. ...(1)

Focus will be (a, 0) and any line passing through (a, 0) may be given by

$$y = m(x - a) \quad ...(2)$$

To get the points of intersection of (1) and (2), we solve them, so putting the value of x from (2) in (1), we get

$$y^2 = 4a\frac{y + am}{m}$$

or $$my^2 - 4ay - 4am = 0. \quad ...(3)$$

If y_1 and y_2 are the roots of (3), then

$$y_1 + y_2 = 4a/m \quad ...(4)$$

and $$y_1 + y_2 = -4a. \quad ...(5)$$

Let (2) intersect (1) in A and B, whose co-ordinates are (x_1, y_1) and (x_2, y_2); then

$$AB = \sqrt{[(x_2 - x_1)^2 + (y_2 - y_1)^2]} \quad ...(6)$$

As (x_1, y_1) and (x_2, y_2) lie on (2), so

$$y_1 = m(x_1 - a) \quad ...(7)$$

$$y_2 = m(x_2 - a) \quad ...(8)$$

Subtracting (7) from (8), we get

$$(y_2 - y_1) = m\,(x_2 - x_1)$$

or $$(x_2 - x_1) = \frac{1}{m}(y_2 - y_1).$$

Putting this will in (4), we get

$$AB = \sqrt{\left[\frac{1}{m}(y_2 - y_1)^2 + (y_2 - y_1)^2\right]} = \sqrt{\left[(y_2 - y_1)^2\left(\frac{1}{m}+1\right)\right]}$$

$$= \sqrt{[\{(y_2 + y_1)^2 - 4y_1y_2\}\ \{(\cot^2 \omega + 1)\}]}$$

[as the angle that (2) makes with x-axis is given to be ω so $m = \tan \omega$]

$$= \sqrt{\left[\left\{\left(\frac{4a}{\tan\omega}\right)^2 - (-16a^2)\right\}\operatorname{cosec}^2\omega\right]} \qquad \text{[by (4) and (5)]}$$

$$= \sqrt{[16a^2\,(\cot^2 \omega + 1)\operatorname{cosec}^2 \omega]}$$

$$= \sqrt{[16a^2 \operatorname{cosec}^2 \omega \operatorname{cosec}^2 \omega]} = 4a \operatorname{cosec}^2 \omega$$ **Proved.**

Again the length perpendicular from vertex (0, 0) on (2)

$$= \frac{am}{\sqrt{(1+m^2)}} = \frac{a\tan\omega}{\sqrt{(1+\tan^2\omega)}} = \frac{a\sin\omega}{\cos\omega\,.\sec\omega} = a\sin\omega.$$ **Proved.**